Advanced Metallization Conference 2009 (AMC 2009)

Advanced Metallization Conference 2009 (AMC 2009)

Proceedings of the Conference held October 13–15, 2009, Baltimore, Maryland, U.S.A. This Conference is affiliated with the Materials Research Society and sponsored by the Department of Continuing Education in Engineering, University Extension, University of California at Berkeley, California, U.S.A.

EDITORS:

Daniel C. Edelstein
IBM Corporation
Yorktown Heights, New York, U.S.A.

Stefan E. Schulz
Chemnitz University of Technology
Chemnitz, Germany

MRS

Materials Research Society
Warrendale, Pennsylvania

Single article reprints from this publication are available through University Microfilms Inc., 300 North Zeeb Road, Ann Arbor, Michigan 48106

CODEN: MRSPDH

Published by:

Materials Research Society
506 Keystone Drive
Warrendale, PA 15086
Telephone (724) 779-3003
Fax (724) 779-8313
Web site: http://www.mrs.org/

Manufactured in the United States of America

CONTENTS

*Invited Paper

MATERIALS AND CHARACTERIZATION

INTEGRATION

*Invited Paper

RELIABILITY AND YIELD

3D INTEGRATION

PREFACE

The 26th session of the Advanced Metallization Conference (AMC 2009) took place at the Marriott Baltimore Inner Harbor at Camden Yards in Baltimore, Maryland, October 13–15, 2009. The conference Co-Chairs were Dr. Daniel C. Edelstein (IBM) and Prof. Dr. Stefan E. Schulz (Chemnitz University of Technology). The Executive Committee for AMC 2009 was responsible for guiding the technical content of the conference and is itself comprised of 24 internationally distinguished researchers from the field of interconnect technology. In Japan, a parallel meeting, the Advanced Metallization Conference 19th Asian Session (ADMETA 2009), was held at The University of Tokyo, Sanjo Conference Hall in Tokyo, Japan, October 19-21, 2009. The Chairman for ADMETA 2009 was Dr. Y. Shimogaki (The University of Tokyo). The meeting was facilitated through the efforts of the conference Executive Committee and members. ADMETA 2009 was co-sponsored by the Japan Society of Applied Physics (JSAP), Silicon Technology Division, and the AMC affiliated with the Materials Research Society.

AMC is a single-session workshop-atmosphere conference featuring the latest R&D and manufacturing results, as well as "real-world" integration and reliability results on the application of metallization and related technologies for advanced ID devices. AMC 2009 hosted scientists and engineers from all over the world, who came to listen and present state-of-the-work in the field of ULSI interconnect technology including metallization, dielectrics, integration, packaging, design, and vertical integration. The conference included 14 invited talks, 26 regular oral presentations, and 23 posters. The Executive Committee carefully reviewed all conference submissions and the presented papers were well received by conference participants. A highlight of the conference was the Keynote speech by Vijay K. Kapur, Ph.D., M.B.A., International Solar Electric Technology, Inc.

The AMC 2010 is scheduled to be held in Albany, New York on October 5–7, 2010 and will be co-chaired by Cindy Goldberg (ST Microelectronics) and Zsolt Tökei (IMEC). ADMETA 2010 is scheduled to be held October 20–22. For more information on AMC 2010 please visit the URL http://www.sematech.org/meetings/announcements/8964/index.htm. For more information on ADMETA, please visit the URL http://www.admeta.org.

Daniel C. Edelstein
Stefan E. Schulz

February 2010

ACKNOWLEDGMENTS

The continued and growing success of both the Advanced Metallization Conference and ADMETA is due to the combined efforts of many individuals and organizations from the global semiconductor community. Of particular note are the contributions of keynote, invited, and oral speakers, poster presenters, short course instructors and panelists who bring the technology to the forefront at this conference. The editors would especially like to thank the active members of the Executive Committees of both the North American and Asian conferences, listed below, for their work in soliciting invited speakers, reviewing abstracts and full papers and serving as session chairs. AMC 2009 received generous support from BASF. Chemnitz University of Technology is gratefully acknowledged for its contribution to cover the publication costs of this proceedings volume. The editors would also like to thank Melody Wells and James Connor of the University of California at Berkeley for valuable help and advice throughout the year.

2009 Advanced Metallization Conference Committee

MATERIALS RESEARCH SOCIETY CONFERENCE PROCEEDINGS

Tungsten and Other Refractory Metals for VLSI Applications, Robert S. Blewer, 1986; ISSN 0886-7860; ISBN 0-931837-32-4

Tungsten and Other Refractory Metals for VLSI Applications II, Eliot K. Broadbent, 1987; ISSN 0886-7860; ISBN 0-931837-66-9

Tungsten and Other Refractory Metals for VLSI Applications III, Victor A. Wells, 1988; ISSN 0886-7860; ISBN 0-931837-84-7

Tungsten and Other Refractory Metals for VLSI Applications IV, Robert S. Blewer, Carol M. McConica, 1989; ISSN 0886-7860; ISBN 0-931837-98-7

Tungsten and Other Advanced Metals for VLSI/ULSI Applications V, S. Simon Wong, Seijiro Furukawa, 1990; ISSN 1048-0854; ISBN 1-55899-068-2

Tungsten and Other Advanced Metals for ULSI Applications in 1990, Gregory C. Smith, Roc Blumenthal, 1991; ISSN 1048-0854; ISBN 1-55899-112-3

Advanced Metallization for ULSI Applications, Virendra V.S. Rana, Rajiv V. Joshi, Iwao Ohdomari, 1992; ISSN 1048-0854; ISBN 1-55899-152-2

Advanced Metallization for ULSI Applications 1992, Timothy S. Cale, Fabio S. Pintchovski, 1993; ISSN 1048-0854; ISBN 1-55899-192-1

Advanced Metallization for ULSI Applications in 1993, David P. Favreau, Yosi Shacham-Diamand, Yasuhiro Horiike, 1994; ISSN 1048-0854; ISBN 1-55899-235-9

Advanced Metallization for ULSI Applications in 1994, Roc Blumenthal, Guido Janssen, 1995; ISSN 1048-0854; ISBN 1-55899-279-0

Advanced Metallization and Interconnect Systems for ULSI Applications in 1995, Russell C. Ellwanger, Shi-Qing Wang 1996; ISSN 1048-0854; ISBN 1-55899-341-X

Advanced Metallization and Interconnect Systems for ULSI Applications in 1996, Robert Havemann, John Schmitz, Hiroshi Komiyama, Kazuo Tsubouchi, 1997; ISSN 1048-0854; ISBN 1-55899-385-1

Advanced Metallization and Interconnect Systems for ULSI Applications in 1997, Robin Cheung, Jeffrey Klein, Kazuo Tsubouchi, Masanori Murakami, Nobuyoshi Kobayashi, 1998; ISSN 1048-0854; ISBN 1-55899-412-2

Advanced Metallization Conference in 1998 (AMC 1998), Gurtej S. Sandhu, Heinrich Koerner, M. Murakami, Y. Yasuda, N. Kobayashi, 1999; ISSN 1048-0854; ISBN 1-55899-484-X

Advanced Metallization Conference 1999 (AMC 1999), Mihal E. Gross, Thomas Gessner, N. Kobayashi, Y. Yasuda, 1999; ISSN 1048-0854; ISBN 1-55899-539-0

Advanced Metallization Conference 2000 (AMC 2000), Dan Edelstein, Girish Dixit, Yukio Yasuda, Takayuki Ohba, 2001; ISSN 1048-0854; ISBN 1-55899-574-9

MATERIALS RESEARCH SOCIETY CONFERENCE PROCEEDINGS

Advanced Metallization Conference 2001 (AMC 2001), Andrew J. McKerrow, Yosi Shacham-Diamand, Shigeaki Zaima, Takayuki Ohba, 2002; ISSN 1048-0854; ISBN 1-55899-670-2

Advanced Metallization Conference 2002 (AMC 2002), Bradley M. Melnick, Timothy S. Cale, Shigeaki Zaima, Tomohiro Ohta, 2003; ISSN 1048-0854; ISBN 1-55899-719-9

Advanced Metallization Conference 2003 (AMC 2003), Gary W. Ray, Tom Smy, Tomohiro Ohta, Manabu Tsujimura, 2004; ISSN 1048-0854; ISBN 1-55899-757-1

Advanced Metallization Conference 2004 (AMC 2004), Darrell Erb, Peter Ramm, Kazuya Masu, Akihiko Osaki, 2005; ISSN 1048-0854; ISBN 1-55899-814-4

Advanced Metallization Conference 2005 (AMC 2005), Sywert H. Brongersma, Thomas C. Taylor, Manabu Tsujimura, Kazuya Masu, 2006; ISSN 1048-0854; ISBN 1-55899-865-9

Advanced Metallization Conference 2006 (AMC 2006), Stephen W. Russell, Michael E. Mills, Akihiko Osaki, Takashi Yoda, 2007; ISSN 1048-0854; ISBN 978-1-55899-947-3

Advanced Metallization Conference 2007 (AMC 2007), Andrew J. McKerrow, Yosi Shacham-Diamand, Shoso Shingubara, Yukihiro Shimogaki, 2008; ISSN 1048-0854; ISBN 978-1-55899-992-3

Advanced Metallization Conference 2008 (AMC 2008), Mehul Naik, Roey Shaviv, Takashi Yoda, Kazuyoshi Ueno, 2009; ISSN 1048-0854; ISBN 978-1-60511-125-4

Advanced Metallization Conference 2009 (AMC 2009), Dan Edelstein, Stefan E. Schulz, 2009; ISSN 1048-0854; ISBN 978-1-60511-218-3

Metallization

A study on Resistivity Increase of Copper Interconnects with the Dimension Comparable to Electron Mean Free Path Utilizing Monte Carlo Simulation

Makoto Wada, Takashi Kurusu, Yosuke Akimoto, Noriaki Matsunaga, Hiroyoshi Tanimoto, Nobutoshi Aoki, Yoshiaki Toyoshima and Hideki Shibata
Center for Semiconductor Research and Development
Semiconductor Company, Toshiba Corporation
8, Shinsugita, Isogo-ku, Yokohama 235-8522, Japan

ABSTRACT

The size effect on the resistivity increase in Cu interconnects has recently become pronounced in the dimension where line width and film thickness approaches the electron mean free path. Understanding of physical mechanisms and an accurate prediction of the size effect on resistivity are required. We investigated the behavior of resistivity increase in Cu interconnects utilizing Monte Carlo simulation. In the dimension where line width and film thickness becomes a dimension comparable to the electron mean free path, the multiple interface scattering occurred around the corners of interconnects and it became a dominant factor for the resistivity increase. Rounding off the corners of interconnects and enhancing the average grain size are effective methods to suppress the resistivity increase. Our Monte Carlo model permits more accurate prediction of the size effect on the resistivity than the theoretical equation does.

INTRODUCTION

The scale of copper interconnects is continually shrinking toward dimensions comparable with the electron mean free path and advancement of the technology node requires further reduction of line width and film thickness. Because of the importance of the size effect on resistivity increase, several researchers investigated the dependence of Cu resistivity on line width and on film thickness. [1]-[6] Steinhögl *et al* theoretically reported the integration model and their model is globally used. [3] [4] However, in our previous work, we observed that the measured resistivity was larger than the value predicted from their theoretical equation in the dimension where line width and film thickness approaches the electron mean free path. [6] We suggested this difference as a possible cause of the electron scattering at the corners of interconnects. Understanding of physical mechanisms and an accurate prediction of the size effect on resistivity are required. In this study, we investigated the behavior of electron scattering in Cu interconnects whose dimensions are comparable to the electron mean free path by using Monte Carlo simulation. In addition, we proposed the methods to suppress the resistivity increase in terms of interface structure and grain boundary structure. (*The electron mean free path is 41.6nm in Cu.)

EXPERIMENTAL

Samples Cu interconnects with line widths between 50 and 500nm have been fabricated using the single damascene process. Ti was used as a barrier metal for Cu interconnects. The samples were annealed at 250 degrees C to stabilize the Cu microstructure. After annealing, film thickness was controlled between 33 and 91nm by CMP. Cu Resistivity was measured at room temperature and calculated from the actual Cu volume after deduction of barrier metal thickness.

Monte Carlo Simulation The Monte Carlo model was constructed based mainly on the work of Kurusu *et al.*[7] Our model considered interface scattering, grain boundary scattering and bulk scattering as shown in Figure 1. Our model contains specularity parameter (*P*) at interface and reflectivity parameter (*R*) at grain boundary as well as the Steinhögl model. We assumed that only electrons near the Fermi surface contribute to the electric current. The conductive electrons move around inside the Cu interconnects with about the Fermi velocity. When electrons in drift motion are scattered, they lose their momentum and reach an equilibrium state. Then electrons are reaccelerated by an external electric field. At any given line width and film thickness, we calculate an average electron drift velocity along the current direction by Monte Carlo simulation. We could evaluate the resistivity form the drift velocity by using Drude's law. [8] Kurusu *et al* presented the details of the Monte Carlo model at SISPAD 2009. [7]

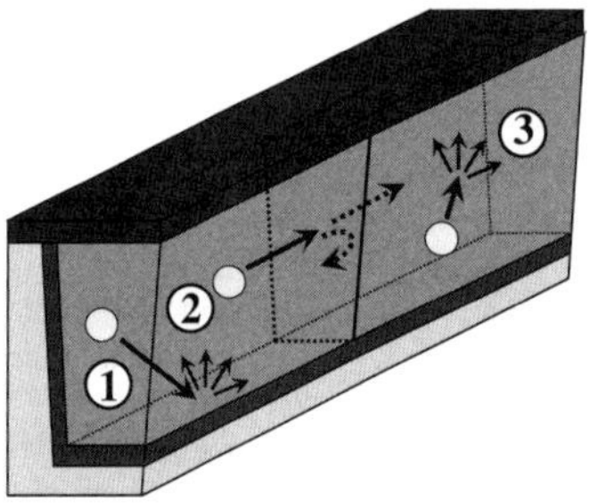

Figure 1. Schematic illustration of the electron scattering in proposed Monte Carlo model.
1. Interface scattering
2. Grain boundary scattering
3. Bulk scattering

RESULTS and DISCUSSION

Behavior of the electron scattering Figure 2 shows the resistivity dependence on line width (a) and on film thickness (b). The same calculation condition was used for the theoretical model and the Monte Carlo model with the parameters P=0 (inelastic scattering), R=0.18 and grain sizes equal to the line width. In terms of both dependences, the Monte Carlo results show better agreement with the experimental data than with the theoretical value. Our Monte Carlo model makes more accurate prediction of the size effect on the resistivity.

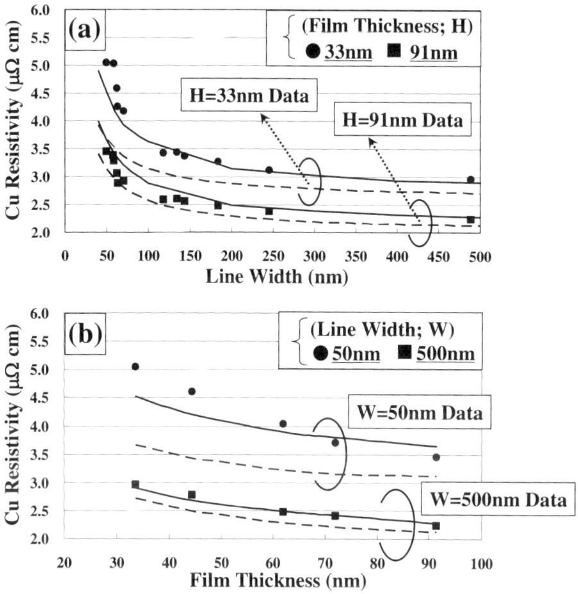

Figure 2. Resistivity data of the Cu interconnects. Marks represent experimental results, solid lines represent Monte Carlo results and dotted lines represent theoretical value.
(a) Resistivity dependence on line width.
(b) Resistivity dependence on film thickness.

In order to understand the differences between the Monte Carlo results and the theoretical model, we focus on the resistivity dependence on film thickness. In this study, Cu grain size did not vary depending on film thickness because the film thickness was controlled by CMP. Therefore, grain boundary resistivity was constant for film thickness and the difference between the Monte Carlo results and the theoretical value was attributed to the interface scattering as shown in Figure 3.

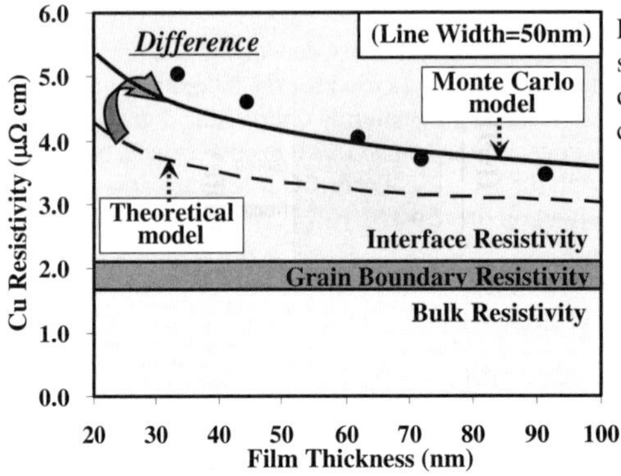

Figure 3. Influence of each electron scattering for resistivity dependence on film thickness. Line width is 50nm constant with film thickness.

Figure 4 shows the electron velocity distributions in the cross-sectional view of interconnects with the film thickness of 33nm (a) and 91nm (b). (Both line widths were the same, 50nm). In both cases, electrons had high drift velocity near the center of interconnects and velocity became slower toward the circumference because of the interface scattering. In particular, large degradation of drift velocity was observed at the corners of interconnects. Figure 5 shows the electron trajectories of motion per picosecond. Inflections of the solid lines indicate the electron scattering. In the case of film thickness of 33nm, scatterings were concentrated at the corners of interconnects and electrons underwent multiple scattering. Multiple scattering decreases both the effective electron mean free path and the electron drift velocity. As a result, further resistivity increase was caused. The theoretical model considered electron scattering as a planar scattering in the width direction and in the thickness direction separately. Therefore the theoretical model underestimates the resistivity equivalent to the corner scattering. In the dimension where line width and film thickness become a dimension comparable to the electron mean free path, corner scattering became a dominant factor for the resistivity increase.

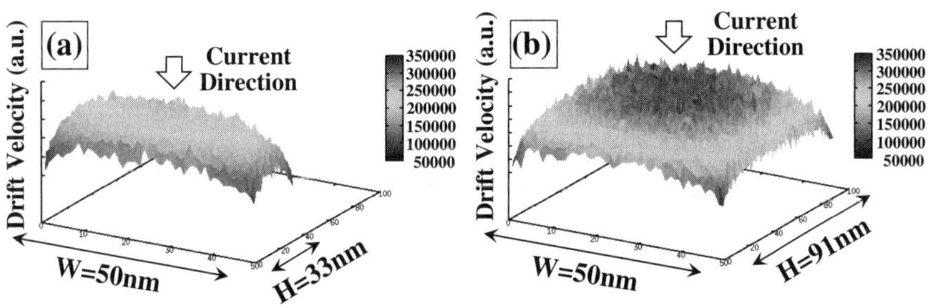

Figure 4. Electron drift velocity distributions in the cross-sectional view of interconnects.
(a) Film thickness is 33nm. (b) Film thickness is 91nm.
Both line widths were the same, 50nm.

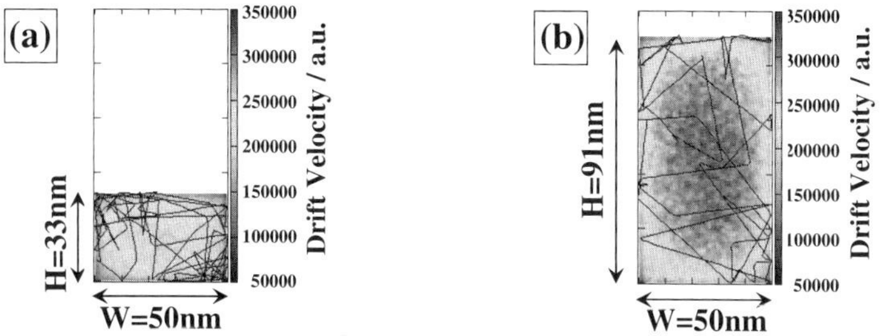

Figure 5. Electron trajectories of motion per picosecond.
Inflections of the solid lines at the circumference represent the interface scattering and the other inflections represent the bulk scattering.
(a) Film thickness is 33nm (b) Film thickness is 91nm
Both line widths were the same, 50nm.

Methods to suppress the resistivity increase Figure 6 shows the resistivity comparison between the rectangular interconnects and the circular interconnects with Monte Carlo simulation. To highlight the influence of interface scattering, bulk scattering and grain boundary scattering were excluded from the calculation. We calculated the resistivity under the assumption that the line width is equal to the film thickness (Aspect Ratio=1). The circular interconnects could keep resistivity lower than the rectangular ones in which the dimensions were below the electron mean free path. This is because the frequency of multiple scattering was reduced by rounding off the corners of interconnects. Figure 7 shows the electron velocity distributions in the 40nm dimension interconnects. It was found that, compared with the rectangular interconnects, the circular ones had larger drift velocity not only at the circumference but also at the center.

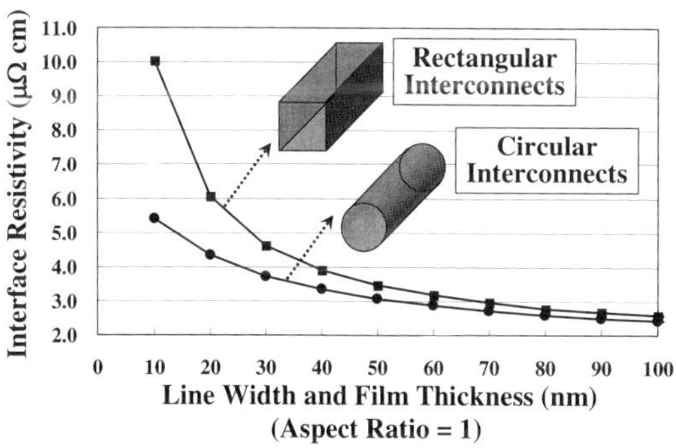

Figure 6. Resistivity comparison between the circular interconnects and the rectangular interconnects with Monte Carlo simulation.
We calculated the resistivity under the assumption that the line width is equal to the film thickness (Aspect Ratio=1).

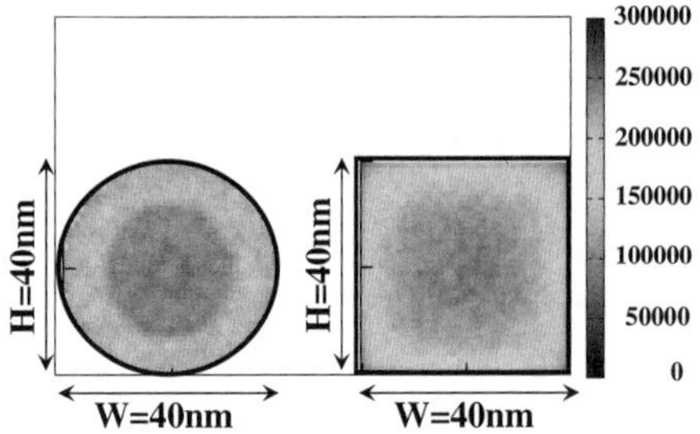

Figure 7. Comparison of the electron drift velocity distributions in the 40nm dimension interconnects between the circular interconnects and the rectangular interconnects.

Figure 8 shows the electrical field distributions in the 40nm interconnects. While the electrical flux lines were concentrated at the corners of rectangular interconnects, those became decentralized in the circular interconnects and the wiring capacitance could be reduced about 15% from the rectangular one. Rounding off the corners is effective for both resistivity reduction and capacitance reduction in the dimensions below the electron mean free path.

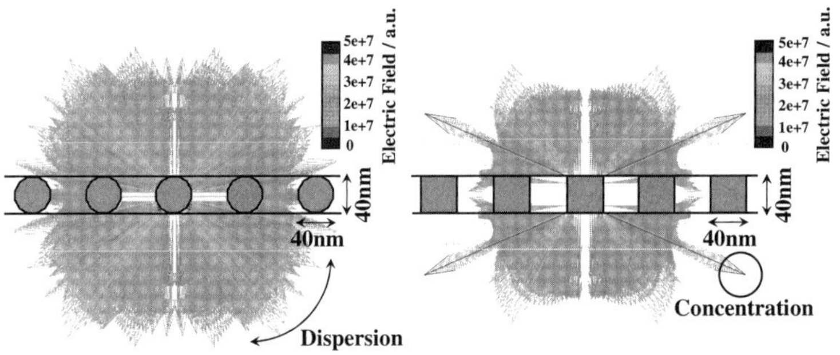

Figure 8. Comparison of the electrical field distributions in the 40nm dimension interconnects between the circular interconnects and the rectangular interconnects. Current was applied only to the central wire.

In the aforementioned discussions, we assumed grain sizes equal to the line width.[4] However, there are actually grain size distributions in Cu microstructure. The grain boundary distances were extracted along the length direction as shown in Figure 9 and the results were presented in Figure 10. Grain boundary distances followed the log-normal distribution and varied considerably in length.

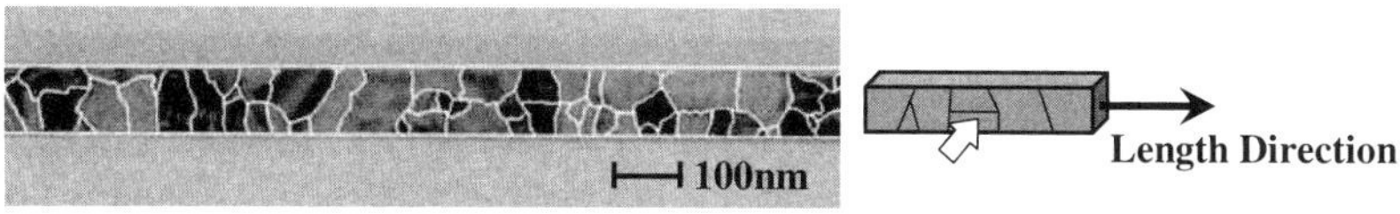

Figure 9. Cross-sectional TEM view image of Cu interconnects along the length direction. The grain boundaries are marked with white lines. The grain boundary distances were extracted along the length direction with respect to every 10nm height. 4.6μm length was measured.

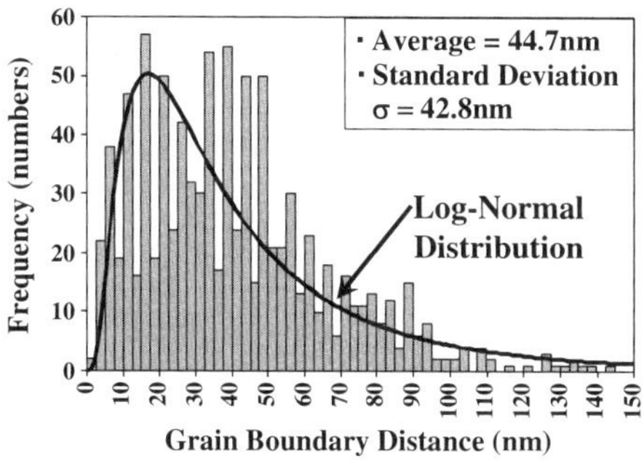

Figure 10. Distribution of grain boundary distances in Cu interconnects along the length direction. Grain boundary distances followed the log-normal distribution with the 44.7nm average grain size and the 42.8nm standard distribution.

Figure 11 shows the resistivity comparison between log-normal and Gaussian cases on which the grain sizes are distributed. While the actual grain size dispersions were represented in log-normal distribution, the less grain size dispersion was assumed in Gaussian distribution as the standard deviation was one-tenth of the average grain size. The average grain size was 44.7nm in both cases. The two resistivities were almost the same, which indicated that the grain size dispersions had little influence on the resistivity increase. In order to suppress the resistivity increase, enhancing the average grain size is effective for the resistivity reduction rather than the control of the grain size dispersion.

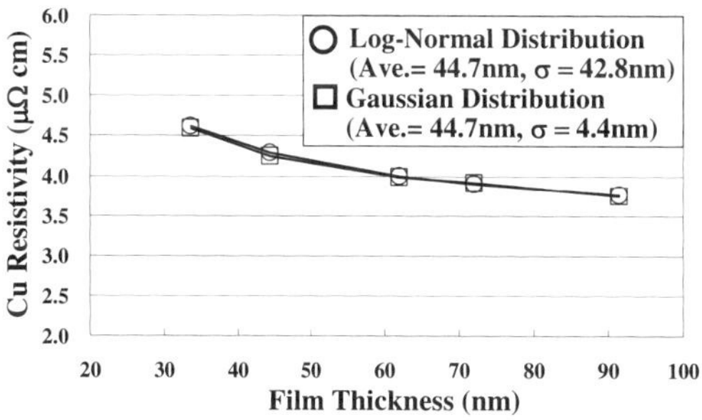

Figure 11. Resistivity comparison between log-normal distribution and Gaussian distribution.

CONCLUSIONS

The behavior of resistivity increase in Cu interconnects was investigated by using the Monte Carlo simulation. The Monte Carlo model permits more accurate prediction of the size effect on the resistivity than the theoretical equation does. In the dimension where line width and film thickness becomes a dimension comparable to the electron mean free path, the multiple interface scattering occurred around the corners of interconnects and it became a dominant factor for the resistivity increase. In terms of the interface scattering, rounding off the corners is effective for both resistivity reduction and capacitance reduction. In terms of the grain boundary scattering, enhancing the average grain size is more effective for the resistivity reduction than the control of the grain size dispersion is.

REFERENCES

1. K. Fuchs, Proc. Cambridge Philos. Soc. **34**, 100 (1938)
2. A. F. Mayadas and M. Shatzkes, Phys. Rev. B **1**, 1382 (1970)
3. W. Steinhogl et al., Proc. of IEEE Semiconductor Process and Devices 2003 (SISPAD 2003), P. 27.
4. W. Steinhogl et al., J. Appl. Phys. **97**, 023706 (2005)
5. T. S. Kuan et al., Proc. of Mat. Res. Soc. 612, P. D7.1.1(2000)
6. M. Wada *et al.,* in Advanced Metallization Conference 2007 (AMC 2007) , edited by A. J. McKerrow, Y. S. Diamand, S. Shingubara and Y. Shimogaki, (Mater. Res. Soc. Symp. Proc. Albany, NY, 2007) P. 245.
7. T. Kurusu *et al.,* Proc. of IEEE Simulation of Semiconductor Process and Devices 2009 (SISPAD 2009), P. 170.
8. N. W. Ashcroft and N. D. Mermin, *Solid State Physics*, (Harcourt College Publishers, 1976) P. 2.

Conference Proceedings AMC XXV © 2010 Materials Research Society

Quantitative analysis of the impact of surface and grain boundary scattering on the resistivity of nanometric Cu films

Tik Sun[1], Bo Yao[1], Andrew P. Warren[1], Katayun Barmak[2], Michael F. Toney[3], Robert E. Peale[4], and Kevin R. Coffey[1,4]
[1]Advanced Materials Processing and Analysis Center, University of Central Florida, 4000 Central Florida Boulevard, Orlando FL 32816
[2]Department of Materials Science and Engineering, Carnegie Mellon University, 5000 Forbes Avenue, Pittsburgh, PA 15213
[3]Stanford Synchrotron Radiation Lightsource, 2575 Sand Hill Rd., Menlo Park, CA 94025
[4]Department of Physics, University of Central Florida, 4000 Central Florida Boulevard, Orlando, FL 32816

ABSTRACT

We report a quantitative analysis of surface and grain boundary scattering in encapsulated Cu thin films where film thickness (27 nm to 158 nm) and grain size (35 nm to 425 nm) were varied so that a series of grain sizes at given film thicknesses were obtained. Two types of samples, namely, $SiO_2/Cu/SiO_2$ and $SiO_2/Ta/Cu/Ta/SiO_2$ were prepared by sub-ambient temperature deposition followed by annealing. Average grain size values were determined from populations of 400 to 1,500 grains per sample using hollow cone dark field transmission electron microscopy imaging. Film thickness and roughnesses of the upper and lower Cu surfaces were determined by specular X-ray scattering. Film resistivities were measured at room temperature and at 4.2 K and were compared with several scattering models in order to quantify the contributions of surface and grain boundary scattering. Grain boundary scattering is found to provide the most significant contribution. However, a comparison of the high and low temperature data indicates that a weak surface scattering contribution to the size effect is likely present. The implications of our results to the future scaling of Cu interconnect resistivity will be discussed.

INTRODUCTION

The classical resistivity size effect, wherein conductors with dimensions on the order of the mean free path of electrons at room temperature (39 nm for Cu) exhibit higher resistivity than bulk conductors, was noted by Thompson[1] in 1901. The phenomenon was first described by the Fuchs and Sondheimer (FS)[2,3] theory of surface scattering. This theory, which was derived from the Boltzmann transport equations, described scattering of conduction electrons from the film surfaces, each such event being defined as either diffuse or specular. In a specular scattering event, a conduction electron incident upon the surface is reflected from the surface with no change in energy or momentum components parallel to the plane of the film. In a diffuse scattering event, the scattering direction is assumed fully random (i.e. isotropic) and momentum in the plane is not conserved. Diffuse surface scattering results in an increase in resistivity in films and lines, $\Delta\rho_{FS}$, given for films by:

$$\Delta\rho_{FS} = \rho_i - \rho_i\left[1 - \left(\frac{3}{2k}\right)(1-p)\int_1^\infty\left(\frac{1}{t^3} - \frac{1}{t^5}\right)\frac{1-\exp(-kt)}{1-p\exp(-kt)}dt\right]^{-1}, \quad (1a)$$

where, $k = h/\lambda$, h is the film thickness, λ is the electron mean free path due to phonon and impurity scattering, and ρ_i is the bulk resistivity of the metal. In the limit of small k, equation

(1a) reduces to the more commonly used equation:

$$\Delta\rho_{FS} = \rho_i\left(\frac{3}{8}\right)\frac{\lambda}{h}(1-p). \qquad (1b)$$

More recently, Rossnagel and Kuan (RK)[4] proposed a semi-empirical extension of the FS model to include surface roughness explicitly.

In 1970, Mayadas and Shatzkes[5] (MS) proposed an alternative resistivity model, which treats grain boundary scattering as the primary mechanism for the resistivity increase with decreasing conductor thickness. Their model assumes a completely columnar grain structure, with all grain boundaries perpendicular to the film. The grain boundaries perpendicular to the film are in turn taken to be either parallel or perpendicular to the current flow. The boundaries parallel to the current flow give rise to specular scattering of the incident conduction electrons, whereas the boundaries perpendicular to the current flow result in either electron reflection with a probability *R*, or transmission with a probability *1-R*. The increase in resistivity given by this model is:

$$\Delta\rho_{MS} = \rho_i - \rho_i\left[1-\frac{3}{2}\alpha+3\alpha^2-3\alpha^3\ln\left(1+\frac{1}{\alpha}\right)\right]^{-1}, \qquad (2a)$$

where $\alpha = \left(\lambda/g\right)R/(1-R)$ and *g* is the grain size. In the limit of small α, equation (2a) reduces to the more commonly used expression: $\Delta\rho_{MS} = \rho_i\left(\frac{3}{2}\right)\left(\frac{\lambda}{g}\right)\left(\frac{R}{1-R}\right)$. (2b)

The attempts to verify the above (or alternative) models of film resistivity have been impacted by the coupling of film thickness and grain size. Surface scattering and grain boundary scattering share a similar geometric dependence, namely that they are inversely proportional to film thickness (*h*) and grain size (*g*), respectively (compare equations 1b and 2b), and that the grain size in a series of similarly processed thin film samples is experimentally observed to scale in proportion to film thickness ($h \propto g$). Hence, these two parameters typically do not have independent variation in polycrystalline thin films[5]. The presence of roughness at the surface may further complicate the identification of surface and

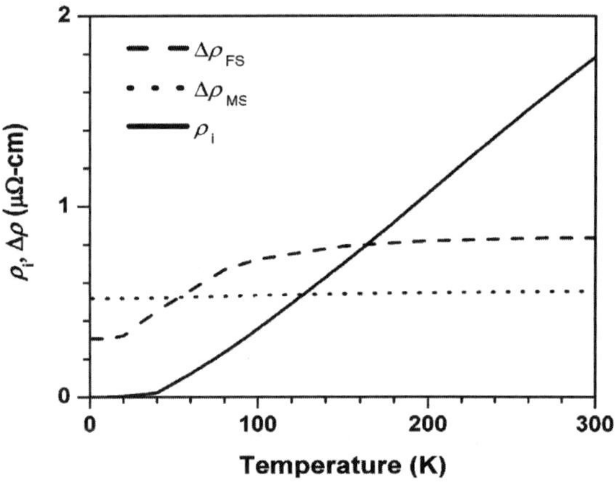

Figure 1 A comparison of the temperature dependence of $\Delta\rho_{FS}$ and $\Delta\rho_{MS}$ (Eqs. 1 and 2) for a hypothetical polycrystalline Cu thin film having a 30 nm grain size and a 30 nm thickness. The temperature dependence of the bulk resistivity of Cu, ρ_i, is also shown.

grain boundary effects. A number of experimental approaches have been used to tackle this problem. One such approach is to measure the resistivity of the polycrystalline thin film samples over a range of temperatures[6,7,8,9,10]. In figure 1, the additional resistivity due to FS model ($\Delta\rho_{FS}$), the additional resistivity due to MS model ($\Delta\rho_{MS}$), and are plotted as a function of temperature for a hypothetical film sample having a 30 nm grain size and a 30 nm thickness, to illustrate their different temperature dependencies. The bulk resistivity of Cu (ρ_i) is also shown in the figure. As can be seen, $\Delta\rho_{FS}$ exhibits a stronger dependence on temperature than does $\Delta\rho_{MS}$ as a result of the strong interaction between the surface scattering and phonon scattering. The increased phonon scattering with increasing temperature allows the fraction of electrons which would have been travelling parallel to the external surfaces to be more frequently redirected towards the surfaces. For grain boundary scattering, the electrons cannot avoid impinging on the boundaries, and, therefore phonon scattering, and thus temperature, has a smaller effect on the resistivity.

In this work, we present both room (296 K, the typical temperature in the lab) and low temperature (4.2 K) resistivity data of a set of samples having substantially independent variations in thickness and grain size. The film thickness ranges from 27 nm to 158 nm and grain size ranges from 35 nm to 425 nm. We also quantify the roughness of the upper and lower Cu film surfaces, and we study two types of scattering interfaces, both SiO_2/Cu/SiO_2 and SiO_2/Ta/Cu/Ta/SiO_2. We compare our data to the well established models for the classical resistivity size effect described above. We show that the MS model (based only on grain boundary scattering) fits our experimental data better than the FS model. We also show that a combination of the FS and MS models using Matthiessen's rule provides a further improved fit.

(a) -2 degree tilt **(b) 0 degree tilt** **(c) 2 degree tilt**

Figure 2 Hollow cone dark field (HCDF) transmission electron micrographs of a 33.6 nm-thick Cu film encapsulated in 20 nm SiO_2 and annealed at 600 °C, imaged at 3 different sample tilts of the same field of view. The three tilts allow the grain boundaries to be identified for hand tracing. The hand-traced boundary network is used for automated measurement of grain size.

EXPERIMENT

SiO_2 encapsulated Cu thin films, with and without Ta barriers, were prepared on Si (100) substrates having a 150 nm-thick layer of thermally grown SiO_2. The processing conditions that were developed aimed for minimal void formation during the post-deposition annealing step that was used obtain a range of grain sizes (greater than film thickness) at a given film thickness. Prior to film deposition, the substrates were RF sputter-cleaned and cooled to -40°C by contact with a liquid nitrogen cooled Cu plate. An underlayer of 20 nm of SiO_2 was RF sputter deposited prior to the Cu film deposition and a 20 nm SiO_2 overlayer was subsequently sputter deposited to form a SiO_2/Cu/SiO_2 structure. The Cu layers were deposited by DC sputter deposition from high purity (99.9999%) Cu targets and had thicknesses in the range of 27 to 158 nm. For films having the SiO_2/Ta/Cu/Ta/SiO_2 structure, a 2 nm Ta layer was DC sputter deposited immediately prior to, and again after, the Cu layer deposition. Samples were annealed at 150 °C and 600 °C for 30 minutes in a tube furnace, and at 400°C for 6 seconds by halogen lamp heating (rapid thermal annealing), all under a reducing gas environment of Ar+3%H_2. The sheet resistance of the samples at 4.2 K was measured by dipping a Van der Pauw geometry four point probe[11] into liquid helium while using a Keithley 2400 Source meter and 2182 nanovoltmeter for data collection. The void area fraction of the samples was quantified using high angle annular dark field (HAADF) imaging in scanning transmission electron microscopy (TEM) mode at relatively low magnifications. The void fractions were found to be between 0 and 2.4%. While 2.4% is unacceptable in the context of a semiconductor interconnect process the resistivity error of this void area fraction is negligibly small. For the grain size measurement, the samples were examined by hollow cone dark field (HCDF) imaging in TEM mode to provide the highest diffraction contrast, as shown in figure 2. By imaging the given field of view at three different sample tilts and thereby varying the diffraction contrast, it was possible to visually identify the grains. Twin boundaries within grains were excluded in the traced boundary network since this type of grain boundary has been reported to have a minimal contribution to sample resistivity[12]. The reported average grain size for each sample is the diameter of the equivalent circle with area equal to the average of the grain areas. The errors on this mean are quoted as 2σ values at a 95% confidence level for the given grain population[13].

To quantify the roughness of the buried Cu/SiO_2 and Cu/Ta interfaces X-ray reflectivity experiments were performed at the Stanford Synchrotron Radiation Lightsource (SSRL). The data are given elsewhere [14]. The Cu film thickness and the roughnesses of the upper and lower Cu/encapsulant layer interfaces were determined by fitting of the reflectivity data to established models[15,16]. These data were collected on the thin film diffraction beam line 2-1. This beam line is equipped with a Huber 2-circle goniometer, a pair of 1 mm slits as the analyzer, and a He filled sample stage was used to decrease the air scattering background. The 1.549Å wavelength X-rays were monochromated with double bounce Si(111) crystals. Two types of

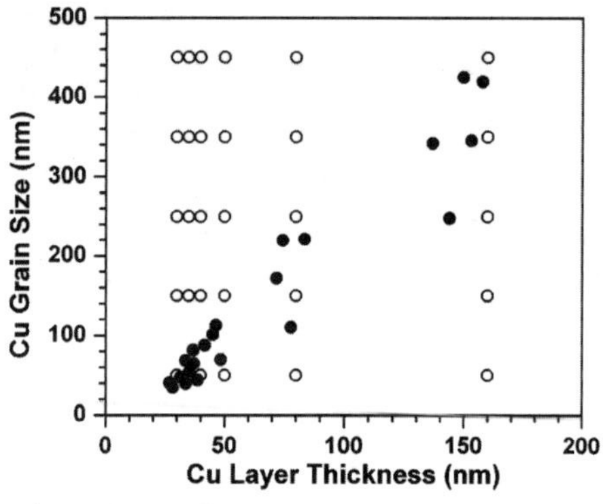

Figure 3 Plot of the average grain sizes of the Cu thin films (filled circles) as a function of Cu layer thickness, showing ~2× variation in grain size for a given Cu layer thickness. Open circles represent an ideal sample set with fully independent variation of grain size and thickness.

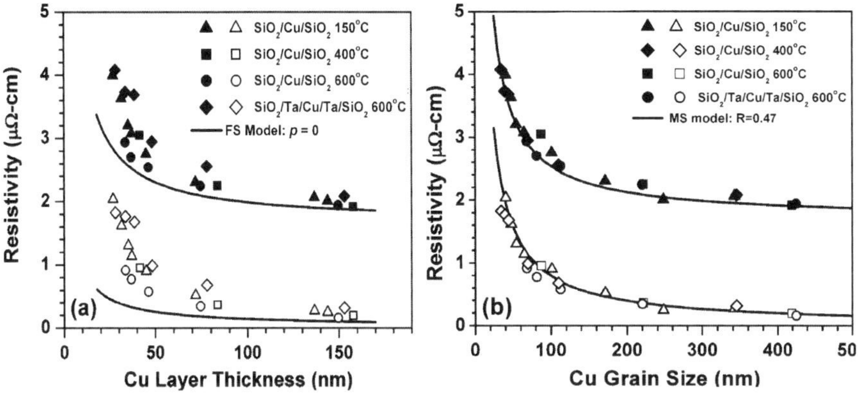

Figure 4 The total resistivity of SiO_2/Cu/SiO_2 and SiO_2/Ta/Cu/Ta/SiO_2 thin films a) as a function of Cu layer thickness, *h*, and b) as a function of Cu layer grain size, *g*. The filled symbols are resistivities measured at room temperature, and the open symbols are resistivities measured at 4.2 K. The solid curve corresponds to the FS model in a) and MS model in b).

scans were performed: specular, where $\omega = 2\theta/2 = \theta$, and off-specular where $\omega = 2\theta/2 \pm 0.15°$. The off-specular scans were used to subtract the contribution of diffusely scattered X-rays to the specular reflection resulting in a purely specular reflectivity pattern. Data were collected from $2\theta = 0.2°$ to 12°, with a step size of 0.02°, 0.01°, or 0.005° depending on film thickness.

DISCUSSION

In order to separate the effect of grain boundary scattering from that of surface scattering, the grain size of the Cu samples should ideally be varied fully independently of film thickness. As can be seen in figure 3, we have produced a wide range of grain sizes for films at each thickness (from ~ 1.2×thickness to ~ 3.1×thickness), although a fully independent variation was not achieved. Nevertheless, with the careful quantification of grain size exercised in this work (error less than ~7 %), this degree of independent variation is sufficient for the separate quantification of the grain boundary and surface scattering effects on resistivity.

In figure 4a, the resistivities of the SiO_2/Cu/SiO_2 and SiO_2/Ta/Cu/Ta/SiO_2 samples are plotted as a function of thickness at both room temperature (RT), typically 296 K in our laboratory, and at 4.2 K. The FS model (using equation 1a) is shown by the lines in the figure and fails to describe the experiment even with the most extreme value for the specularity coefficient ($p = 0$). Within the SiO_2/Cu/SiO_2 samples, it can be seen that for samples of similar thickness the resistivity decreases with increasing annealing temperature. Further, the Cu samples with Ta inter-layers are found to have consistently higher resistivities than those without Ta. The presence of samples with different resistivities at each thickness indicates that surface scattering alone can not explain the resistivity size effect in these Cu thin films. Figure 4b is a plot of the room and low temperature resistivity as a function of grain size. It is immediately evident from the figure that the additional variations in resistivity associated with the annealing temperature or the presence of Ta are no longer present, but are instead accounted for by sample grain size. Figure 4b compares both the room temperature and low temperature experimental data with the predictions of the MS model of equation 2a with a reflection coefficient of $R = 0.47$ and no other variable parameters. The MS model fits the data well. However, it can be observed that experimental resistivities at room temperature tend to be generally higher than the MS model prediction while the resistivities at 4.2 K are lower, i.e., the temperature dependence of the resistivity is not well described. This may reflect limitations inherent in the MS model (e.g., assumption of purely specular scattering from parallel grain boundaries or the use of a temperature independent reflection coefficient) or this may indicate the presence of a surface scattering contribution to the resistivity, such as that given by the FS model, which has significant temperature dependence (figure 1).

To assess the limits of our experimental data in understanding the resistivity size effect, the data at both temperatures are compared to the models previously described above. The sum of the residual squared error (SSE), $\sum(\rho_{exp} - \rho_{model})^2$, where differences between the calculated resistivity from a model is subtracted from the experimentally measured resistivity and squared, and where the summation is then taken over all of the forty four experimental resistivity values in figure 4, is used as a measure of the model's ability to describe the experiment. Table 1 lists the parameters used in fitting the models to the experimental data, the optimum values of those parameters found within the physical limits allowed by each model, e.g., $0<p<1$, and the sum of the squared residual error for each these models: the FS model, the MS model, the FS+MS model, the FS+MS model with "Caps" (i.e., FS+MS model with different specularity coefficients for Cu/SiO_2 and Cu/Ta/SiO_2 interfaces), and the RK+MS model. In the fitting of FS+MS models, two different assumptions regarding the surface scattering were considered. In one case, identical scattering from the Cu/SiO_2 and the Cu/Ta interfaces was presumed and a single specularity parameter was used for the combined data. In the second case, two separate specularity parameters were used for the Cu/SiO_2 and for the Cu/Ta interfaces and this model variation is identified as "Caps." At room temperature, a fixed bulk resistivity value of 1.7 μΩ–cm and an electron mean free path of 39 nm were used for all models. At 4.2 K, a fixed bulk resistivity of 0.002 μΩ–cm and an electron mean free path of 33 μm were used[9]. To further compare models having a similar minimum residual squared error, the F-test[17] (confidence level of 95%) was used to determine if the apparent improvement in fit was significant.

Table 1 Summary of classical size effect models, the sum of the residual squared errors (SSE) for all forty four data points of figure 4, and optimized model parameters.

Model Name	**SSE ($\mu\Omega^2$-cm^2)**	**Model Parameters**
FS[2,3]	22.9	$p = 0$
MS[5]	0.96	$R = 0.47$
FS+MS[14]	0.48	$p = 0.52 \quad R = 0.43$
FS+MS (Caps)[14]	0.48	$P_{SiO_2} = 0.51 \quad P_{Ta} = 0.55 \quad R = 0.43$
RK+MS[4]	0.43	$p = 0.68 \quad R = 0.43 \quad n = 18.2$

The large summed squared error of 22.9 $\mu\Omega^2$-cm² observed with the FS model when fitting our experimental data indicates that it clearly fails to describe the experiment, even with the most extreme value allowed for the specularity coefficient ($p = 0$). As is also seen in figure 4, the MS grain boundary scattering model (figure 4b) is a dramatic improvement over the FS surface scattering model (figure 4a). However, a considerable summed squared error of 0.96 $\mu\Omega^2$-cm² remains due, in part, to the inability of the MS model to account for the experimental drop in $\Delta\rho$ with temperature. The FS+MS model (using a simple Matthiessen's rule addition of surface and grain boundary scattering effects)[14] with a summed squared error of 0.48 $\mu\Omega^2$-cm² provides a statistically significant improved fit over the MS model. Based on this improved fit, we conclude that thickness dependence is present in the experimental data. Table 1 gives the grain boundary scattering reflection coefficient of 0.43 and surface scattering specularity coefficient of 0.52 for the combined FS+MS model. This intermediate value for the specularity coefficient is in contrast to the many previous works, wherein a surface specularity coefficient of zero was assumed, which simply maximized the surface scattering contribution to the resistivity increase[7,9].

To further explore the role of the interface properties on surface scattering, the FS+MS model was extended to include two specularity coefficients for the two interfaces we used (p_{SiO_2} and p_{Ta}, for Cu/SiO_2 and Cu/Ta interfaces, respectively). Surprisingly, this did not improve the residual summed squared error, nor did it change the grain boundary reflection coefficient from 0.43 $\mu\Omega^2$-cm². This confirms that, in spite of the higher resistivities observed for the SiO_2/Ta/Cu/Ta/SiO_2 samples, little of the resistivity increase can be attributed to differences in the surface scattering of conduction electrons at the Cu/Ta and Cu/SiO_2 interfaces. Rather the resistivity increase is a simple consequence of the smaller grain size present in the SiO_2/Ta/Cu/Ta/SiO_2 samples.

The role of interfacial roughness was further explored by comparing our data to the RK+MS model (a simple Matthiessen's rule addition of the MS and RK models[4]) which reduces the summed squared residual error from that of the FS+MS (0.48 $\mu\Omega^2$-cm²) to 0.43 $\mu\Omega^2$-cm². While this reduction in error justifies additional efforts to relate surface roughness to resistivity, it is worth noting that the improvement observed was not statistically significant.

CONCLUSIONS

In summary, we carefully studied the contributions of surface scattering and grain boundary scattering to the resistivity increase observed with reduction of conductor thickness in polycrystalline Cu thin films through quantitative measurement of the primary experimental variables and comparison of the data to accepted models. We found that grain boundary scattering was the dominant scattering mechanism and provided an unambiguous experimental verification of the grain boundary scattering model of Mayadas and Shatzkes[5]. We were additionally able to provide a measurement of the surface scattering for our samples and found it to be well described by the Fuchs-Sondheimer model with an intermediate value, $p = 0.52$, for the surface specularity coefficient. This value is in contrast to prior reports where fully diffuse scattering surfaces (coefficient of zero) were claimed. The surface scattering from Cu/SiO_2 and Cu/Ta interfaces was found to be indistinguishable. Our data indicates that grain size will have an important role in scaling Cu interconnects to future products.

ACKNOWLEDGMENTS

We thank A. D. Darbal, V. Kumar, and S. Roberts for their assistance with the grain size data, D. C. Berry for rapid thermal annealing, C. Fredricksen for his assistance with the low temperature measurements, and our colleagues for many fruitful discussions. We gratefully acknowledge the financial support of the Semiconductor Research Corporation, Task 1292.008, and partial support from the MRSEC program of the NSF under DMR-0520425. Portions of this research were carried out at the Stanford Synchrotron Radiation Lightsource (SSRL), a national user facility operated by Stanford University on behalf of the US Department of Energy, Office of Basic Energy Sciences.

REFERENCES

1 J. J. Thomson, Proc. Cambridge Philos. Soc. **11**, 120 (1901).
2 K. Fuchs, Proc. Cambridge Philos. Soc. **34**, 100 (1938).
3 E. H. Sondheimer, Adv. Phys. **1**. 1, (1952).
4 S. M. Rossnagel and T. S. Kuan, J. Vac. Sci. Technol. B **22,** 240 (2004).
5 A. F. Mayadas and M. Shatzkes, Phys Rev. B **1**, 1382 (1970).
6 W. Steinhögl, G. Schindler, G. Steinlesberger, and M. Engelhardt, Phys. Rev. B **66,** 075414, (2002).
7 H. Marom, J. Mullin, and M. Eizenberg, Phys. Rev. B **74**, 045411 (2006).
8 W. Wu, S. H. Brongersma, M.Van Hove, and K. Maex, Appl. Phys. Lett. **84,** 2838 (2004).
9 W. Zhang, S.H. Brongersma, Z. Li, D. Li, O. Richard, and K. Maex, J. Appl. Phys. **101,** 063703 (2007)
10 P.M. T. M. van Attekum, P. H. Woerlee, G. C. Verkade, and A. A. M. Hoeben, Phys. Rev. B **29**, 645 (1984)
11 L. J. Van der Pauw, Philips Res. Rep. **13,** 1 (1958).
12 L. Lu, Y. Shen, X. Chen, L. Qian, and K. Lu, Science **304**, 422 (2004).
13 D. T. Carpenter, J. M. Rickman, and K. Barmak, J. Appl. Phys. **84**, 5843 (1998).
14 T. Sun, B. Yao, A. P. Warren, K. Barmak, M. F. Toney, R. E. Peale, and K. R. Coffey, Phys. Rev. B **79**, 041402(R) (2009).
15 M. Worminton, C. Panaccione, K. M. Matney, and D. K. Bowen, Phil. Tran. R. Soc. London A **357**, 2827 (1999).
16 S. K. Sinha, E. B. Sirota, S. Garoff, and H. B. Stanley, Phys. Rev. B **38**, 2297 (1988).
17 G.Casella and R.L. Berger, Statistical Inference (Duxbury, Pacific Grove, 2002), p. 534.

Conference Proceedings AMC XXV © 2010 Materials Research Society

Origins of Large Variation in Ductility of Thin Metal films on Polymer Substrates

Teng Li[1, 2, *], Benoit Michaux[1], Zhao Zhang[1]

[1]Department of Mechanical Engineering, University of Maryland, College Park, MD 20742
[2]Maryland NanoCenter, University of Maryland, College Park, MD 20742
[*]Corresponding author. Tel: 301-405-0364, Fax: 301-314-9477, Email: LiT@umd.edu

ABSTRACT

Recent developments of flexible electronics require that thin metal films grown on polymer substrates be deformable. It has been reported that, when a laminate of a thin metal film (~100 nm thick) on a polymer substrate (10s~100s microns thick) is stretched, the metal film may rupture at strains ranging from 1~2 percent up to 50 percent. The mechanistic understanding of this large variation in the ductility of thin metal films on polymer substrates has been uncertain. To reveal the underpinning origins for the large variation in ductility, we report a systematic study of the governing mechanisms by considering two completing failure processes: debonding-assisted necking and grain boundary cracking of the thin metal films on polymer substrates.

INTRODUCTION

Rising interests in flexible electronics in recent years have been motivated by the promising future of its potential applications including paper-like displays, printable thin-film solar cells, and skin-like smart prostheses [1-3]. Such flexible devices will undergo large and cyclic stretches, bends and twists during service, therefore requires large deformability of the device components (e.g., metal interconnects). Flexible electronic devices are often made as organic/inorganic hybrids. For example, polymers are often used as the substrate material, replacing the rigid substrates (e.g., silicon) in traditional microelectronic devices. While polymers are compliant and can sustain large deformation, thin films of most electronic materials (e.g., metals, semiconductors, and dielectrics) are brittle and thus fracture at small strains (less than a few per cent). The mechanical failure of these thin films of electronic materials leads to functional breakdown of flexible devices, therefore poses substantial challenges to the future success of flexible electronics technology. This paper reports a systematic study via computational modeling on the failure mechanisms of thin metal films on polymer substrates. These thin metal films are widely used as interconnects and electrodes in flexible devices.

It has been often reported that thin metal films, freestanding or polymer-supported, rupture at small strains (< 2%) [4-7]. On the other hand, it has also been reported that some polymer-supported metal films can sustain large strains (up to or beyond 50%)[8-11]. The cause for the substantial difference in rupture strain has been uncertain, but recent modeling and experimental studies may have shed light on the subject [11, 12]. The rupture of freestanding thin metal films under tension often initiates from deformation instability, such as necking or shear banding. Once the instability (e.g., a neck) sets in, it leads to failure of the metal films without constraint, thus often results in small rupture strains. By contrast, it has been shown by both simulations and experiments that, a thin metal film well bonded to a polymer substrate can deform uniformly to a large tensile strain (tens of per cent) due to the geometric constraint of the substrate [8, 10, 13]. Li and Suo further modeled the co-evolution of the necking and the debonding of the thin metal films on polymer substrates, and found that the interfacial adhesion plays a crucial role in determining the deformability of such thin metal films [12]. This prediction has recently been verified by the tensile tests of thin Cu films on polyimide substrates, in which Cu films well bonded on polyimide substrates can sustain tensile strain above 50%, while those poorly bonded on polyimide substrates fracture at strains about 2% [10]. Recent experiments also revealed the mixing of ductile necking and brittle inter-granular cracking of the thin metal films on polymer substrates under tension [8]. Further simulations on the inter-granular cracking of thin metal films on polymer substrates showed that the rupture strain is also modulated by the grain boundary adhesion and the grain size of the thin metal films [14].

So far, the results from the experiments and simulations on the tensile failure of thin metal films on polymer substrates are suggestive for understanding the large variation of the metal films ductility, but not yet systematic. For example, the competing ductile necking and brittle grain boundary cracking has been observed in experiments [8, 10, 11]. Existing simulation studies, however, consider these two failure mechanisms separately, by assuming either perfect grain boundary adhesion or perfect interfacial adhesion [12, 14-16]. It is desirable to conduct systematic studies to explore the interplay between the two failure mechanisms from which the parameters underpinning the ductility of thin metal films on polymer substrates can be quantitatively determined. To address this issue, we report here systematic simulations of the tensile behaviors of thin metal films on polymer substrates in which various combinations of grain boundary adhesion and interfacial adhesion are explored.

SIMULATION MODEL

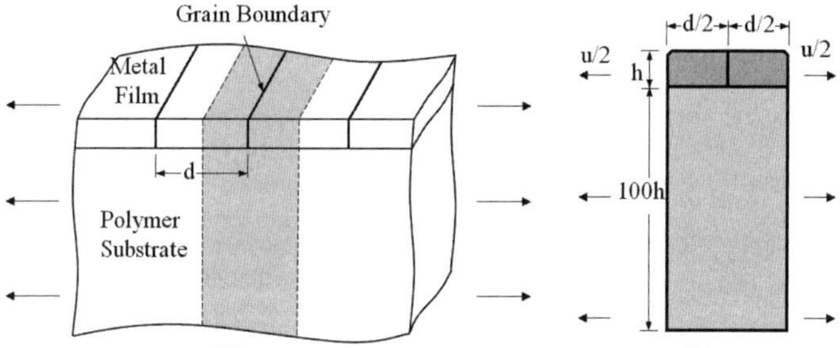

(a) (b)

Figure 1. (a) Schematics of a blanket thin metal film on a polymer substrate, subject to tension; (b) The unit cell used in the simulation model (the shaded area in (a)).

Figure 1 depicts the simulation model. Under tension, a thin blanket metal film well bonded to a thick polymer substrate is subject to grain boundary cracking (Fig. 1a). For simplification, the following assumptions are made: there is only one grain along the film thickness direction (e.g., a very thin metal film with columnar grains); all grains have the same size *d* along the tensile direction; the intergranular cracking occurs along grain boundaries perpendicular to the tensile direction (all other grain boundaries are not subject to cracking, thus are not considered in the simulation). The limitations of the above assumptions are to be discussed later in the paper. The metal/polymer laminate is taken to deform under the plain strain conditions. Taking advantage of symmetry we model only a unit cell of the laminate, consisting of two halves of adjacent grains, the grain boundary in-between and the substrate underneath (Fig. 1b). In the simulation model, the film is a layer of thickness *h*, and the substrate is a block of thickness 100*h* and length *d=40h*. The horizontal displacement is set to be zero along the centerline of the laminate, and set to be $u/2$ along both sides of the laminate. The quantity u/d will be called the applied strain. To initiate non-uniform deformation of the metal film, Two V-shaped notch, 0.2*h* wide and 0.02*h* deep, are placed at the centers of the top surface of the two neighboring grains (i.e., two top corners in Fig. 2b) to introduce imperfection.

Both the metal grains and the polymer are modeled as elastic-plastic solids. Under uniaxial tension, the true stress σ and the natural strain ε follow the relation

$$\sigma = \begin{cases} \sigma = E\varepsilon, & \varepsilon \le \sigma_Y / E \\ \sigma_Y \left(\dfrac{\varepsilon}{\sigma_Y / E} \right)^N, & \varepsilon > \sigma_Y / E \end{cases}, \quad (1)$$

where *E* is Young's modulus, *N* the hardening exponent, and σ_Y the yield strength. In the simulations, the following values are used: $E = 100GPa$, $N = 0.02$, and $\sigma_Y = 100MPa$ for the metal; and $E = 8GPa$, $N = 0.5$ and $\sigma_Y = 50MPa$ for the polymer.

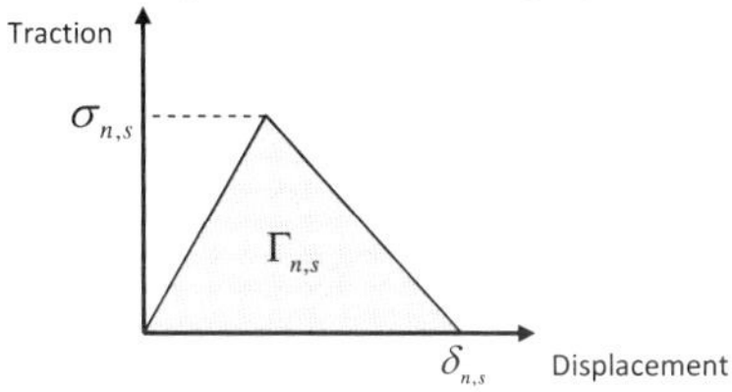

Figure 2. The traction-separation laws used to model the grain boundary and the metal/polymer interface.

Both the grain boundary and the metal/polymer interface are modeled as an array of nonlinear springs, each of which is characterized by a tensile and a shear traction-displacement law, with six parameters: interfacial tensile strength σ_n and shear strength σ_s, critical opening displacement δ_n and sliding displacement δ_s, and the areas under the traction-displacement curves Γ_n and Γ_s (i.e., the normal and shear adhesion energy of the grain boundary, respectively), as illustrated in Fig. 2. We assume that $\sigma_n = \sigma_s$, $\delta_n = \delta_s$ and $\Gamma_n = \Gamma_s$ for both the grain boundary and the metal/polymer interface. In the rest of the paper, we will use $\sigma_{n,s}^{GB}$, $\delta_{n,s}^{GB}$, and $\Gamma_{n,s}^{GB}$ to denote the parameters of the traction-displacement law for the grain boundary, and $\sigma_{n,s}^{INT}$, $\delta_{n,s}^{INT}$, and $\Gamma_{n,s}^{INT}$ for the metal/polymer interface. The grain boundary and the metal/polymer interface are meshed with four-node cohesive elements sharing nodes with the neighboring elements in the film and the substrate.

RESULTS

To explore the interplay between the interfacial adhesion and grain boundary adhesion on the ductility of thin metal films, we first assume a low grain boundary strength and vary the interfacial adhesion, and then assume a high grain boundary strength and vary the interfacial adhesion. In each case, we determine the rupture strain and the dominating failure mode of the thin metal films on polymer substrates.

Table 1 lists the values of $\sigma_{n,s}^{INT}$, $\sigma_{n,s}^{INT}/\delta_{n,s}^{INT}$ (interfacial stiffness), and $\Gamma_{n,s}^{INT}$ used in the simulation to characterize the weak, intermediate and strong interfacial adhesion, respectively. These values are representative of a metal/polymer interface with various adhesion qualities.

Table 1: Interfacial adhesion parameters used in simulations

Interfacial adhesion	$\sigma_{n,s}^{INT}$	$\sigma_{n,s}^{INT}/\delta_{n,s}^{INT}$	$\Gamma_{n,s}^{INT}$
Weak	1 MPa	0.2 MPa/nm	0.05 J/m^2
Intermediate	50 MPa	150 MPa/nm	0.3 J/m^2
Strong	100 MPa	200 MPa/nm	0.5 J/ m^2

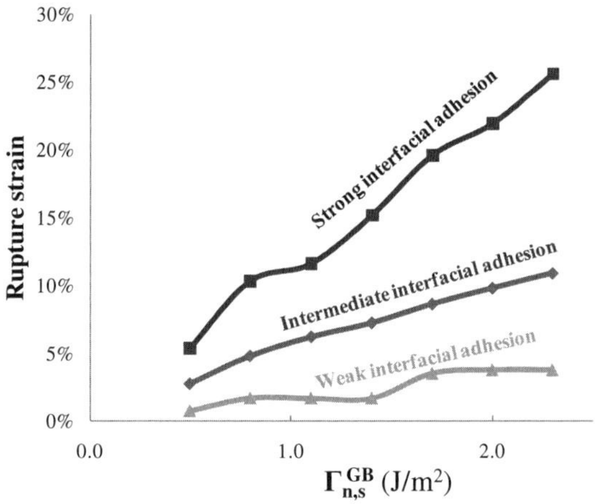

Figure 3. Rupture strain vs. grain boundary adhesion energy $\Gamma_{n,s}^{GB}$ for various interfacial adhesion qualities. Here grain boundary strength is low ($\sigma_{n,s}^{GB}$=50MPa).

Figure 3 plots the rupture strain of the thin metal films on polymer substrates as a function of grain boundary adhesion energy $\Gamma_{n,s}^{GB}$ for weak, intermediate and strong interfacial adhesion. Here, $\sigma_{n,s}^{GB}$=50MPa, $\sigma_{n,s}^{GB}/\delta_{n,s}^{GB}$ = 100MPa/nm, denoting a low grain boundary strength. As shown in Fig. 3, for a given interfacial adhesion quality, the rupture strain increases as the grain boundary adhesion energy increases, in a roughly linear manner. For a given grain boundary adhesion energy $\Gamma_{n,s}^{GB}$, the stronger the interfacial adhesion, the higher the rupture strain. For example, with a strong interfacial adhesion and $\Gamma_{n,s}^{GB}$=2.3 J/m^2, the rupture strain can be as high as 26%.

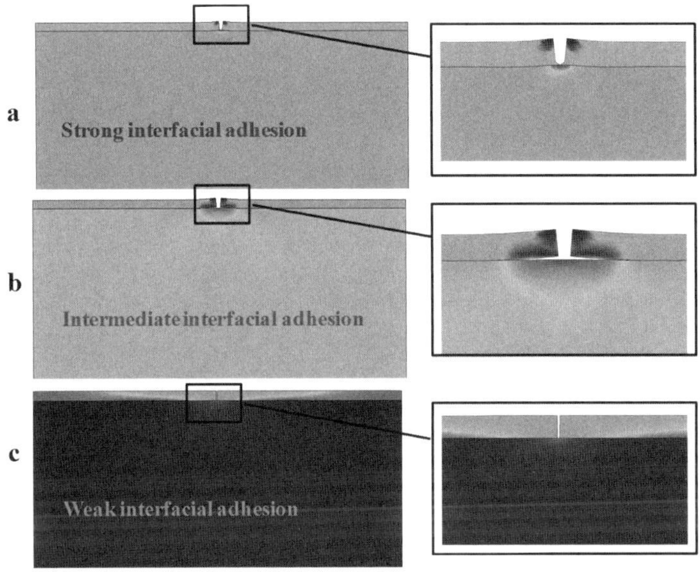

Figure 4. Deformed shapes of thin metal films on polymer substrates at rupture for various interfacial adhesion qualities. Insets show close-up of the grain boundary cracking. Note the difference of failure characteristics for various interfacial adhesion qualities. Shades denote von Mises stress. Only partial substrates are shown.

To further understand the trend of the rupture strains depicted in Fig. 3, Fig. 4 plots the deformed shapes of the thin metal films with low grain boundary strength on polymer substrates at tensile rupture, with strong, intermediate, and weak interfacial adhesions, respectively. The color shades represent the von Mises stress level. When the interfacial adhesion is strong, the thin metal film ruptures by grain boundary cracking while the metal/polymer interface remains bonded even though the stress level in the polymer near the grain boundary crack tip is rather high. The constraint from the substrate allows the thin metal film to deform uniformly to a relatively large strain until the grain boundary crack eventually opens up and ruptures the metal film. When the interfacial adhesion is intermediate, the high stress concentration at the interface near the grain boundary causes local debonding. The locally debonded thin metal film facilitates the grain boundary crack opening, leading to a relatively low rupture strain. When the interfacial adhesion is weak, the interfacial debonding occurs at small tensile strain and can easily propagate through the whole interface, leaving the thin metal film nearly freestanding. This becomes evident as the nearly zero stress level in the metal film shown in Fig. 4c. The lack of substrate constraint results in a rather low rupture strain of the thin metal film due to grain boundary cracking. In summary, when the grain boundary strength is low, the thin metal film on polymer substrates rupture by grain boundary cracking, and the rupture strain is modulated by the interfacial adhesion.

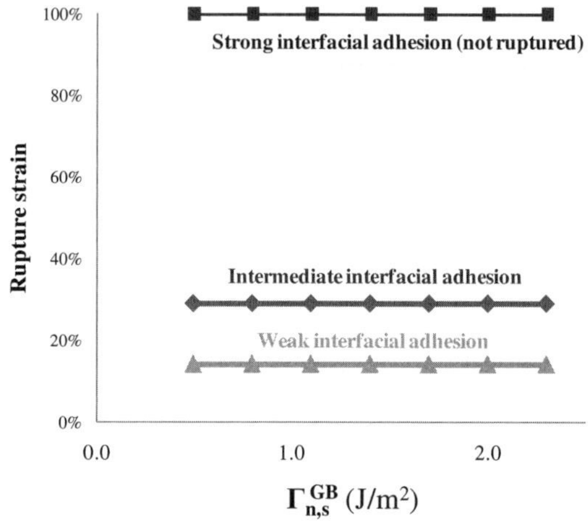

Figure 5. Rupture strain vs. grain boundary adhesion energy $\Gamma_{n,s}^{GB}$ for various interfacial adhesion qualities. Here grain boundary strength is high ($\sigma_{n,s}^{GB}$=150MPa).

Figure 5 plots the rupture strain of the thin metal film on polymer substrates as a function of grain boundary adhesion energy $\Gamma_{n,s}^{GB}$ for weak, intermediate and strong interfacial adhesion. Here, $\sigma_{n,s}^{GB}$=150MPa, $\sigma_{n,s}^{GB}/\delta_{n,s}^{GB}=$100MPa/nm, denoting a high grain boundary strength. As shown in Fig. 3, the rupture strain increases as the interfacial adhesion increases, and is independent of grain boundary adhesion energy. Figure 6 further plots the deformed shapes of the thin metal film with high grain boundary strength on polymer substrates at tensile rupture, with strong, intermediate, and weak interfacial adhesions, respectively. The color shades represent the von Mises stress level. When the interfacial adhesion is strong, the thin metal film can deform uniformly to a very large strain (e.g., 100% as shown in Fig. 5a) without rupture due to the strong constraint from the polymer substrate. When the interfacial adhesion is intermediate, multiple necks appear in the thin metal film before it eventually ruptures. Since plastic deformation of the thin metal film is volume-reserved, each neck (i.e., local thinning) results in a local elongation, adding up to an intermediate rupture strain. When the interfacial adhesion is weak, the interfacial debonding occurs at small tensile strain and can easily propagate through the whole interface, leaving the thin metal film nearly freestanding. This is also evident as the nearly zero stress level in the metal film shown in Fig. 6c. The freestanding thin metal film ruptures by forming a single neck near the imperfection location in each grain, leading to a rather low rupture strain. In summary, when the grain boundary is strong, the thin metal films on polymer substrates rupture by necking, and the rupture strain is also further modulated by the interfacial adhesion. These results agree with those from earlier modeling and experimental studies [8, 10, 12].

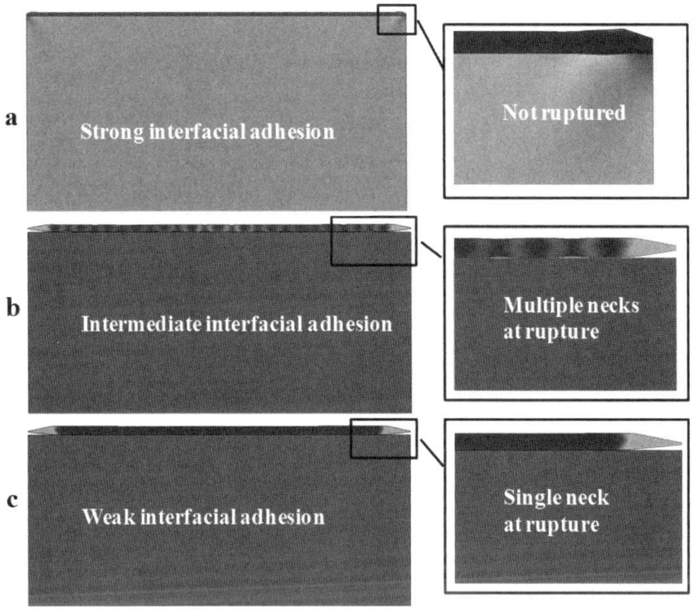

Figure 6. Deformed shapes of thin metal film on polymer substrates at rupture for various interfacial adhesion qualities. Insets show close-up of thin metal film necking. Note the difference of failure characteristics for various interfacial adhesion qualities. Color shades denote von Mises stress (Red: high stress; blue: low stress). Only partial substrates are shown.

CONCLUSIONS

We study two competing failure mechanisms of thin metal films on polymer substrates, through comprehensive computational modeling. We show that, when the grain boundaries of the metal film are weak, the thin metal film tends to rupture by grain boundary cracking. The rupture strains are modulated by the metal/polymer interfacial adhesion, but in general the deformability of such thin metal film is small. When the grain boundaries of the metal film are strong, the thin metal film mainly ruptures by necking. There are large variations of the rupture strain, depending on the interfacial adhesion quality. The results from this study suggest that, to achieve a large deformability of thin metal films on polymer substrates, it is desirable to have strong grain boundaries of the metal films and strong metal/polymer interfacial adhesion.

REFERENCES

1 G. P. Crawford, *Flexible flat panel displays* (John Wiley & Sons, Chichester ; Hoboken, NJ, 2005).

2 C. J. Brabec, *Solar Energy Materials and Solar Cells* **83**, 273 (2004).

3 S. Wagner, S. Lacour, J. Jones, P. Hsu, J. Sturm, T. Li, and Z. Suo, *Physica E* **25**, 326 (2004).

4 D. Pashley, *Procdings of the Royal Society of London A- Mathematical and Physical Sciences* **255**, 218 (1960).

5 H. B. Huang and F. Spaepen, *Acta Materialia* **48**, 3261 (2000).

6 S. L. Chiu, J. Leu, and P. S. Ho, *Journal of Applied Physics* **76**, 5136 (1994).

7 B. E. Alaca, M. T. A. Saif, and H. Sehitoglu, *Acta Materialia* **50**, 1197 (2002).

8 Y. Xiang, T. Li, Z. Suo, and J. Vlassak, *Applied Physics Letters* **87**, 161910 (2005).

9 R. M. Niu, G. Liu, C. Wang, G. Zhang, X. D. Ding, and J. Sun, *Applied Physics Letters* **90**, 161907 (2007).

10 N. S. Lu, X. Wang, Z. G. Suo, and J. Vlassak, Applied Physics Letters **91,** 221909 (2007).

11 N. Lu, X. Wang, Z. Suo, and J. Vlassak, *Journal of Materials Research* **24**, 379 (2009).

12 T. Li and Z. Suo, *International Journal of Solids and Structures* **44**, 1696 (2007).

13 T. Li, Z. Huang, Z. Xi, S. Lacour, S. Wagner, and Z. Suo, *Mechanics of Materials* **37**, 261 (2005).

14 Z. Zhang and T. Li, *Scripta Materialia* **59**, 862 (2008).

15 T. Li, Z. Huang, Z. Suo, S. Lacour, and S. Wagner, *Applied Physics Letters* **85**, 3435 (2004).

16 T. Li and Z. Suo, *International Journal of Solids and Structures* **43**, 2351 (2006).

Conference Proceedings AMC XXV © 2010 Materials Research Society

Nucleation effects of Tungsten Chemical Vapor Deposition on B_2H_6 Pre-treated Titanium Nitride for sub-45 nm contacts

Avraham Rozenblat[1], Rotem Drori[1], Maor Rotlain[1], Yosi Shacham-Diamand[2] and Dror Horvitz[1]
1. Numonyx Israel Ltd., Qiryat-Gat 82109, Israel
2. Electrical Engineering, Tel Aviv University, Tel Aviv 69978, Israel

ABSTRACT

Characteristics of tungsten nucleation layer are strongly governed by surface pretreatment and by the early stages of deposition known to affect the electrical properties of the film. In this work we investigated the effect of B_2H_6 treatment on the initial stages of tungsten chemical vapor deposition (CVD). Chemically sensitive x-ray photoelectron spectroscopy (XPS) was employed to investigate the surface reactions between B_2H_6 and air exposed (technical) TiN and the reaction between WF_6 molecules and B_2H_6 treated TiN. We show that the reaction of B_2H_6 results in reduction of TiO_2 and formation of TiB_2 and BN species. It is also shown that the BN act as a template site for WF_6 reaction. The surface coverage of W-CVD process on B_2H_6 treated TiN was quantitatively measured by XPS for various CVD pulse durations and characterized by AFM and HRTEM. Results suggest that W CVD on B_2H_6 treated TiN, follows the Stranski-Krastanov (S-K) growth mode, similar to SiH_4 treatment but with significantly shorter incubation time.

INTRODUCTION

Tungsten-CVD is an important process to make interconnection in advanced integrated circuit devices. Common contact fill technology is composed of three sequential layers, which differently contributing to contact overall conductivity: 1) Titanium Nitride (TiN) adhesion layer. 2) Tungsten nucleation layer using SiH_4 reduction of WF_6 and 3) W-plug fill by H_2 reduction of WF_6 (W-CVD). W nucleation layer is used due to its ability to promote faster deposition on the conventional barrier layer, TiN, with little incubation time. W-CVD fill is used due to its good conductivity [1-4]. As device dimensions continue to decrease below 45nm with aspect ratio above 1:10, incomplete deposition inside the contacts/plug holes and increase in contact resistance is becoming a crucial issue. Furthermore, W nucleation layer characteristics are strongly governed by the early stages of nucleation on TiN which in turn, is found to affect the properties of W-CVD film [2]. This has motivated researchers to study the nucleation mechanism in these CVD processes. One of the focus topics of nucleation studies has been the clarification of an incubation time at the initial stages of deposition. Nucleation of tungsten-CVD on TiN surface is known to take place in Stranski–Krastanov (S-K) mode i.e. W starts to grow slowly on the TiN surfaces up to about one-monolayer coverage, next W islands nuclei start growing rapidly [3,4]. Surface pretreatment of TiN was found to affect the incubation time and deposited layer properties. Pretreatment by SiH_4 on TiN substrate before deposition of W-nucleation layer, is a common used method. Several studies have been carried out to investigate SiH_4 pretreatment. Jones *et al.* and Herner *et al* observed the existence of Si at the interface between W film and TiN substrate [5,6]. Lantz et al. observed that SiH_4 pretreatment on clean and oxidized TiN accelerates the subsequent WF_6 reaction [8]. Recent tungsten CVD technology uses B_2H_6 as surface pretreatment of TiN [9] and as a reduction agent for WF_6 nucleation layer [10-12]. Kim *et al.* showed that B_2H_6 pretreatment decreased the incubation time of W-ALD over SiO_2 and drastically improved step coverage [9]. However, the chemical reaction between B_2H_6 and the TiN surface and the effect of B_2H_6 treatment on the kinetics of W-CVD, was not investigated. In this study the effect of B_2H_6 surface treatment on the initial stages of W-CVD nucleation was investigated. The surface reaction between B_2H_6 and technical TiN was studied initially. Then the reaction of WF_6 molecules on B_2H_6 treated TiN was analyzed and finely the surface coverage of W CVD process on B_2H_6 treated TiN was measured for various W-CVD pulse durations. Surface chemistry and coverage were analyzed by x-ray photoelectron spectroscopy (XPS). Surface morphology was investigated by atomic force microscopy (AFM) and high resolution transmitted electron microscopy (HRTEM).

EXPERIMENT

Sample preparation

A cold-wall CVD system (Novellus Concept 2 Altus) was used to deposit W on air exposed TiN/Ti substrates. Process in the deposition chamber was performed at a pressure of 40 torr at temperature of 300°c. Three experimental routs were performed and analyzed: 1- Surface treatment on technical TiN for five seconds of B_2H_6 at Ar atmosphere, 2- B_2H_6 treated TiN was exposed to WF_6 at Ar atmosphere for 0.5-6 seconds. 3- Five seconds pretreatment of either B_2H_6 at Ar atmosphere or SiH_4 at H_2/Ar atmosphere, followed by W-CVD deposition from a gas mixture of WF_6, H_2 and Ar at durations varied from 0.5 s to 55 s.

XPS analysis

Analysis performed with Thermo Microlab350 system. The photoelectrons are excited by Mg Kα source. Samples were analyzed at a take-off angle of 90º.High resolution spectra were taken with 20 eV pass energy and 0.05 eV step size. Information on the chemical bonding was obtained by deconvoluting the spectra by curve fitting method [16]. Due to different charging effects between samples, the spectra was shifted according to the position of N 1s to 396.7eV, reported for TiN-CVD [14]

AFM analysis

We used the AFM Dimension Tapping Mode™ 5000SPM, produced by Veeco Digital instruments. The measurements were performed with the Tapping Mode model using a Si probe. The tip height is 10÷15μm, Side angle 220, TSB range 5÷25 μm. All samples were examined in scanning areas of 1 μm × 1 μm.

TEM analysis

TEM analysis was done in FEI Tecnai F20 equipped with Fischione HAADF STEM detector and EDAX EDS. All analysis was performed under accelerating voltage of 200kV. HAADF STEM images were taken with camera length (CL) of 80 mm for Z contrast and CL of 490 mm for diffraction contrast, spot size is 2A. For EDS analysis, dwell time of 15 sec was used with α-tilt of 3 degrees towards the detector.

RESULTS

Surface analysis of B_2H_6 treated TiN

Figure 1a shows the Ti 2p photoelectron peak taken from the TiN surface after B_2H_6 treatment. The high resolution Ti 2p spectrum envelop is composed of Ti $2p_{3/2}$ and the Ti $2p_{1/2}$ spin-orbit. Our discussion will refer to the Ti $2p_{3/2}$ component that has lower binding energy (b.e.). The Ti $2p_{3/2}$ is decomposed to three main chemical states. The lower binding energy peak, at 455.2 eV (#1) is related to TiN or TiB_2 bonds. The highest binding energy at 458.3 eV (#3) is related to TiO_2 bonding and the middle peak, at 456.6 eV (#2) indicate intermediate oxidation in the form of TiO_x. The differences in chemical states of the titanium after B_2H_6 treatment are examined by the figure 1b. Before surfacc treatment, the Ti 2p spectrum has a broad peak with significant contribution of TiO_2 species located at 458.3 eV [13,14]. After exposure to B_2H_6 the oxide contributions (TiO_2 and TiO_x) are significantly reduced. In addition, the low energy peak is slightly shifted toward lower binding energy indicating on the formation of TiB_2 bonds. A shift of the peak by <1eV between TiN (455±0.2) and TiB_2 (454.3-454.6) states was previously reported on a Ti-N-B layers analysis [15,16]. It should be noted that the slight deviation in the absolute binding energy between previous reported results and ours current spectrum can be explained by differences in spectrum calibration and by our choice to shift the spectra according the N1s peak location (396.7eV) [14].

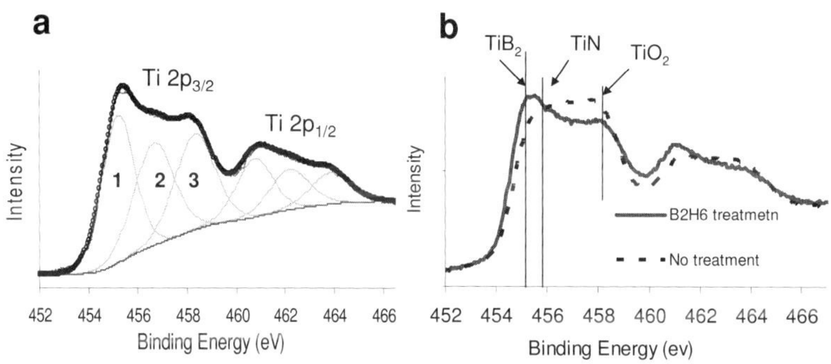

Figure 1: Ti 2p photoelectron peak: a- Decomposition of the Ti 2p spectrum after B_2H_6 surface treatment. b- Comparison of Ti 2p spectrum before treatment and after B_2H_6 treatments.

Figure 2a shows comparison of the O 1s spectra taken form TiN before and after B_2H_6 treatments. Before treatment the broad O 1s distribution contains three contributions situated at 529.6, 531.2 and 532.8. The peaks at 529.6 and 531.2 eV are related to O-N and TiO_2 species respectively [13,16]. After treatment, a sharp decrease is seen at the 529.6 eV and 531.2 eV contributions, whereas an increase is seen at 533 eV peak which relates to B_2O_3 bonds [16]. Examination of the B 1s peak is shown in figure 2b. The spectra show two peaks indicating the presence of TiB_2 and BN species at 187.8 and 191.6 eV respectively [15,16].

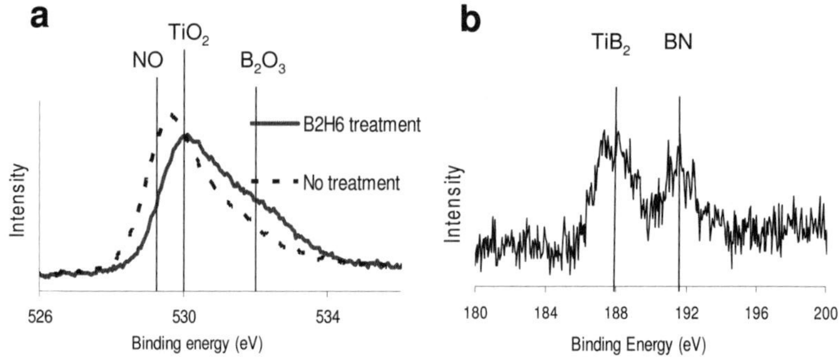

Figure 2: a- Comparison of O 1s spectrum before and after B_2H_6 treatments. b- B 1s spectra of the TiN surface after treated with B_2H_6.

Table 1 shows a summary of the content of elements on TiN, as result from the treatment. It is noticed that B_2H_6 treatment reduces the amount of oxygen attached to the surface. A sharp decrease in the TiO_2 peak (Ti (#3)) and significant increases in the TiN/TiB_2 contribution, Ti (#1), are seen.

Table 1: XPS elements quantification (at %) of TiN before and after B_2H_6 treatments. Ti contributions are marked 1-3, according to the fitting of Figure 1a.

Treatment\ at%	Ti (1)	Ti (2)	Ti (3)	O	N	C	Si	B
TiN	15.4	14.8	15.4	45.5	27.0	8.3	-	-
TiN + B_2H_6	17.3	13.9	12.1	20.5	25.3	6.4	-	3.1

Surface analysis of WF_6 reaction on B_2H_6 treated TiN

We have examined the early stages of tungsten nucleation and surface reaction on diborane-treated TiN by exposing the film to WF_6 molecules. Figure 3a shows the W 4d spectrum taken after 0.5s, 1.5s and 3s of WF_6 flow. The W 4d peak, located at 246.5eV and 258.7eV, indicates that the W is found in enriched, oxidized state [17]. No chemical change was seen in peak location for different exposure time. Figure 3b shows the evolution of the B 1s spectrum. The B-N peak, situated at 191.6 eV is decomposed immediately after 0.5s of WF_6 while the TiB_2 peak located at 187.8 eV is reduced dramatically after 0.5s but later stabilizes on constant amount. The total change of boron amount is presented in figure 4a. Boron is reduced and then saturates on low constant amount related to TiB_2 bond. The overall surface composition change is shown in figure 4b. The amount of W, F and O increased and then saturates after one second of deposition while, as deposition proceeds, Ti and N are reduced. The reduction in Ti is attributed to reduction in the Ti-oxide species (not shown here). The reduction of WF_6 by TiO_x and desorption of TiF_{4-y} was reported by other authors [3,7, 17] yet the reduction of nitrogen spices was not reported and will be discussed here after.

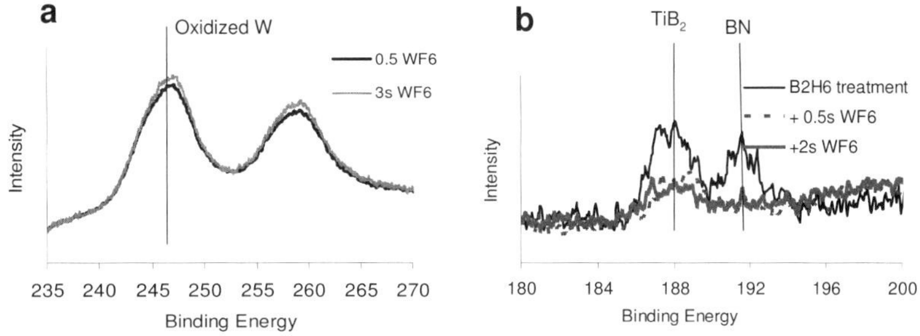

Figure 3: XPS spectrum of TiN surface treated with B_2H_6 and then exposed to WF_6 molecules for 1-6 seconds (at 300°c). a- W 4d spectra overlay after 0.5 and 3 seconds of WF_6 flow. b- Change in B 1s spectrum after B_2H_6 and after 0.5s, 2s of WF_6 flow.

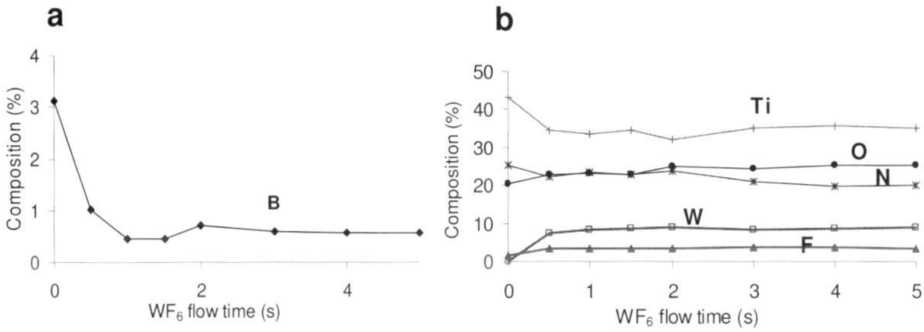

Figure 4: Element composition determined by XPS of the TiN film treated with B_2H_6 and then exposed to WF_6 for 1-6 seconds (at 300°c). a- Quantification of Boron. b- Composition of Ti, O, N, W and F.

Surface coverage of W CVD on B_2H_6 treated TiN for various deposition times

The kinetics of deposition on TiN surface is shown in figure 5 by the amount of W at% vs. deposition time. The behavior of the W amount can be taken as simple representation of the change in layer thickness [3]. In figure 5, both lines for the no-treatment and SiH_4 treatment exhibit similar growth rate with saturation around ~20 s. When the diborane is used, the amount of W on the surface increases significantly immediately after 0.5 seconds (4.5 at%) compared with no B_2H_6 treatments (3 at%). It was also seen that the earlier saturation occurred around 10 s and then increased again. This behavior agrees with previous results for WF_6 deposition on TiN [3]. The final increase corresponds to formation and growth of 3D islands. AFM was used to examine surface morphology during nucleation [Figure 6]. The root mean square (RMS) of the morphology measurements show that the SiH_4 treated sample had RMS of 0.6 nm with maximum peak of 5nm, similar to the results we got on untreated TiN (not shown here). B_2H_6 treated sample (Figure 6b) show that 3D island were formed with height of 12nm and RMS of

0.9 nm. Figure 7 shows the TEM-STEM analysis of the B_2H_6 treated sample followed by 10s of CVD. Figure 7a is a BF-TEM image of cross section sample shows the thin partial coverage of the W deposition over the TiN. HADF-STEM was used on plane view sample in order to verify the chemical composition of the 3D islands [Figure 7b]. STEM image was taken with short, 80mm CL, to achieve maximum mass contrast. Bright spots on the image indicate on differences in mass between background surface and the spots, correspond to W 3D islands. Figure 7c presents the EDS spectra taken from points 1 and 2 in the STEM area. Spectra result verifies that the bright spots show higher amount of W compare to background surface.

.

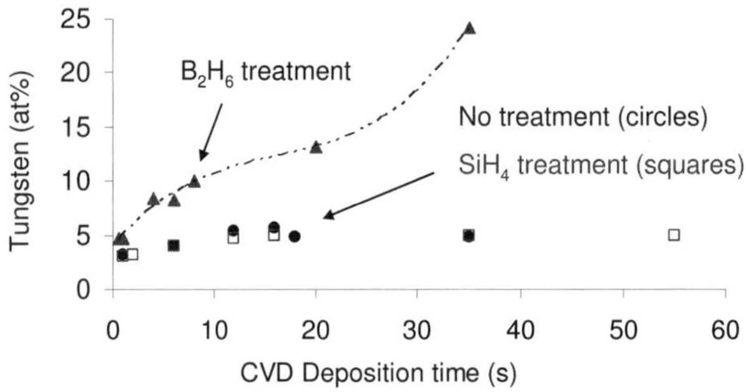

Figure 5: XPS quantification of W at% on the TiN surface vs. time of CVD (WF_6+H_2 at T= 300°C).

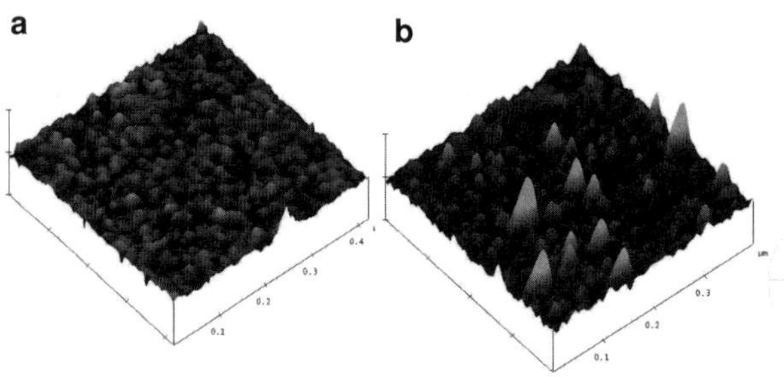

Figure 6: 0.5x0.5 μm of AFM image. a- SiH_4 treated TiN followed by 16 s CVD. b- B_2H_6 treated TiN followed by 10 s CVD.

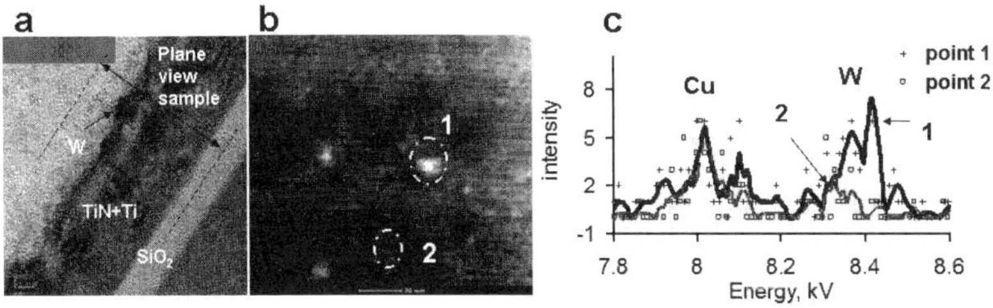

Figure 7: a- BF-TEM on cross section and b- HADF-STEM image taken with 80mm CL on plane section of the B_2H_6 treated TiN followed by 10s CVD. Bright spots indicate on mass difference between surface and spot. c- EDS spectra taken from points 1 and 2 on STEM area. Cu peak in both spectra is emitted from sample holder.

DISCUSSION

In this study we investigate the effect of surface treatment on the kinetics of tungsten CVD process. We show that the reaction of B_2H_6 and technical TiN results in reduction of TiO_2 and TiO_x bonds. Reduction of oxide by SiH_4 treatment on TiN, which resulted in formation of Si-N bond on the surface have been reported previously [5-8,17]. However, when B_2H_6 treatment is used the TiO_2 reduction results in formation of TiB_2 and BN species on the surface [Figure 2]. The reaction of WF_6 molecules was examined on a B_2H_6 treated TiN. After short exposure time, at 300°C, both tungsten and fluorine were detected on the surface [Figure 4a]. The thickness of deposited tungsten increased and then saturated after 1s at ~9 at%. The fluorine amount on the surface found to follow the same behavior. Chemical analysis of the deposited tungsten layer found tungsten to be in enriched chemical state [Figure 3a] indicating on the formation of WF_y or WF_xO_y [317]. Maragon *et al.* reported that W nucleation accompanied the reduction of the titanium oxide layer on titanium nitride substrates [7]. Indeed, our result shows that the W deposition resulted in desorption of Ti-oxide species but also by the reduction of BN [Figure 3]. The saturation seen during deposition of WF_6 [Figure 4b] results from the inability of WF_6 to adsorb onto WF_x therefore, for W deposition to proceed, WF_6 must be reduced by SiH_4 or H_2. These saturation behaviors characterize the CVD process when the growth mode changes from thin mono-layer to 3D, auto catalysis mode [3,4]. Our measurements show that the transition from monolayer to 3D mode with B_2H_6 treatment is significantly faster and occurs with twice the W coverage than without B_2H_6 treatment. Figures 4, 5 show that WF_6 exposure of B_2H_6 treated TiN either with or w/o additional reduction agent, results in mono-layer saturation at ~10 at%. Yet for TiN without B_2H_6 treatment this mono-layer saturates at ~5 at%
In addition, diborane pretreatment result in significant decrease in incubation time between the first and the second layers [Figure 5]. We confirmed by AFM and HADF-STEM that the second sudden increase corresponds to 3D island formation.

CONCLUSION

This study investigated the mechanism of W nucleation on B_2H_6 treated TiN. It was shown that B_2H_6 treatment results in reduction of TiO_2 and TiO_x and in formation of TiB_2 and BN species on the surface. When WF_6 molecules are introduced to the surface, a deposition of WF_xO_y takes place by the desorption of Ti-oxide and also by BN species. The additional BN is believed to act as a template for the WF_6 reaction and by that accelerates the nucleation reaction. Under WF_6 reaction the deposition is known to reach saturation, yet for B_2H_6 treated TiN the amount of W was found to be twice that of technical TiN or SiH_4 treated TiN. W-CVD process on B_2H_6 treated TiN follows the S-K growth mode, similar to SiH_4 treatment, but with significantly shorter incubation time.

ACKNOWLEDGMENTS

The Authors would like to thank Shai Haimson, Tal Bronshtein, Rena Razilov, Tanya Roif, Shai Perets and Daniel Fishman.

REFERENCES

1. T. Luoh , C.T. Su, T.H. Yang, K.C. Chen, C.Y. Lu, Microelectron. Eng, 85, 8, 1739-1747, 2008.
2. S.H. Kim, J.T. Kim, N. Kwak, J. Kim, T.S. Yoon, H. Sohn, J. Vac. Sci. Technol. B 25, 5, 2007.
3. Y. Kajikawa, T. Tsumura, S. Noda, H. Komiyama and Y. Shimogaki, Jpn. J. Appl. Phys., 43, 6B, 3945-3950, 2004
4. S. Noda, T. Tsumura, J. Fukuhara, T. Yoda, H. Komoiyama and Y. Shimogaki, Jpn. J. Appl. Phys., 43, 10, 6974-6977, 2004.
5. C. M. Jones, J. Zhao and T. Yuan: Appl. Surf. Sci. 165, 154, 2000.
6. S. B. Herner, S. A. Desai, A. Mak and S. G. Ghanayem, Electrochem. Solid-State Lett. 2, 398, 1999.
7. M. S. Maragon, G. Queirolo and C. Savoia, Appl. Surf. Sci. 91, 157, 1995.
8. S. L. Lantz, A.E. Bell, W.K. Ford and D. Danielson: J. Vac. Sci. & Technol. A 12, 1032, 1994.
9. S.H. Kim, E.S. Hwang, B.M. Kim, J.W. Lee, H.J. Sun, T.E. Hong, J.K. Kim, H. Sohn, J. Kim and T. S. Yoon Electrochem. & Solid-State Lett., 8 ,10, C155-C159, 2005.
10. S.H. Kim, S.J. Yeom, N. Kwak, and H. Sohn ,J. Electrochem. Soc., 155, 2, D148-D154, 2008.
11. S. Smith, B.J. Walgenwitz, O.J.H. Bonnin, R.H. Braspenning, C. Alliance, conference proceedings AMC XIX, 2004, MRS, pp.367-371,
12. I.C. Rho, C.H. Kim, H.S.Lee, J.M. Oh, I.C. Ryu, J.W. Chung, S.W. Lim, K.B. Rouh, Y.S. Sohn and H.S. Kang, Conference Proceedings AMC XXIII, 2008, MRS, pp. 589-593.
13. P.-Y. Jouan, M.-C. Peignon, Ch. Cardinaud and G. Lempericre, Appl. Surf. Sci., 68, 595-603, 1993.
14. J. Zhao, E. Gene Garza, K. Lam, C.M. Jones, Appl. Surf. Sci. 158, 246–251, 2000.
15. C. Ruby, R. Ott, F. Huang, M. L. Weaver and J. A. Barnard, Surf. Interface Anal. **29**, 823–828, 2000.
16. Y.H. Lu, Z.F. Zhou, P. Sit, Y.G. Shen, K.Y. Li, H. Chen, Surface & Coatings Technology 187, 98– 105, 2004.
17. M.L. Yu, K.Y. Ahn and R.V. Joshi, J. Appl. Phys, 67, 2, 1990.

Evaluation of Selective Copper Deposition (Selectroplating)

Rashid Mavliev[1] and Jeff Gambino[2]

[1]IPGRIP LLC., 502 E. John St., Carson City, NV, 89706
[2]IBM Microelectronics, 1000 River Street, Essex Junction, VT, 05452

ABSTRACT

In this paper we present results on a new technology for selective deposition of copper - Selectroplating. This technology is based on a selective chemical modification of field areas of a wafer before electrochemical deposition (ECD). A protective agent deposited on the field area is strong enough to survive conventional plating conditions. Selective copper deposition has been demonstrated on coupons and 200 mm test wafers for trenches as narrow as 40 nm and shallow and deep (TSV) structures. Sheet resistance measurements suggest that the selective plating process does not degrade the copper resistivity. The results show the basic feasibility of the process and are very promising for reducing manufacturing costs and enabling new materials.

INTRODUCTION

Dual damascene copper is the dominant interconnect technology for advanced semiconductor chip manufacturing [1]. In a traditional process, trenches in the dielectric are filled by electrochemical deposition (ECD) of copper, followed by chemical mechanical planarization (CMP) for removal of the metal from field areas of the wafer. CMP has been an enabling technology for the use of copper in damascene technology. It has rapidly become one of the most important and widespread processes. Overall, the total number of CMP operations is expected to grow 67% from 0.9 billion unit operations in 2007 to 1.5 billion in 2011 [2]. The aggregated cost of the CMP process for the semiconductor industry exceeds $1 billion per year (Gartner, 2007). CMP cost reduction and process enabling are major factors for recent interest in selective electrochemical deposition processes (ECD) [3]. This is especially true for global interconnects and inductors on RF and mixed signal technology, where the Cu wire thickness can be as much as 6 um [4-6].

An "ideal" ECD process would result in complete filling of trenches and via with high quality copper (without compromising the line resistivity), with no deposition of copper in the field areas. One method to achieve this is to "poison" the field area, prohibiting copper deposition, with a self-assembled monolayer (SAM) of alkanthiols [7-9]. However, formation of SAMs on Cu using alkanethiole monolayers has poor reproducibility, because of copper surface oxidation when copper is exposed to air [10]. In addition, formation of SAMs on copper surfaces from solution may take up to 1 h or more [11]. Finally, according to Love et al. [12], the structures formed on copper surfaces are not well understood yet. All of this makes selective copper deposition using SAMs unpractical for application on an industrial scale.

A new method for selective deposition (Selectroplating) of metal into the trenches has been developed using protective films in the nanometer scale range using thiol class substances. The selective surface modification process takes place after depositing the physical vapor deposition (PVD) seed layer and before the ECD process. The method consists of selective chemical modification of the seed layer with a surface modification agent, followed by a conventional electroplating process.

Figure 1 shows the conventional process flow for single damascene interconnect fabrication compared to the process flow using selective plating.

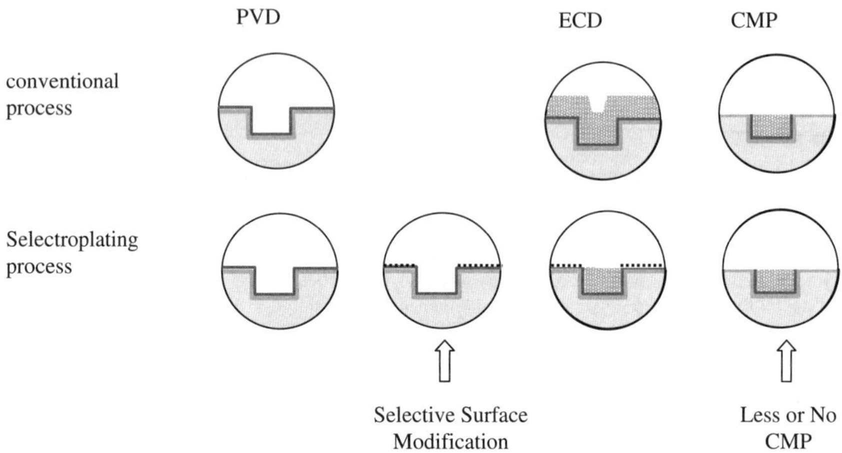

Figure 1. Comparison of conventional and selective fabrication processes.

EXPERIMENT

Chemical modification is achieved by the use of a solid flat applicator covered with a thin film (a few nanometers in thickness) of the inhibiting agent. The applicator is put into contact with the top surface of the wafer without affecting the cavities of the sample. The application and surface modification processes are fast (a few seconds) and take place under atmospheric conditions. The modified areas of the seed layer become non-conductive and thus prevent metal deposition on the field. Electrochemical deposition takes place only in the recesses, which are not inhibited with the inhibiting agent. Preliminary experiments have been run on wafer coupons with trenches of 40 nm to 100 μm widths, and on 200 mm wafers using a 130 nm node process technology [13]. The Selectroplating agent can withstand plating bath chemistries up to one hour, making the technology feasible for through-silicon via (TSV) applications.

The sheet resistance of Cu interconnects formed with selective deposition was measured on 200 mm wafers with a Sematech 854 pattern. After deposition of barrier and seed layers, test wafers were treated for selective surface modification, then plated with Cu, and finally polished. The fabrication of the test structures used conventional processes (including conventional ECD), except for the surface modification step. Approximately 50% of one wafer surface was not modified with the inhibiting agent, allowing that area to be used as a baseline for non-selective deposition.

RESULTS

Standard Interconnects

The effect of selectivity can be clearly observed in the FIB cross sections presented in Figure 2. Figure 2a is a cross-sectional SEM image after conventional copper ECD. The wafer surface in Figures 2b and 2c was processed with the inhibiting agent prior to ECD, resulting in minimal copper deposition on the field region. Figure 2a, indicates that with the non-selective process, the copper ECD is just enough to fill the 400 nm deep trench in wide features (the blue line represents the top of the trench).

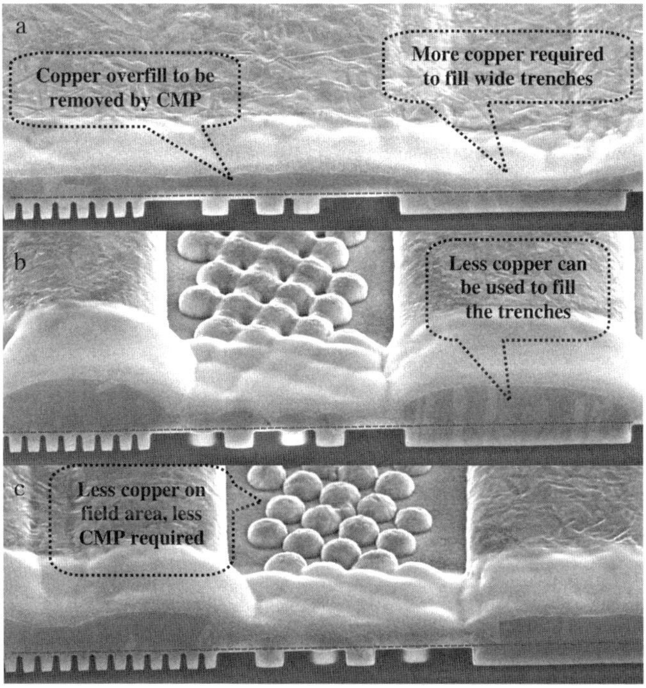

Figure 2. FIB cross sections of wafers without (a) and with Selectroplating in the center (b) and at the edge of the wafer (c).

By limiting the surface area of copper deposition with the inhibiting agent (while keeping the same ECD parameters), the Cu plated in the trenches is thicker, resulting in overdeposition of copper (see Fig. 2b –c). Hence, the use of the inhibiting agent reduces the time and material required for the ECD process. No systematic differences were observed for Selectroplating in the center and at the edge of the wafer (see figures 2b and 2c).

Figure 3. FIB cross sections of narrow trenches (a) without and (b) with Selectroplating in the center and (c) at the edge of the wafer. Small voids at top cusps are due to plating overgrowth on inhibited areas, as opposed to undesired defects.

Figure 3 shows a detailed picture of a FIB cross section revealing the structure of copper deposition in small trenches. Copper overplating of closely spaced trenches takes place in both selective and non-selective deposition. The quality of the Cu fill in the trenches appears to be similar in both cases, with no obvious voids in the Cu inside the trench due to the presence of the inhibiting agent. The black areas above the trenches in Figures 3b and 3c are a manifestation of inhibiting agent. The inhibiting agent is embedded in the structure due to lateral overgrowth of the Cu. When deposited copper significantly overfills the trench, it is no longer confined, and starts to grow in all the directions forming a "mushroom" shape, which can be seen in Figures 2b and also in Figure 4b.

Deep trenches and Through-Silicon Vias (TSV)

The overplating and the consequent long CMP polishing of the copper from field area is more severe in the case of global interconnects used in the last layers of metallization, in TSVs or for MEMS structures. An example of global interconnect trench structures filled with selective plating are presented in figures 4a and 4b. Figure 4a shows a top-down SEM of 3 um deep trench structures on a 200 mm wafer after selective surface modification followed by conventional ECD of 1.5 um of copper (the deposited copper amount was estimated based on full wafer coverage typical in conventional ECD). Figure 4a is FIB cross section of narrow and wide lines. One can see that the trenches are overfilled, even though the copper plating time was reduced by half compared to the conventional (non-selective) plating process. Figures 4a and 4b demonstrate the good selectivity of the process – minimal deposition occurs on field areas, whereas the trenches are filled completely.

Overplating takes place in narrow trenches and on the edges of wide lines, indicating the strong effect of accelerator additive in plating bath.

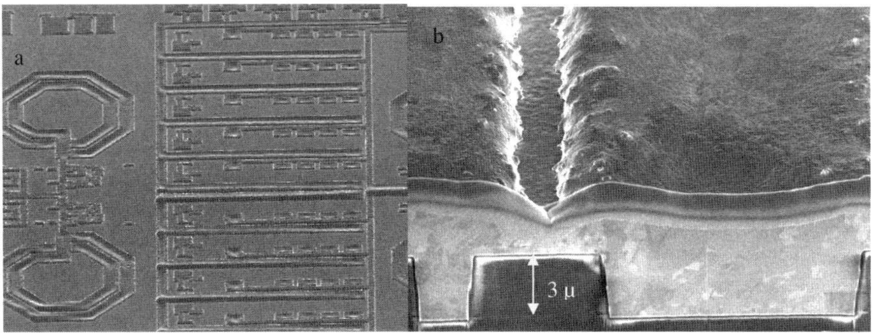

Figure 4. SEM picture (a) and FIB cross sections of deep trenches (b) filled using a selective Cu plating process. Deposition of copper takes place only in trenches, with minimal deposition on the field regions.

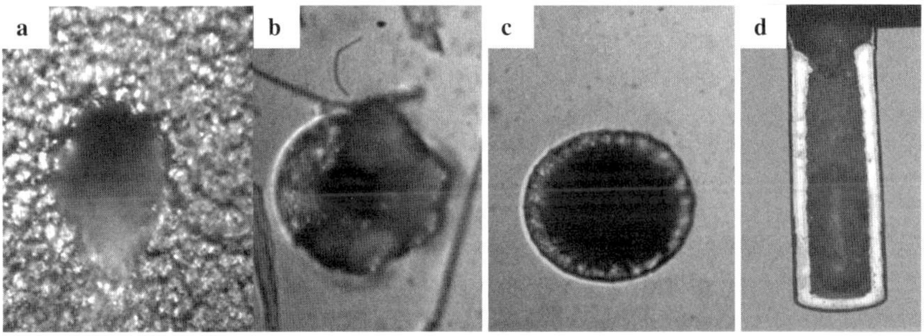

Figure 5. Optical images (a-c) and SEM cross-sections (d) of deep (200 μm) TSV structures without (a) and with (b-d) Selectroplating process. Deposition time was 60 min (a-b) and 15 min (c-d).

Figure 5 presents experimental data for the selective plating process applied to 200 μm deep TSVs. Small coupons (~2") of TSV wafers were subject to surface modification, followed by electrochemical plating in a beaker. The electroplating solution (WA-30 from Technic Inc.) and plating current (~20 mA/cm^2) were not optimized, resulting in high roughness of deposited copper. In some experiments, an ultrasonic bath was used to agitate the electrolyte during the long (~1 h) deposition process. There is no copper deposition on the field area after a 1 hour plating process, indicating that the protective film has a strong bond to the copper surface (which is very important for TSV applications).

Sheet resistance measurements

The resistance of 10 μm wide lines was measured on all wafers with > 90% yield, and with similar values of the resistance in the die with selective versus non-selective Cu deposition (see Figure 6). Approximately 10% or less of dies have shorts, detailed analysis of die maps shows that short are caused by underpolishing on wafer edge.

The yield for sheet resistance was lower on narrow lines (line width of 0.18 μm or 0.25 μm) lines, on the order 10-35 % yield. However, the sheet resistance is similar for both types of die (ie. selective Cu versus non-selective Cu). Smaller test features are more sensitive to contamination and process imperfection resulting in lower yield. The results suggest that the selective plating process does not degrade the resistivity of the Cu.

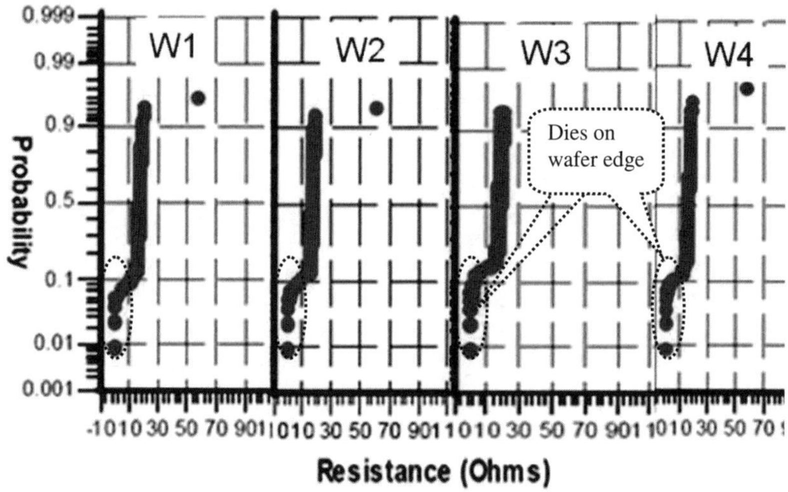

Figure 6. Resistance distribution for 10 μm lines for 4 wafers (W1-W4). Dies on the wafer edge are underpolished, resulting in "shorts".

CONCLUSIONS

Evaluation of Selectroplating technology is presented based on a comparison of the selective plating process to conventional plating. By applying a plating inhibitor on the field regions, selective Cu plating in trenches is achieved while still using conventional plating technologies. The plating inhibitor withstands the plating bath conditions for up to one hour, so it is compatible with thick Cu deposition, such as that used for global interconnects or TSVs. The sheet resistance of Cu interconnects fabricated with the selective process is similar to those fabricated with a conventional process, suggesting that the inhibiting agent does not affect the resistivity of Cu. The use of selective plating can reduce the reliance on CMP for clearing metal off the field regions as well as reduce the time and material used for ECD process. The results indicate that Selectroplating is a promising technology for interconnect processes.

REFERENCES

1. D. Edelstein, J. Heidenreich, R. Goldblatt, W. Cote, C. Uzoh, N. Lusting, P. Roper, T.McDevitt, W.Motsiff, A. Simon, J. Dukovic, R. Wachnik, H. Rathore, R. Schulz, L. Su, S. Luce, and J. Slattery, in Proc. IEEE Electron Device Meeting 1997, (IEEE, Piscataway, NJ, 1997), pp. 773-776.
2. M. Corbett, *Solid State Technology* April, 2008, p.52.
3. C. Uzoh and D.Edelstein, US Patent 6,251,528 (2001); C. Uzoh and D.Edelstein, US Patent 6,297,140 (2001); R. Volant et al., US Pat. 6368484, (2002); Chung et al., US patent 6605534 (2003); B. Basol, US Patent 6,921,551 (2005); C. Barns; US Patent 7,238,610 (2007).
4. J. Gambino, T. Stamper, H. Trombley, S. Luce, F. Allen, C. Weinstein, B. Reuter, M. Dunbar, V. Samek, P. McLaughlin*, T. Kane* in Materials, Technology and Reliability for Advanced Interconnects and Low-k Dielectrics, edited by A. J. McKerrow, J. (Jim) Leu, O. Kraft, T. Kikkawa (Mat. Res. Soc. Symp. Proc. Vol. 766, San-Francisco, 2003) pp. E5.5-9 .
5. J. Dunn et al., IBM J. Res. Dev., **47**, 101 (2003).
6. C.H. Chen, C.S. Chuang, C.P. Chao, J.F. Kuan, C.L. Chang, S.H. Wang, H.M. Hsu, W.Y. Lien, Y.C. Tsai, H.C. Lin, C.C. Wu, C.F. Huang, S.M. Chen, P.M. Tseng, C.W. Chen, C.C. Ku, T. Y. Lin, C.F. Chang, H.J. Lin, M.R. Tsai, S. Chen, C.F. Chen, M.Y. Wei, Y.J. Wang, J.C.H. Lin, W.M. Chen, C.C. Chang, M.C. King, C.M. Huang, C.T. Lin, J.C. Guo, G.J. Chern, D.D. Tang, J.Y.C. Sun, in Proc. IEEE Electron Device Meeting 2003, (IEEE, Piscataway, NJ, 2003), pp. 39-42.
7. Younan Xia, Enoch Kim, Milan Mrksich, and George M. Whitesides *Chem. Mater.*, ***8***, 601-603 (1996).
8. Laibinis, P. E.; Whitesides, G. M. *J. Am. Chem. Soc.*, **114**, 9022. (1992).
9. Soo-Kil Kim and Jae Jeong Kim *Electrochemical and Solid-State Letters*, **7** (9) C101 (2004).
10. Myung M. Sung, Kiwhan Sung, Chang G. Kim, Sun S. Lee, and Yunsoo Kim. *J. Phys. Chem. B*, **104**, 2273-2277 (2000).
11. Mirjana Metikoš-Huković, Ranko Babić, Željka Petrović, and Damir Posavec. *Journal of The Electrochemical Society*, **154**, 2 C138-C143 (2007).
12. J. Christopher Love, Lara A. Estroff, Jennah K. Kriebel, Ralph G. Nuzzo and George M. Whitesides*, Chem. Rev., **105**, 1103-1169 (2005).
13. A.K. Stamper, C. Adams, X. Chen, C. Christiansen, E. Cooney, W. Cote, J. Gambino, J. Gill, S. Luce, T. McDevitt, B. Porth, T. Spooner, A. Winslow, R. Wistrom, in Advanced Metallization Conference 2002 (AMC 2002), edited by B.M. Melnick, T.S. Cale, S. Zaima, T. Ohta Proc., (Mat. Res. Soc. Conf. Proc. V-18, 2003), pp. 485-492.

Low-k and ULK Dielectrics

Conference Proceedings AMC XXV © 2010 Materials Research Society

Silicon Precursor Development for Advanced Barrier Integration

Laura Matz*, Anupama Mallikarjunan, Raymond N. Vrtis, Andrew D. Johnson, Xuezhong Jiang, Mark L. O'Neill[2]
[1]Air Products and Chemicals, Inc., 7201 Hamilton Blvd, M/S R4203, Allentown, PA 18104
[2]Air Products and Chemicals, Inc., 1969 Palomar Oaks Way, Carlsbad, CA, 92011

ABSTRACT

For back end of line (BEOL) interconnects, maintaining device reliability while scaling the dielectric stack is a continual challenge. In this paper, the ability to improve the dielectric barrier film properties based on precursor engineering is demonstrated. Two examples for improvement are provided with precursors for SiCO and SiCN. In the case of SiCO, the precursor was optimized for maximizing the density value for a given dielectric constant without compromising the electrical properties. In the SiCN case, improvements in UV stability were established through precursor optimization relative to baseline 3MS-SiCN films. The basic film properties are important to understand, but performance is also contingent upon the interface engineering for advanced generations targeting improvements in electromigration lifetime. Further precursor development targeted improvements to Cu-SiCN interface engineering through identification of an adhesion promoter chemistry. The benefits of this precursor, including selectivity to Cu are discussed.

INTRODUCTION

For back end of line (BEOL) interconnects, maintaining device reliability while scaling the dielectric stack is a continual challenge. New barrier dielectrics have seen less frequent adoption owing to the stringent interconnect integration requirements, such as good copper-barrier adhesion for electromigration, stable electrical properties for time dependent dielectric breakdown (TDDB), and etch selectivity. Current industry standard films incorporate trimethylsilane (3MS) or tetramethylsilane (4MS) for depositing SiCN and SiCO dielectric films; with capacitance improvements being made through reduced film thickness, reduction of trench level etch stops in dual damascene integration and bilayer stack approaches with thin SiCN interfaces. The most advanced devices in manufacturing are using interlevel dielectrics (ILD) films with dielectric constant (k) values from 2.5-2.0[1,2], enabling improvements in capacitance metrics. The majority of ILD processes now implement a porogen/organosilicate glass co-deposition process that requires an ultraviolet curing process post deposition to remove the porogen in the OSG framework. The UV process poses challenges for the underlying barrier film exposure as a result of the transparent nature of porous OSG. In addition to capacitance needs for the barrier dielectric process, adhesion promotion to copper has also become an issue in advanced technology nodes due to dimensional scaling effects on electromigration performance. Traditionally, copper adhesion (primary metric for electromigration) has either been enhanced by reducing gas plasma pre-treatments[3] or alternative metallic interfaces. As these approaches reach the limits, focus is now being placed on k reductions of the barrier dielectric films as well as integration challenges introduced by the adoption of porous OSG materials. In this study, the benefit of precursor engineering to address barrier dielectric integration challenges is presented, both for bilayer films as well as single layer SiCN films.

EXPERIMENTAL

All depositions were performed using a 200 mm P5000 Applied Materials PECVD DXZ chamber with direct liquid injection. With the exception of 3MS results, all other precursors were liquid precursors with varied delivery temperatures dependent on the precursor's boiling point. A mercury probe was utilized for all film measurements where dielectric constant, electrical breakdown field and leakage are presented. Modulus and hardness were measured by nanoindentation and bonding properties of the dielectric films were

*matzlm@airproducts.com,

analyzed with a Nicolet transmission FTIR tool. All density measurements were performed with x-ray reflectivity using a 2-layer model. Experiments reported with ultraviolet stress change utilized a sweeping broadband UV source and film stress was measured before and after UV exposure for the reported exposure time. Electrical test structures comprised of M1 Cu lines with OSG inter-level dielectric (ILD) and a dense organo-silicate glass hardmask were evaluated after M1 CMP and CAP (copper adhesion promoter) processing with 50Å silicon nitride in order to evaluate the effect of CAP on resistance and leakage on Cu single damascene structures.

RESULTS

SiCO Precursor Optimization

Traditionally, process optimization has enabled reductions in k values and improvements in barrier properties such as film density and hermeticity[4] with SiCN films demonstrating k values of 4.8-5.5 and SiCO films with k values of 4.2-4.8. An alternative approach is to engineer the precursor to affect the desired network, targeting improved film structures. In Figure 1, precursor screening results of various oxygen-containing silicon precursors for improved SiCO films is shown. The target for this screening was to maintain film density (nominal 1.9-2.0g/cc for 3MS SiCN and SiCO) while reducing the k value to less than 4.2. In the case of a 3MS process to form dense barriers, the nitrogen or oxygen doping is controlled through the addition of reactant gases where carbon dioxide or oxygen can be used for SiCO and ammonia can be used for SiCN films. As the thickness is scaled, the continuity of the films becomes critical to the reliability. In this study, the development effort was focused on how to improve the precursor efficiency in the plasma for maximizing the desired film properties. In the case of SiCO, oxygen-containing silicon precursors were tested where the oxygen is bonded directly to the silicon rather than incorporation as a dopant.

Our metric for comparison was to evaluate the density versus k value trends for a given precursor over a range of test conditions, with the goal being higher density for the same k value. As observed in Figure 1, initial testing of LKB (lower k barriers) 200, 300, and 400 showed differences in the density to k value relationships, even when identical process conditions were tested. More specifically, the LKB-200 and 400 differed in the type of carbon that was incorporated in the film. Although the 200 and 400 films have similar k value ranges, the density is improved with the structure of 400 due to modifications to the network bonding. The LKB 300 and 400 precursors varied in the oxygen to carbon ratio. Out of this initial structural comparison, LKB-500 was identified to have optimal structure based on the appropriate tuning of carbon (k reduction) and oxygen content (higher density).

While carbon introduction can reduce k, it may negatively affect electrical properties. As demonstrated in Figure 2, the IV ramp characteristics of two LKB-500 SiCO films with k values of 3.7 and 4.0 are shown to have lower leakage and higher electrical breakdown characteristics relative to a standard 3MS-SiCN film (k=5.1). Interestingly, the IV curve for the k=3.7 film shows a lower leakage at low electrical fields, even though the film incorporates more carbon in the film. In a 32nm integration utilizing a 300 Å bilayer stack (200 Å SiCO / 100 Å SiCN), replacing the SiCO material with LKB-500 film at k=3.7 would reduce the capacitance by ~5% when compared to a monolayer SiCN 300Å film. This improvement can be achieved without a significant reduction in the overall density of the bilayer film.

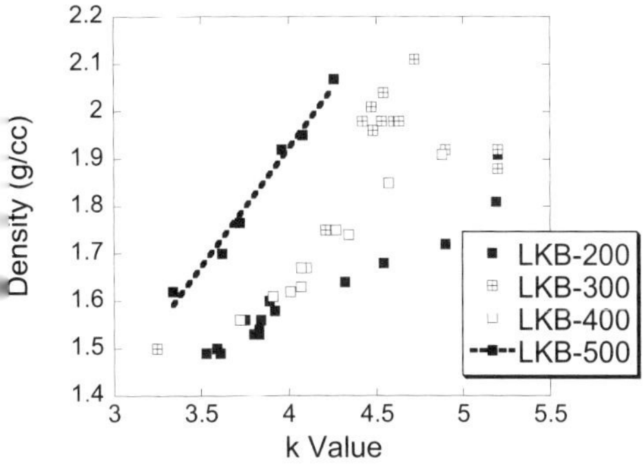

Figure 1: Oxygen-containing silicon precursor screening for highest density with reduced k values. LKB-500 structure optimized for efficient k reduction

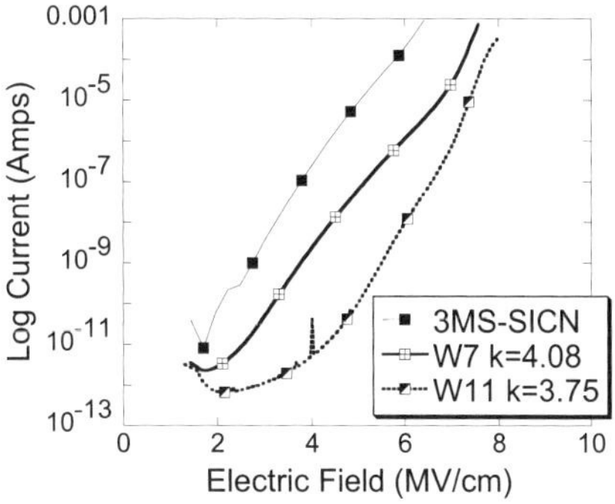

Figure 2: IV Characteristics for LKB-500 films compared with 3MS-SiCN

One challenge in PECVD is the ability to predict the optimum process conditions for rapid film development. In the LKB-500 deposition process that is demonstrated above, the best conditions found were generated through reaction of the primary dopant gas, hydrogen, which provided control of the chemical bonding and film properties. In Figure 3, the FTIR comparison of a 3MS-SiCN film and LKB-500 (k=4.08) film are presented. Shifts in the Si-H bonding are due to differences in the network bonding. For the SiCO films, the primary bonding observed is networked Si-C (e. g. silicon carbide at ~800cm^{-1}), Si-H, Si-O, Si-CH_3 and C-H_x (around 2800-3000cm^{-1}). In Figure 4, the correlation between Si-CH_3 bonding and refractive index demonstrates the ability to easily predict the film properties. For this series of films, the modulation of k value originates from the conversion of Si-CH_3 bonds to Si-O and networked Si-C species in the backbone, producing a higher refractive index that directly correlated to higher k values. The Si-CH_3 quantification is based on the integration of the Si-CH_3 stretch at 1270cm^{-1} from FTIR analysis.

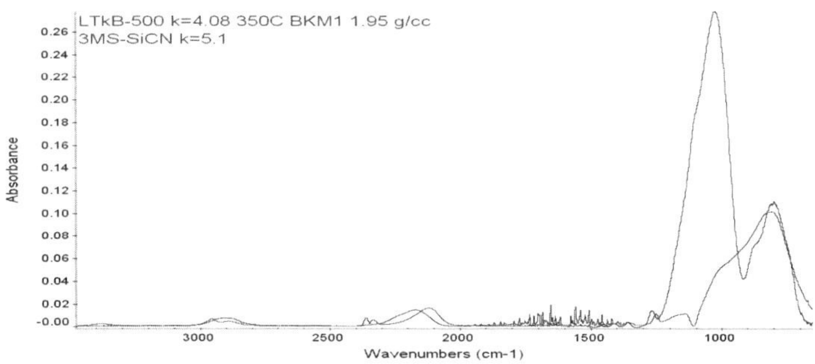

Figure 3: FTIR comparison of 3MS-SiCN film and LKB-500 SiCO film properties

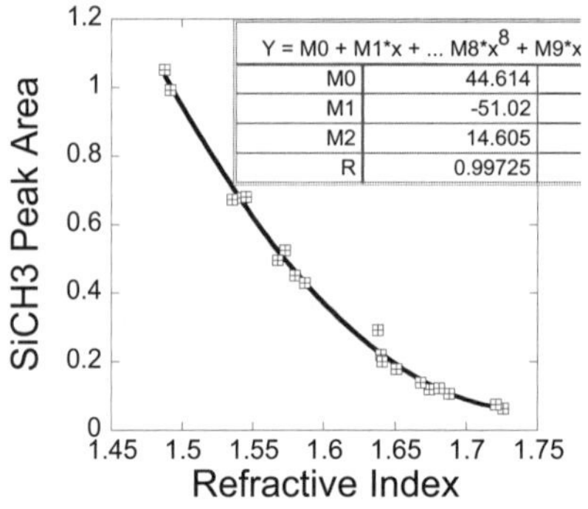

Figure 4: Correlation between refractive index and $SiCH_3$ bonding within SiCO films deposited using LKB -500

For other applications such as displays there is a need to reduce the temperature of the deposition process while maintaining the barrier properties. In Figure 5, the influence of temperature on LKB-500 film properties under the same deposition conditions is demonstrated. The refractive index is reduced significantly with lower temperature, indicating more Si-CH_3 in the film (based on the correlation in Figure 4) and reduced dielectric constants. By increasing the ratio of H_2:LKB-500 in the plasma at lower deposition temperatures, RI, k value and densities were achieved similar to those demonstrated at higher temperatures. Optimization of the H_2:LKB-500 ratio for each temperature produced films of similar density and k value within a deposition temperature range of 150 °C-350 °C as demonstrated in Figure 6. The ability to easily tune and control the film properties under PECVD conditions originates from incorporating the desired bonding in the precursor. Control of the fragmentation of the precursor in the plasma was obtained by using H_2 with the precursor, which appears to target carbon containing portions of the precursor. This example demonstrates the value in designing the precursor for better control of the PECVD process and film structure.

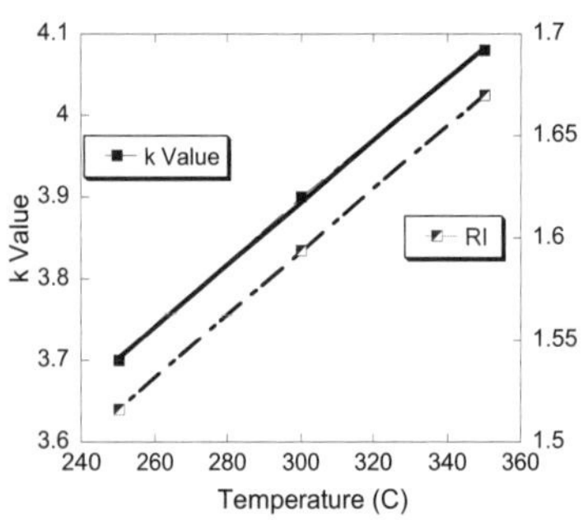

Figure 5: Effect of deposition temperature on RI and k value modulation of LKB-500 film properties under same PECVD reactor conditions

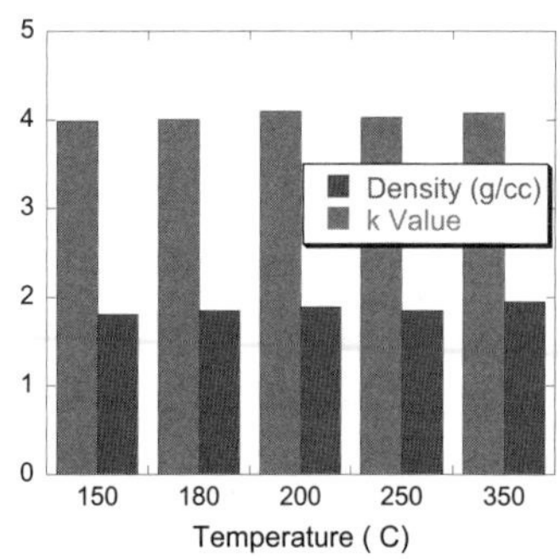

Figure 6: Density and k value for LKB-500 films deposited from deposition temperatures of 350 °C-150 °C.

SiCN Precursor Optimization- UV Stability

Similar to SiCO films, incorporation of nitrogen bonding into the silicon precursor may enable more efficient deposition in SiCN films. In Figure 7, the comparison of a 3MS-SiCN DOE with two alternative nitrogen-containing silicon precursors is presented (designated as BASICN precursors). BASICN 100 and 200 differ significantly in the nitrogen bonding and as such, demonstrate different k versus density trends. The higher density for BASICN 200 suggests closer to the optimum nitrogen structure within the precursor. In comparing the structures for BASICN 100 and 200, the 200 molecule contains the same number of nitrogens but a higher carbon content in the precursor. Initially, it might be expected that more carbon would create a less dense film, however the carbon bonding and corresponding nitrogen environment seem to play a strong role in the density trends. When looking at the IR analysis for the BASICN 200 in Figure 8 there is evidence of NH_x species in the IR at ~3400cm^{-1} even though the deposition process does not utilize ammonia or another external nitrogen source. Interestingly, XPS for a 3MS-SiCN film and BASICN 200 film both demonstrate similar carbon levels in the film, but the level of nitrogen is much greater with the BASICN 200 as reported in Table 1.

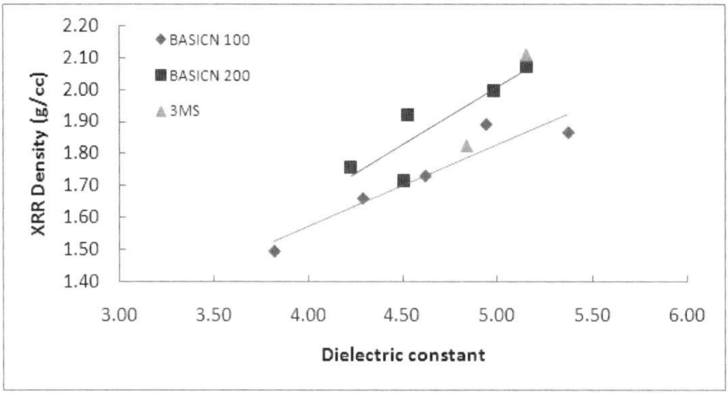

Figure 7: Dielectric constant versus density trends for 2 nitrogen-containing silicon precursors compared to 3MS films

K=5 Films	%C	%N	%O	%Si	N/Si	C/Si
3MS/NH3	40.1	8.6	4.3	47.1	0.18	0.85
BASICN 200	41.9	19	2	37.1	0.51	1.13

Table 1: Comparison of atomic percent composition measured by XPS for 3MS-SiCN and BASICN 200 films. Although carbon incorporation is similar; nitrogen incorporation is greater with BASICN 200.

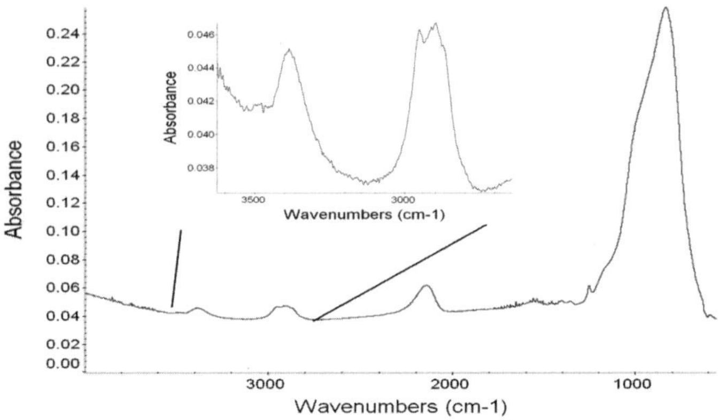

Figure 8: FTIR of BASICN 200 film k=5.0 deposited at 350C

K=5 Films	%C	%N	%O	%Si	N/Si	C/Si
3MS/NH3	40.1	8.6	4.3	47.1	0.18	0.85
BASICN 200	41.9	19	2	37.1	0.51	1.13

Table 1: Comparison of atomic percent composition measured by XPS for 3MS-SiCN and BASICN 20 films. Although carbon incorporation is similar; nitrogen incorporation is greater with BASICN 200.

Post-deposition ultraviolet (UV) curing associated with porous low k adoption may lead to fil stability issues with existing SiCN films. The transparency of the low k films permits the UV radiation t interact with the underlying films in addition to influencing the properties of the low k itself. Internal testir of 3MS-SiCN films exposed to UV curing showed that the films undergo a stress change from compressive t tensile, as demonstrated in Figure 9. In further testing it has been found that 3MS-SiCN films with varie process conditions demonstrate different sensitivity levels to UV exposure, but all films tested eventual result in tensile film stress. Review of the film bonding with 3MS films demonstrates both Si-CH_3 and Si- reductions with UV exposure, although it appears that the sensitivity of the films is not necessarily due to the presence of these bonds in the film but rather the influence of the surrounding network. In contrast to 3MS, BASICN 200 films demonstrate a wide compressive stress window as presented in Figure 10. By varying the deposition conditions in the process, the film stress for BASICN 200 can be varied from 200-800MPa compressive stress demonstrating significantly higher compressive stress values than any comparable experiments performed with 3MS films. For the BASICN 200 films the more compressive films are also higher k value films with an upper limit for this deposition process of k ~ 5.5. Review of the structural bonding in the BASICN 200 film shows that the higher stress films contain more NH_x species and greater Si-N bonding, as eluded to based from the backbone peak at 1000cm^{-1} shifting to higher wavenumbers. This bonding in the film typically leads to higher k values due to the polarity of the Si-N bonding.

Upon extended direct UV cure of 12 minutes, the BASICN 200 films remain compressive as shown n Figure 11. Extensive DOE testing for 3MS films were similarly performed but conditions were not found hat enabled the films to retain compressive film stress. In the BASICN 200 films, there is some stress change that occurs upon UV exposure however, the high initial stress values minimize the effect and enable compressive film stress to exist even after long exposure times. Comparison of film bonding properties to understand this effect showed that several conditions for BASICN 200 demonstrated little film property change. This has not been true for all films that undergo stress change with UV exposure. In the case of the LKB-500 films doped with nitrogen, although the films were high density, the Si-CH_3 and Si-H bonding was reduced and silanol groups were observed in the IR analysis post UV cure. The unique properties of BASICN 200 films are attributed to the enhanced stability of the film bonding given the incorporation of network components into the precursor. This demonstrates the value of structural tuning rather than solely process optimization to meet integration needs.

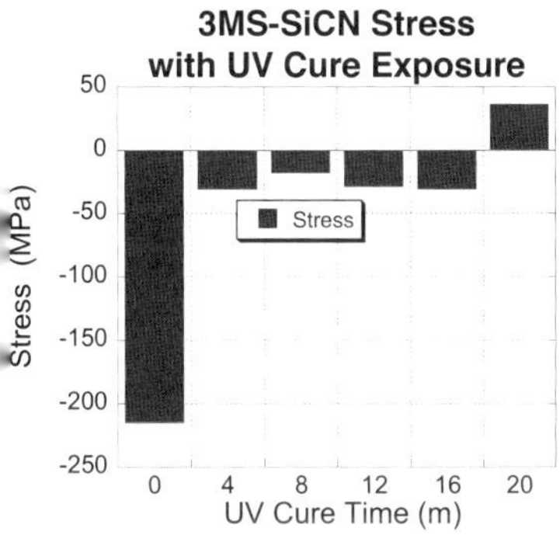

Figure 9: Influence of UV cure time on 3MS-SiCN Stress change

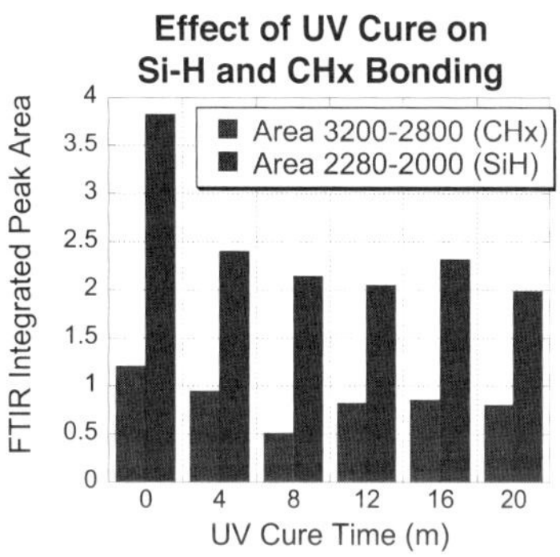

Figure 10: Effect of UV exposure on 3MS-SiCN bonding properties for CH_x region and Si-H region of FTIR

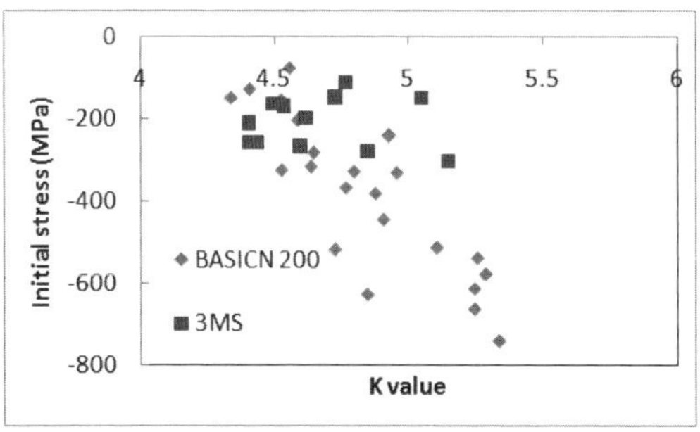

Figure 11: As-deposited film stress range for 3MS SiCN films compared to similar conditions for BASICN 200

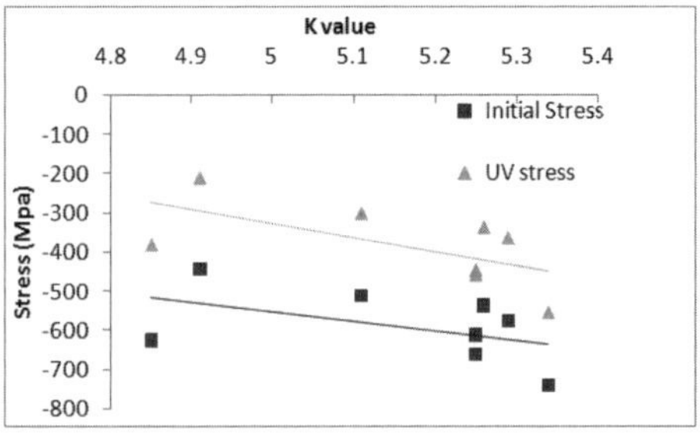

Figure 12: Comparison of initial film stress and film stress after 12 minute UV exposures

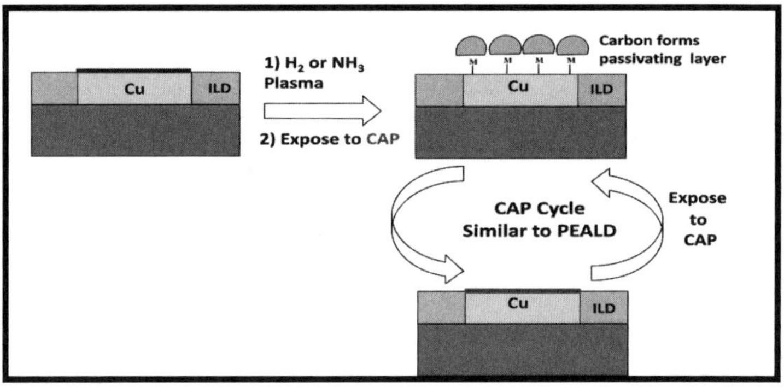

Figure 13: Diagram of CAP adhesion promotion process

Cu-SiCN Adhesion Promotion with CAP precursor

The final integration challenge for a barrier dielectric process is the copper and SiCN interface layer, which is one of the primary regions for early electromigration failure[3]. There have been several studies demonstrating the improvement of electromigration resistance resulting from the introduction of a metal layer between the copper interconnect line and barrier dielectrics (typically SiCN), such as electroless CoWP, CVD Co[7], and CVD Ru[8] between the copper and the barrier dielectric. However, it has proven difficult to maintain the selective deposition of these materials. In contrast to chemical vapor deposition precursors, it is believed that a vapor phase stable organometallic precursor such as (copper adhesion promoter) CAP[5] can enable atomic layer deposition at the copper surface without causing leakage issues within the porous low k films. This effect originates from high surface selectivity of the CAP precursor to adsorb to the Cu surface but not on the ULK or damaged ULK surface. We have previously demonstrated increased adhesion with CAP at the Cu/SiCN interface with little impact to blanket ULK and Cu properties (leakage and sheet resistance, respectively)[5]. Other processes that have good selectivity to low k films have been reported such as copper silicidation or CuGe formation, however these processes typically suffer from increased resistance due to Si or Ge diffusion into the bulk Cu lines. This effect presents a challenge as resistance requirements for advanced technology nodes are already under pressure. The scaling effects for Cu and low k films at the

2nm and 22nm nodes highlight that an adhesion promoting process requires good selectivity and little olubility in the bulk Cu.

Prior reports for the CAP process have demonstrated that blanket resistivity on Cu showed no vidence of increasing resistance. In Figure 14, single level metal electrical testing of the CAP precursor is resented. Unlike other selective processes such as $CuSiN_x$ formation using SiH4 which leads to an increase n Rs, the CAP process shows no impact to the line resistance even with extended CAP (7X CAP) exposure. he higher line Rs seen with the H_2 pretreatment in Figure 14 is suspected to be the result of an unoptimized ilicon nitride deposition process where the SiH_4 / NH_3 gases are allowed to stabilize flow before striking lasma thus forming CuSi on the surface. It is interesting that the resistance results are reduced with the CAP rocess, suggesting that the presence of CAP at the copper interface may potentially minimize silicon iffusion from the seasoning and copper reduction steps. For metallic interfaces such as CoWP, the major ssue has related to leakage challenges between minimum pitch electrical lines. In contrast, the example hown below with both short (1X) and long (7X) CAP exposures demonstrates little influence on the line to ine leakage. From this data set, the 7X CAP split actually shows the lowest level of leakage, providing an dditional degree of evidence in the CAP selectivity to copper rather than any appreciable deposition on the ielectric surface. As in the prior examples of precursor optimization for SiCN and SiCO films, tuning the recursor structure is an effective means to optimize materials properties and address new integration hurdles. he next generation of interconnects will require refined film properties, pushing the limits of interconnect eliability and requiring new approaches such as precursor optimization to address these challenges.

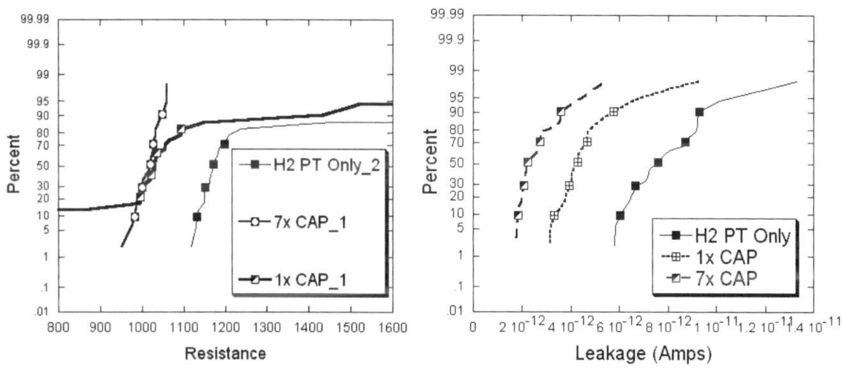

Figure 14: Effect of CAP process exposure to leakage and resistance in single damascene copper lines. H_2 pretreatment only process is compared to short (1XCAP) and long (7XCAP) exposures to understand the influence of CAP exposure on copper structures.

REFERENCES

[1] D. Restaino, S.Molis, A.Grill, V.Patel, M.W.Lane, T.Cheng, A.Demos, S.M.Gates, D.Kioussis, N.Klymko, H.Liu, S.Nguyen, V.Nguyen, J.C.Patel, S.Reiter, H.Shobha, J.Widodo, K.Yim, D.Edelstein, in Advanced Metallization Conference Proceedings, edited by A.J.McKerrow, Y.Shacham-Diamand, S.Shingubara, and Y.Shimogaki, (Materials Research Society Conference, Albany, NY) p. 419-426.

[2] R.N.Vrtis et al., U.S.Patent # 6846515, 25 January 2005.
[3] A.K. Stamper, H.Baks, E.Cooney, L.Gignac, J.Gill, CKHu, T.Kane, E.Liniger, Y.Y.Wang, J.Wynne in Advanced Metallization Conference Proccedings XXI, edited by S.Brongersma, T.C.Taylor, M.Tsujimura, and K.Masu, (Materials Research Society Conference, Colorado Springs, CO) pp. 727-733.
[4] L.Matz, T.Tsui, E.R.Engbrecht, K.Taylor, G.Haase, S.Ajmera, R.Kuan, A.Griffin, R.Kraft, A.McKerrow in Advanced Metallization Conference Proccedings XXI, edited by S.Brongersma, T.C.Taylor, M.Tsujimura, and K.Masu, (Materials Research Society Conference, Colorado Springs, CO) pp. 437-443.
[5] R. Vrtis, L.Matz, M.L.O'Neill, MRS Proceedings, 2008

Conference Proceedings AMC XXV © 2010 Materials Research Society

Restoration of plasma damaged porous ultra low-k SiOCH films: A coating process with UV activation versus a vapor phase process with thermal activation.

Thomas Oszinda[1], Matthias Schaller[2], Susanne Leppack[2], Stefan E. Schulz[3]
[1] Fraunhofer CNT, Königsbrückerstraße 180, 01099 Dresden, Germany
[2] GLOBALFOUNDRIES Dresden Module Two GmbH & Co KG, Dresden, Germany
[3] TU Chemnitz and Fraunhofer ENAS, Chemnitz, Germany

ABSTRACT

Two processes to repair plasma damaged porous ultra low-κ SiOCH dielectric were studied. One process is based on a spin on approach in combination with UV radiation. The second process uses a vapor phase reaction chamber, while the silylation reaction is thermally activated. Comparison of both processes was achieved by applying four different silylation agents. After processing of blanket wafers, the restoration performance was characterized by different analytical techniques like Fourier transformed infrared and Auger electron spectroscopy as well as contact angle and mercury probe measurements. Electrical tests on patterned wafers were carried out only for the spin on UV process. Both repair approaches show variable restoration performance, while the vapor phase process seems to be favorable. It is suggested that the reason for this advantage is related to a continuous chemical delivery and a better diffusion behavior of chemicals at elevated temperatures. Restoration of the surface free energy and partial recovery of the κ-value was obtained for both processes. This suggests that the supported energy, i.e. by UV or high temperature, is sufficient to activate the silylation reaction.

INTRODUCTION

Ultra low-κ dielectric (ULK) layers face a lot of difficulties especially with respect to plasma processing after patterning of such layers [1]. The requirement of a κ-value of smaller than 2.5 [2] is achieved by the introduction of carbon and porosity into SiO_2 layers. It is reported that during plasma processes the ULK film is damaged, which is mainly related to carbon loss [3]. This is caused by smaller dissociation energy for Si-C bonds (~ 447 kJ/mol) compared to Si-O bonds (~ 800 kJ/mol) [4]. It was observed that during the carbon depletion dangling bonds occur [5], which form new Si-O or Si-OH bonds [6] after exposing the damaged film to atmosphere or during subsequent plasma processing. Hence, an unfavorable degradation of the dielectric properties occurs, i.e. increasing κ-value [7]. For ULK layer restoration, silylation based repair processes are proposed [8]. During the repair process mainly Si–OH bonds are replaced by $Si–O–Si(CH_3)_x$, using e.g. supercritical CO_2 [9,10], vapor phase [11,12] or spin on processes. The aim of the restoration treatment is to replace polar silanol groups by less polar tri-, di- or monomethylsilyl groups. Many factors influence the silylation reaction like kind of reactive group and number of reactive groups in the repair chemical as well as steric hindrance effects [8]. Due to steric hindrance, chemicals with one reactive group like Hexamethyldisilazane (HMDS) are not satisfactory. Those chemicals do not react with all existing silanol groups, since one trimethylsilygroup can block neighboring silanol groups [13]. Chemicals with two or three reactive groups tend to polymerize, which is stronger for chemicals with three reactive groups. Therefore, Nitta et al. [8] suggested chemicals with two reactive groups as most promising for silylation based repair approaches. In this study HMDS as a baseline chemical and three other chemicals having two reactive groups were used for this reason.

EXPERIMENT

500nm chemical vapor deposited ultra low-κ (ULK) SiOCH dielectric films were fabricated on blanket silicon wafers with an initial porosity of ~ 25% and an average pore diameter of ~ 1.5 nm. These layers were treated by a photo resist removal plasma process for 60 seconds, using H_2 as the main gas component. This plasma process was found to leave the pore structure open [14], which is needed for the diffusion of repair chemical into the damage region. Furthermore, patterned wafers were prepared using typical 45 nm production processes at GLOBALFOUNDRIES. To restore the damage, pieces of blanket wafers were treated under two different process conditions. Process A uses a coating based spin on treatment to deposit the repair chemical for 1 min on the wafer. Afterwards the wafer is exposed to UV radiation. The broad band UV radiation was applied also for one minute without and with thermal support, provided by an infrared flash array. For process A, a Semitool™ Raider platform was equipped with a coating chamber and an UV system for 300 mm wafers. In process B the repair chemicals were carried into a closed reaction chamber by using nitrogen gas during the whole activation process. Starting at ~ 60 °C the final temperature of > 300 °C was reached after more then 30 minutes. Afterwards, the temperature was hold for one more minute before cooling. For patterned wafers only process A was applied.

Four different repair chemicals were used to study the effect of both processes. Besides HMDS with one reactive and three methyl groups, also other repair chemicals having two reactive and two methyl groups were used [15]. Repair chemical two, three and four (RC2, RC3 and RC4, respectively) are helpful in reducing problems of steric hindrance during the repair process. The sizes (width to length) of these repair chemicals are ~ 0.56 nm to 0.67 nm (HMDS or RC1), ~ 0.50 nm to 0.61 nm (RC2), ~ 0.53 nm to 0.64 nm (RC3) and 0.78 nm to 0.94 nm (RC4). More details, like formula or on the chemicals can't Blanket wafers were characterized using FTIR (in transmission), AES carbon depth profiling, contact angle and mercury probe measurements. For characterization of the electrical performance test structures in the metal level N, where the restoration process was applied, were used to determine the impact of the different repair chemical. RC values were measured after processing of metal level N and after processing of the next metal level N+1.

RESULTS AND DISCUSSION

<u>Investigation of structural changes by using FTIR spectra</u>

From measured transmission spectra absorbance spectra have been derived. For a better analysis difference spectra were computed by subtracting the reference spectrum from the respective absorbance spectrum. Difference spectra of damaged and differently treated ULK films are shown in Fig. 1. All difference spectra numbered with "1" (Ref. - 1, A - 1 and B - 1) indicate vibration modes related to silanol groups and physically adsorbed water. The UV only treatment shows an increase of these features, while all other UV assisted treatments in Fig. 1 (Ref. - 1 and A - 1) do not show appreciable changes. In comparison repair process B shows for all repair chemicals a decrease in this FTIR region. The difference spectra numbered with "2"

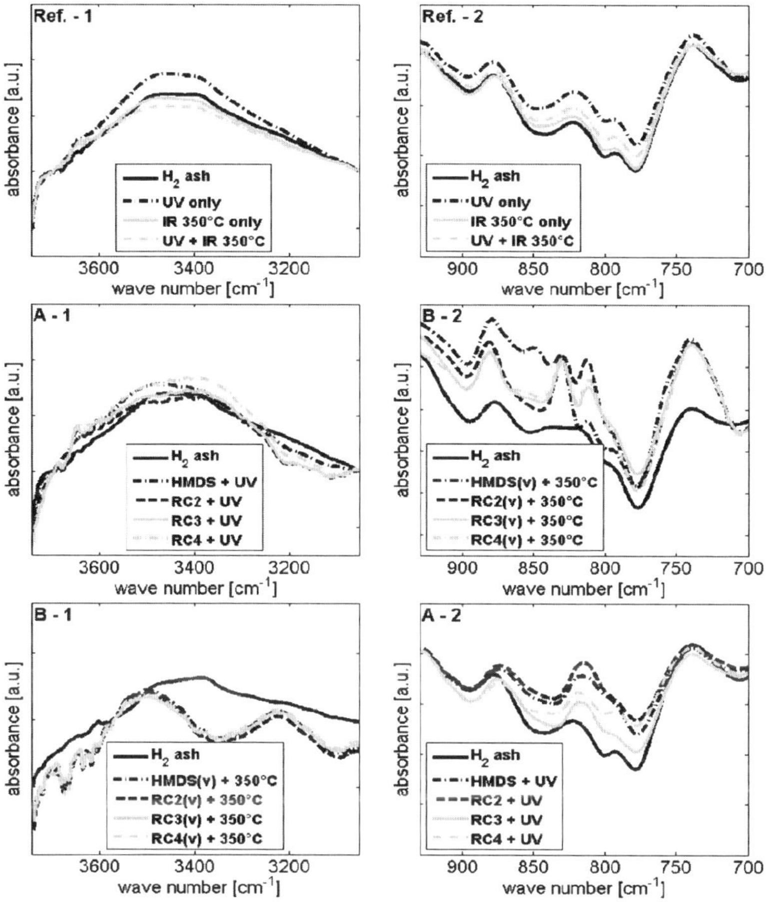

Figure 1. FTIR difference spectra of ULK changes due to UV treatment (Ref.), repair chemical coating and UV treatment (A) and repair chemical vapor treatment at 350 °C (B). FTIR regions are 3000 cm^{-1} to 3700 cm^{-1} for silanol and physical adsorbed water vibration modes (1) and 700 cm^{-1} to 950 cm^{-1} for Si-C related vibration modes (2).

(Ref. - 2, A - 2 and B - 2) show Si-C related vibration modes. Due to the H_2 ash these modes are reduced over a broad region. After restoration, Si-C related peaks increase after process B, while no specific change for the UV only treatment and process A can be found. In process A only a broad increase of Si-C related vibration modes is visible. This suggests the formation of a dielectric layer with a few monolayer thicknesses and a less ordered crystalline structure. Applying process B, for all repair chemicals three peak changes (Si-C/C-H_3 at ~ 850 cm^{-1}, at

~ 830 cm^{-1} and Si-O-Si at ~ 730 cm^{-1}) are visible. HMDS shows a further feature at ~ 880 cm^{-1} while all chemicals with two reactive groups show an additional peak at ~ 800cm^{-1}. Probably tri- and dimethylsilyl groups result in a different silylation reaction (reaction with one or two silanol groups at the same time). These specific and clear FTIR peaks support the conclusion of a more pronounced silylation reaction using a vapor phase repair process (B) compared to a spin on process (A).

Carbon depth profiling using Auger electron spectroscopy

For carbon depth profiles the sputtering time was converted into depth scales, as described earlier [16]. Fig. 2 shows the AES carbon depth profiles for the same samples. For the UV, IR and UV/IR only treatments (Ref.) the carbon depletion increases compared to the initial damage state. After plasma processing the damage region has a less strongly ordered and less linked bond

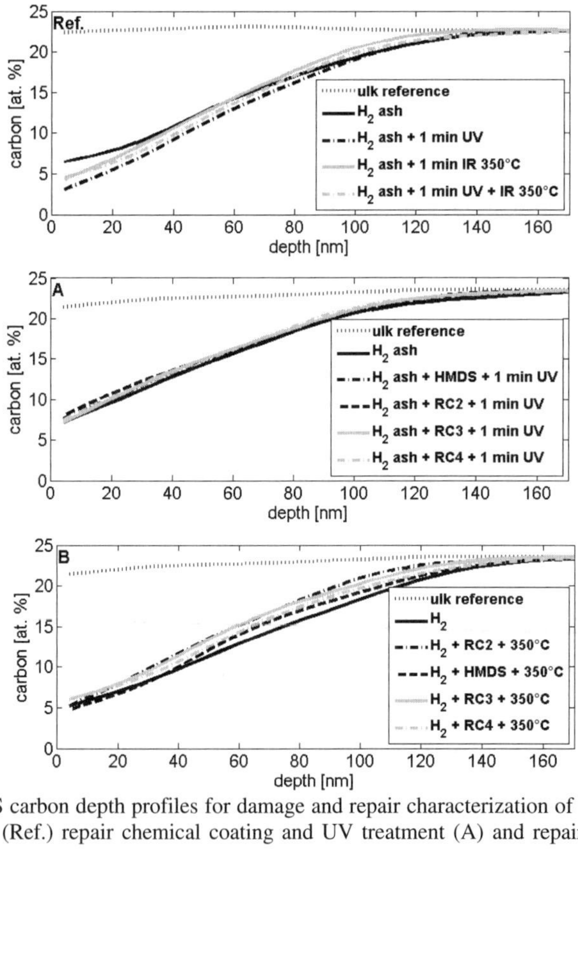

Figure 2. AES carbon depth profiles for damage and repair characterization of ULK layers after UV treatment (Ref.) repair chemical coating and UV treatment (A) and repair chemical vapor treatment (C).

structure. In consequence, an additional energy feeding leads into further release of carbon, especially for the surface near damage region. Applying process A no clear increase of carbon in the bulk-near and the surface-near damage region can be found. But the chemicals seem to inhibit further carbon depletion. After process B, the carbon content was increased in comparison to the damaged film showing difference for the respective repair chemicals. RC2 seems to be most effective. This might be due to its small size and hence, a favorable diffusion behavior. In the surface-near damage region the repair effect is much smaller than in the bulk. Near the surface the loss of Si-C leads to less Si-OH but to more Si-O-Si bonds. These cross-linked bonds cannot be restored using silylation processes explaining the low restoration rate. In the bulk-near damage region a restoration of more then 50% could be achieved, supporting this assumption. In the bulk-near damage region the impact of plasma species is less pronounced and therefore, the carbon is less frequently removed. Hence, neighboring silanol groups which can form new Si-O-Si bonds, also occur less frequently.

Surface Free Energy from contact angle measurements

The surface free energy (SFE) of the investigated samples is illustrated in Fig. 3. Evaluation of the SFE for damaged ULK surfaces with its polar and dispersive contribution was published recently [14]. The dispersive ULK surface is first transformed into a polar surface after H_2 plasma treatment. The polarity of the surface does not change after exposure to UV, IR or UV/IR radiation (Ref. in Fig. 3). After repair process A and B the surface free energy was decreased to < 40mN/m which is comparable to the SFE of pristine ULK. Here process A and B show similar trends in restoring the surface free energy. Hence, it can be assumed that both UV and thermal activation are in principle appropriate for silylation based repair processes. The ratio of polar to dispersive part of the SFE depends mainly on the repair chemical and not on the kind of activation, i.e. UV or thermal treatment. HMDS and RC3 are most efficient in restoring the hydrophobic character of the ULK surface. RC2 and RC4 are both chemicals having reactive groups with oxygen atoms. Probably, the restoration at the surface goes along with a formation of a thin dielectric layer. Incorporation of oxygen into this layer would then lead into a surface free energy with a higher polar contribution.

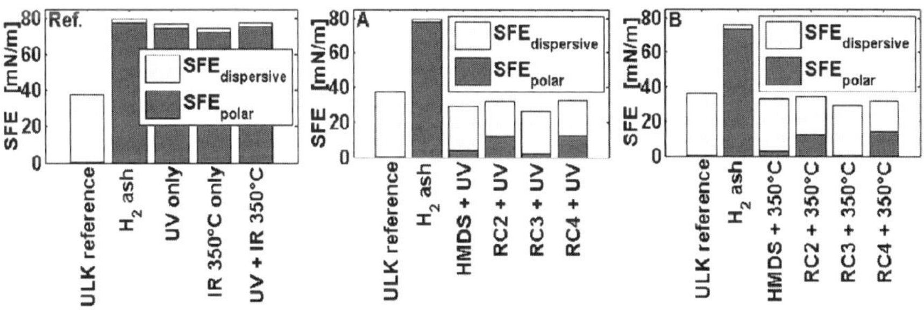

Figure 3. Surface free energies (SFE) from contact angle measurements of ULK layer surfaces after UV treatment (Ref.), repair chemical coating and UV treatment (A) and repair chemical vapor treatment (B).

K-value and leakage current from mercury probe measurements

The κ-value and the leakage current of all samples are plotted in Fig. 4 and Fig. 5, respectively. After UV and thermal treatment (Ref.) the κ-value increases slightly due to UV or IR treatment and strongly for UV in combination with IR. The leakage current does not change for UV only and IR only and decreases for UV/IR treatment. The energy feeding due to the combination of UV and IR probably leads to a densification, which would explain the κ-value increase and the leakage current decrease. UV/IR probably can cause a reduction of lattice vacancies and change of the pore size orientation as a result of a structural rearrangement under formation of new Si-O-Si bonds. The energy support using UV/IR radiation is higher and therefore promotes this rearrangement much stronger. Process A in comparison, provides a decrease of κ-value and leakage current by applying UV radiation after treating the layers with repair chemical, see Fig. 4(B) and Fig. 5(B), respectively. Interestingly, this improvement occurs although no substantial carbon restoration has been observed in FTIR and AES measurements. RC4, which has a high boiling point, shows the best κ-value improvement. This chemical does evaporate decelerated (T_B > 170 °C) during the UV process (T_{End} ~ 170 °C). Hence, the capping by formation of a thin dielectric layer is much likely, which can cause this κ-value improvement [17, 18]. The most efficient leakage improvement was found for RC2 and RC4, which cannot be explained so far. The κ-value improvement after process B is more realistic. As shown in FTIR difference spectra and AES depth profiles an incorporation of carbon takes place. The restoration

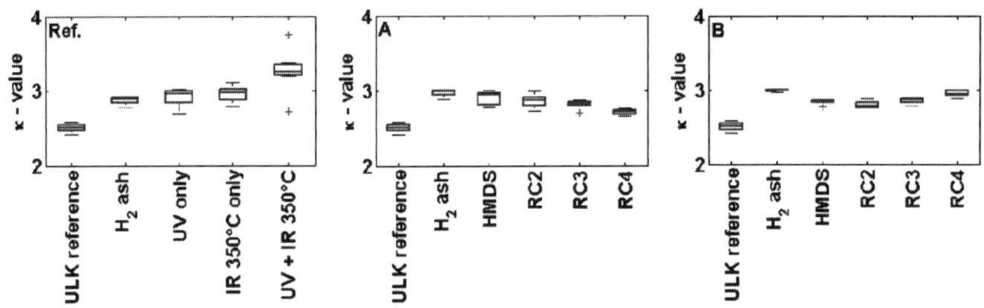

Figure 4. K-value from mercury probe measurements of ULK layers after UV treatment (Ref.), repair chemical coating and UV treatment (A) and repair chemical vapor treatment (B).

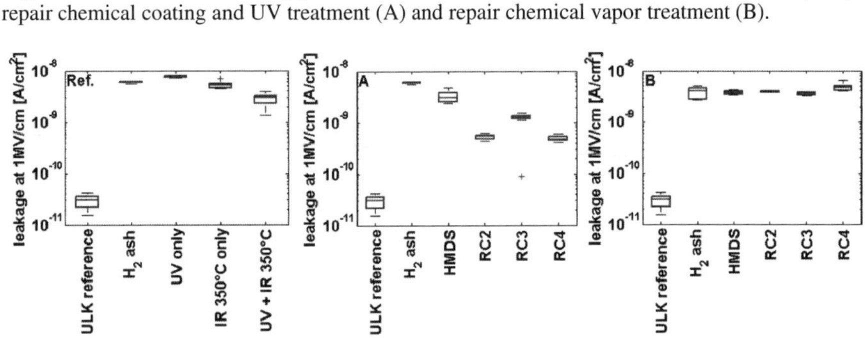

Figure 5. Leakage from mercury probe measurements of ULK layer surfaces after UV treatment (Ref.), repair chemical coating and UV treatment (A) and repair chemical vapor treatment (B).

can be considered as a "true" restoration. HMDS, RC2 and RC3 provide a κ-value improvement of ~ 0.2 (equal to ~ 40%). All chemicals have a relatively high vapor pressure and a low boiling point. Hence, an adequate amount of chemicals is provided during the whole vapor phase process. Most efficient in κ-value restoration is RC2. This correlates with the highest amount of incorporated carbon found for RC2 as illustrated in Fig.2 B. RC4 has a very small vapor pressure and is the largest molecule. Its size is only slightly smaller then the average pore size in the damage near surface region (~ 1.1 to 1.2 nm) [19]. It can be assumed that less molecules of RC4 are delivered by the carrier gas and probably less chemical diffuses into the damage region. Hence, κ-value recovery for RC4 was found to be small. The leakage current does not really change for all chemicals, which is surprising and cannot be explained up to know.

As κ-value improvement was found even for the spin on process without incorporation of carbon at the same time, different effects seem to be important. Carbon incorporation and formation of a very thin dielectric film can explain the reduction of the κ-value. The assumption of a layer formation is supported by the fact that the applied chemicals are similar or the same as the ones used for ULK CVD processes. To distinguish the two contributions to the κ-value decrease is difficult, in particular for the vapor process, where we found evidence for both.

Resistivity capacitance (RC) product

RC measurements were performed only for process A. In Fig. 4 the RC behavior is shown. After the repair experiments the wafers were shipped to GLOBALFOUNDRIES. There standard metallization processes were performed. First the metal process at the level of repair (level N)

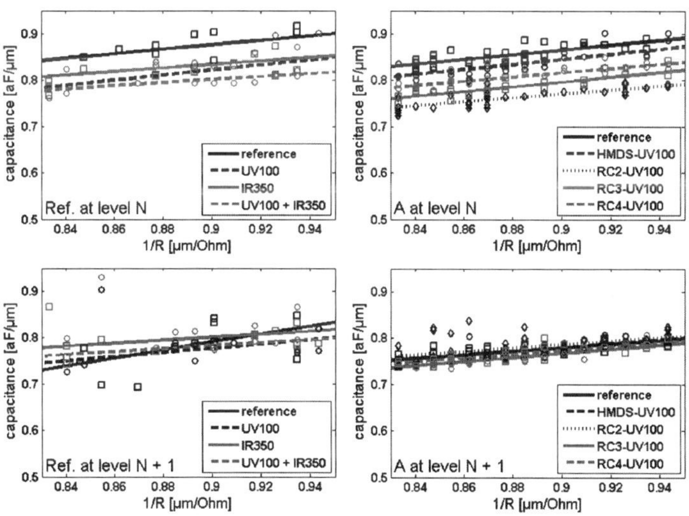

Figure 6. RC change of patterned structures after UV treatment (Ref.) and coating based repair processes in combination with UV (A). RC behavior were measured after restoration and processing of the same metal level (level N) and after processing of the next level (level N+1)

was carried out and RC measured. Afterwards the complete process for the next metal level (level N+1) was done. Again the first metal level N was measured to study the remaining RC improvement at this level. An improvement of RC values has been achieved by restoration process A, after complete processing of metal level N. HMDS shows the lowest and RC2 the highest impact. The thermally assisted UV and the UV only treatments also show improvement in the RC plot. This can be related to desorption of water and to conversion of Si-OH in new Si-O-Si bonds. After complete processing of the next metal level N+1, all RC values provide the same amount of RC improvement, see Tab. 1. In both cases even the reference shows the same amount of RC improvement compared to the treated samples. The difference of both references (reference of the experiment with chemicals is worse to those without chemicals) is ~ 4% at the level N, while no difference remains after level N+1 processing. This explains why the improvement in both experiments shows a difference of about 4%. During the deposition of the next level ULK a UV curing step at a temperature of > 400 °C is applied for several minutes. Hence, it can be assumed that water can further be desorbed and remaining silanol groups can be transformed into new Si-O-Si bonds. In consequence, the rearrangement is promoted and probably the pore size is reduced further. The RC behavior improves for all in the same way. The chemicals do not show a real or stable restoration, which is confirmed by results from FTIR, AES etc. for blanket wafers. RC behavior investigations for the vapor phase process were not possible so far due to a missing test arrangement. An appropriate chamber is in construction and results will be shown in future.

Table 1. Improvements of the capacitance at a fixed resistance for the level of restoration after complete processing of the same metal level N and the next metal level N+1

treatment	C improvement (at level N)	C improvement (at level N+1)
Reference (trench/via ash only)	0 %	~ 8 % / ~ 12 %
1 min UV	~ 4 %	~ 4 %
1 min IR 350°C	~ 7 %	~ 8 %
1 min UV + IR 350°C	~ 8 %	~ 8 %
HMDS + 1 min UV	~ 3 %	~ 12 %
RC2 + 1min UV	~ 12 %	~ 11 %
RC3 + 1min UV	~ 7 %	~ 12 %
RC4 + 1min UV	~ 5 %	~ 12 %

CONCLUSIONS

In this study two different repair approaches for plasma damaged ULK layers were compared. It was found that a vapor phase process is advantageous compared to a coating based processes. For silylation based repair processes pore size reduction, formation of adsorption places (mainly silanol groups) due to plasma treatment and steric hindrance play a key role. In consequence, the diffusion of repair chemical can be reduced or even inhibited. This is more pronounced at low temperature. Hence, it can be assumed that during the spin on process only a minor quantity of repair chemical was diffusing into the damage region. The vapor process delivers the repair chemical during the whole activation process (> 30min). In comparison, the spin on process only provides the chemical for a short time (1min), before the activation step. Therefore, only for the vapor phase process a "true" restoration of the damage zone was found. The most efficient restoration was obtained by RC2, which can be linked to the small size and a relatively high vapor pressure. To activate the silylation reaction both thermal and UV processes seem to be appropriate. For both a comparable surface restoration was achieved, which can be considered as independent on diffusion and time of exposure. Furthermore a κ-value restoration and a broad increase of Si-C related FTIR vibration modes without carbon incorporation were found for the spin on approach. A formation of a thin dielectric layer in competition to the silylation could be one explanation. It has also been shown that the next metal level processing has a positive effect on the RC behavior in the metal level below. It is suggested that this is mainly caused by the post-ULK deposition UV curing step.

ACKNOWLEDGMENTS

The work described in this publication has been funded in line with the technology funding for regional development (ERDF) of the European Union and by funds of the Free State of Saxony.

REFERENCES

1. K. Maex, M.R. Baklanov, D. Shamiryan, F. Iacopi, S.H. Brongersma, and Z.S. Yanovitskaya, *Appl. Phys. Rev.* **93**, 8793 (2003).
2. International Technology Roadmap for Semiconductors, SEMATECH
3. K. Kurihara, T. Ono, K. Kohmura, H. Tanaka, and N. Fujii, *J. Appl. Phys.* **101**, 113301 (2007).
4. *CRC Handbook of Chemistry and Physics*, **86**th edition, edited by D.R. Lide (Taylor and Francis, Boca Raton, 2006)
5. A.M. Hoyas, J. Schuhmacher, C.M. Wehlan, M.R. Baklanov, L. Carbonell, J.P. Celis, and K. Maex, *J. Vac. Sci. Technol. B* **23**, 1551 (2005).
6. T. Furusawa, D. Ryuzaki, R. Yoneyama, Y. Homma, and K. Hinode, *Electrochem. Solid-State Lett.* **4**, G31 (2001).
7. K. Yonekura, S. Sakamori, K. Goto, M. Matsuura, N. Fujiwara, and M. Yoneda, *J. Vac. Sci. Technol.* B **22**, 548 (2004).
8. S.V. Nitta, S. Purushothaman, N. Chakrapani, O.Rodriguez, N. Klymko, E. Todd Ryan, G. Bonilla, S. Cohen, S. Molis, K. McCullough, *Proc. of Advanced Metalization Conference 2005*, 325 (2005).
9. B. Lahlouh, J.A. Lubguban, G. Sivaraman, R. Gale, S. Gangopadhyay, *Electrochem. Solid-State Lett.* **7**, G338 (2004).
10. B. Xie, A.J. Muscat, *Microelectron. Eng.* **76**, 52 (2004).
11. T.C. Chang, Y.S. Mor, P.T. Liu, T.M. Tsai, C.W. Chen, C.J. Chu, F.M. Pan, W. Lur, S.M. Sze, *J. Electrochem. Soc.* **149**, F145 (2002).
12. T. Rajagopalan, B. Lahlouh, J.A. Lubguban, N. Biswas, S. Gangopadhyay, J. Sun, D.H. Huang, S.L. Simon, D. Toma, R. Butler, *Surf. Sci.* **252**, 6323 (2006).
13. P.G. Clark, K.K. Christenson, *Proc. of IEEE/SEMI Advanced Semiconductor Manufacturing Conference and Workshop*, 1 (2005).

14. T. Oszinda, M. Schaller, D. Fischer, C. Walshand S.E. Schulze, *Microelectron. Eng.* **87**, 457 (2009)
15. N.P. Hacker, M. Thomas and J.S. Drage, United States Patent No.: US 7,029,826 B2, 2006
16. T. Oszinda, V. Beyer, M. Schaller, D. Fischer, C. Bartsch and S.E. Schulze, *J. Vac. Sci. Technol. B* **27**, 521 (2009).
17. A.P. Singh, D.D. Gandhi, E. Lipp, M. Eizenberg and G. Ramanath, *J. Appl. Phys.* **100**, 14504 (2006)
18. A.P. Singh, D.D. Gandhi, B. Singh, E. Simonyi, E.G. Liniger, S. V. Nitta, M.W. Lane and G. Ramanath, *Applied Phys. Lett.* **94**, 093502 (2009)
19. T. Oszinda, M. Schaller, D. Fischer and S.E. Schulze, *Proc. of IEEE International Interconnect Technology Conference*, 83 (2009).

Conference Proceedings AMC XXV

Fabrication of Porogen Residue Free Ultra Low-*k* PECVD Material by Subsequent H_2-afterglow Plasma Treatment and UV Curing

A.M. Urbanowicz[1,a], K. Vanstreels, P. Verdonck, D. Shamiryan, M. Cremel[2], S. De Gendt[1] and M.R. Baklanov

IMEC, Heverlee, Belgium
[1] also at Dept. Chem, Katholieke Univ Leuven, Heverlee, Belgium
[2] also at Dept. Mat. Eng., INSA de Lyon, Villeurbanne, France

[a)] Corresponding author: Tel: +32 16281469, Fax: +32 16281214, E-mail: urbano@imec.be

ABSTRACT

In this work we report a new curing procedure of a plasma enhanced chemical vapor deposited (PECVD) low-*k* film. The new curing allows us to achieve a mechanically robust and hydrophobic ultra low-*k* film (k < 1.8). The method is based on a subsequent treatment of deposited films in an afterglow of H_2-based plasma and UV assisted thermal curing. The organic porogen removal by the H_2-based plasma afterglow with the following UV-curing allows us to produce the porogen-residue-free ultra low-*k* films with porosity higher than 50% and high elastic modulus of ~ 5 GPa.

INTRODUCTION

The ITRS scaling of ultra-large-scale integrated circuits requires mechanically robust materials with low *k*-value. Low-*k* materials recently used in the Cu/low-*k* integration scheme have *k*-values between 2.5 and 3.0. One of the limiting factors in further reduction of *k*-value is mechanical robustness, since major way to decrease *k*-value is increasing the material porosity [1]. The PECVD low-*k* deposition of ultra low-*k* films uses a porogen-based approach [2]. The matrix material is deposited by oxidation of alkylsilanes in a plasma-enhanced chemical vapor deposition (PECVD) process. The porogen molecules, usually cyclic hydrocarbons, are introduced into a SiOCH film by co-deposition with the matrix material. To create porosity, the porogen is removed from the films using UV-assisted-thermal curing [3]. The porogen molecules are photo-dissociated by UV light with the formation of volatile hydrocarbons and nonvolatile carbon-rich residues (porogen residue). The effect of the porogen residue on optical, chemical and mechanical properties of PECVD films has recently been reported [4,5]. Degradation of the mechanical properties as a result of porogen residue removal from advanced PECVD low-*k* films (*k*<2.3) during a photo mask ash has been recognized as an integration challenge [5]. Therefore, it is necessary to optimize the curing process of low-*k* films in order to avoid porogen residue formation.

In this work we investigate a new PECVD curing approach that includes two steps. The first step is the complete removal of porogen by thermo-chemical treatment. The second step is improvement of mechanical properties by conventional UV-curing. The porogen removal prior to UV-curing prevents cross-linking of porogen inside the Si-based low-*k* skeleton. Therefore, subsequent creation of porogen residues in the low-*k* film skeleton is avoided. However, there is a narrow process window for selective removal of the C-based porogen from the C-containing SiOCH skeleton. In order to keep the dielectric constant low, the removal of organic-porogen has to be performed without the Si-CH_3 bond scission from the SiOCH skeleton of the PECVD deposited low-*k* material. The removal of organic porogen without skeleton damage (Si-CH_3 bonds scission) is found to be possible by annealing of the low-*k* film in H_2-based plasma afterglow. The porogen removal efficiency of the H_2-plasma afterglow and its effects on SiCOH films is compared to the conventional UV-curing effect.

EXPERIMENTAL

Materials and experimental procedure. The matrix material was co-deposited with sacrificial porogen by PECVD on 300 mm Si wafers with thicknesses of 65 nm, 120 nm and 190 nm. Then the films were treated with four combinations of the He/H_2-downstream-plasma (DSP) treatments and UV curing: UV, He/H_2-DSP, He/H_2-DSP + UV, UV + He/H_2-DSP. The He/H_2-DSP treatments were performed at a wafer temperature of 280 °C using 350 s of the He/H_2 20:1 DSP treatment in a 300 mm asher. The UV-curing was performed in nitrogen ambient at temperature close to 430 °C. The new curing procedure was performed using a narrow-band 172 nm UV-source. In order to compare porogen removal efficiency of He/H_2-DSP with the different wavelengths of UV irradiation, an additional experiment with a broadband UV-source with the wavelengths higher than 200 nm was performed.

Instrumentation. The surface hydrophobic properties before and after the plasma treatments were evaluated using water contact angle measurements (WCA). Optical properties were determined by spectroscopic ellipsometry (SE) at an incidence angle of 70 deg. The result were fitted by a single and a double layer optical models using the Marquardt-Levenberg algorithm. The optical models were constructed as described in the literature [6,8]. The depth of modification and the optical properties of 190 nm films were estimated using a double layer SE model. The bottom layer was assumed to have optical properties of the as deposited film, while the optical characteristics of the top modified layer were determined by fitting. The mass change related to plasma treatments was measured by mass balance metrology on 300 mm wafers (Metrix: Mentor SF3). The open porosity and pore size distributions were evaluated using ellipsometric porosimetry (EP) [7]. Mechanical properties, Young's modulus and hardness of the low-*k* dielectric films were measured using a nano indenter XP system (MTS Systems Corporation) with a dynamic contact module (DCM) and a continuous stiffness measurement (CSM) option under the constant strain rate condition. A standard three-sided pyramid diamond indenter tip (Berkovich) was used for the indentation experiments. As the indenter tip is pressed into each sample, both depth of penetration (h) and the applied load (P) are monitored.

RESULTS AND DISCUSSION

I. Porogen removal by H_2-based plasma afterglow versus porogen removal by UV-curing.

In order to investigate the removal efficiency of porogen residue in H_2-based plasma afterglow, we studied the effects of 350 s He/H_2 downstream plasma (DSP) at 280 °C and UV treatments at 350°C on as-deposited PECVD films of 60 nm. After all the treatments the films remained hydrophobic. The water contact angles with surfaces for all the films were approximately 90 degree.

The results obtained by UV-SE show that the conventional thermo-assisted UV curing (photo-chemical) results in creation of porogen residues in contrast to He/H_2-DSP (H radical treatment). According to the recent results reported by Marsik et al. [8], the absorption bands located between 200 and 300 nm are related to the presence of amorphous-carbon-like porogen residue and porogen. The porogen residues are characterized by absorption band of 4.3 eV (~275 nm) related to the sp^2 orbital from C=C. Therefore, the difference in the absorption spectra of these films is mainly related to different amounts of the porogen or the porogen residues embedded in low-*k* skeleton.

Figure 1 shows the optical properties of the as deposited matrix-porogen film, the He/H_2-DSP treated film and the UV-cured film. The extinction coefficient of the as deposited film (with porogen) is the highest due to the highest porogen content. The treatments involving He/H_2-DSP results in complete removal of the porogen and porogen residues and the final absorption spectra are becoming similar to the UV spectra of the low-*k* matrix material [8]. On the contrary, the standard UV-curing processes using narrow band (~172 nm) or broadband (> 200 nm) UV-source results in the porogen residues creation. The extinction coefficient has a maximum at 275 nm which marks amorphous-carbon-like porogen residues. Furthermore, the relative changes in refractive indices of the as deposited film and treated ones are proportional to the porosity increase. The He/H_2-DSP treatments result in the highest RI reduction (the porosity increase) due to efficient porogen removal.

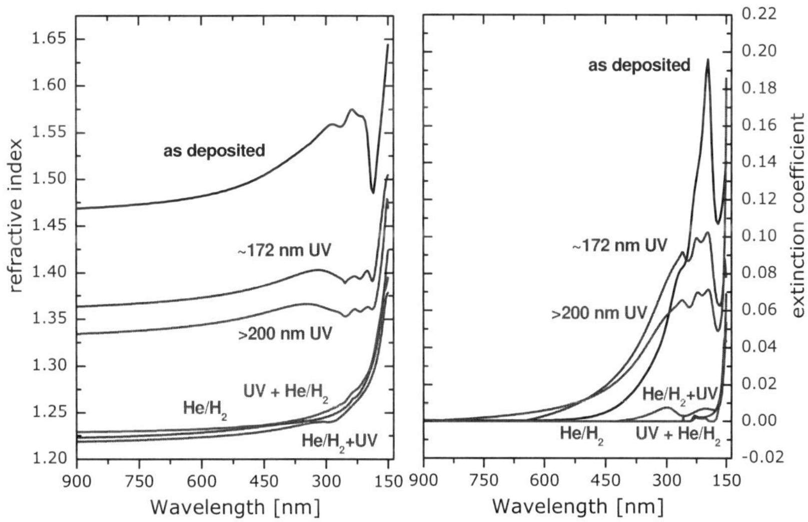

Figure 1. Optical properties of low-*k* films as measured by UVSE: as deposited film (with porogen); as deposited low-*k* films cured with UV light wavelength of 172 nm (narrow band) and > 200 nm (broad band), He/H_2-DSP and two combined He/H_2-DSP and 172 nm UV cures.

In order to prove that the amount of the organic residues is lower after the He/H_2-DSP than after the UV-curing, mass balance metrology was used. We measured the mass loss of 300 mm wafers with 60 nm films treated with the He/H_2-DSP and/or 172 nm UV, as shown in Figure 2. The error bars represent the mass measurement error due to backside cleaning. The percentage number reflects approximated mass loss of the film after the treatments. The results show that the conventional 172 nm UV-curing process leaves approximately 46 % more mass in comparison to the He/H_2-DSP treatment. The results agree with the UV-SE data (see the change of the extinction coefficient).

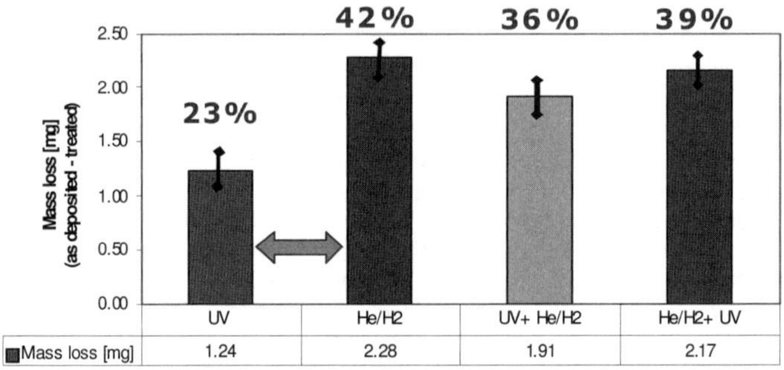

Figure 2. The mass loss after UV, He/H_2 and combined He/H_2 + UV treatments as measured by mass balance on 300 mm wafers.

II. Mechanism and depth of organic removal by He/H_2-DSP plasma.

We presume that the porogen removal mechanism at high temperatures (around 300 °C) is similar to the photo mask removal mechanism in He/H_2-DSP plasma [9]. Heating the low-*k* in H_2-based plasma atmosphere supports the evaporation of the porogen residue. Hydrogen atoms promote cracking of the high mass porogen or the porogen residue, generating smaller C-H polymer chains of lower vapor pressure. However, the organics that need to be removed using He/H_2-DSP are embedded into the porous low-*k* skeleton. Therefore the penetration depth of the H radicals into porous low-*k* skeleton determines the effective depth of porogen removal. The penetration depth of H radicals is limited by their loss in low-*k* pores as a result of recombination on low-*k* pore walls or chemical reaction with porogen or porogen residues. The time dependent depth of penetration of H radicals for different low-*k* films was reported in literature [5]. It was found that it saturates logarithmically with time. In order to determine the depth of H radicals' penetration we treated the as deposited film (with porogen) of 190 nm time of 350 s of He/H_2-DSP. Next, we measured the thickness of the porogen depleted layer by UV-SE using double-layer SE model. The depth of porogen removal was determined to be about 160 nm. Therefore the thickness limit to achieve the uniform films using subsequent He/H_2-DSP and UV-curing is about 120 nm - 140 nm. Thicker film fabrication should involve subsequent film deposition combined with He/H_2-DSP curing. The UV curing should be then performed as a last step due to much higher penetration depth of UV-light (>172 nm) as compared to H radicals.

III. Effect of combined H_2-based plasma afterglow and UV-curing on mechanical properties of the obtained films.

In order to study the effect of porogen or porogen residues on mechanical properties of the obtained films, we measured Young's modulus (YM) and open porosity of the as deposited film, the UV-treated film and the films after combined He/H_2-DSP and UV treatments. The mechanical properties of the low-*k* films were evaluated using Nano Indentation (NI) [10]. Since the film thicknesses were relatively small, only a relative comparison was possible due to Si substrate effect [11].

Figure 3 shows YM versus open porosity. The drop in YM and increase of open porosity after all treatments is due to the porogen or the porogen residues removal. As we reported previously [5] the He/H_2-DSP treatment of conventionally UV-cured film (UV + He/H_2) results in the porosity increase accompanied with mechanical properties reduction due to the porogen residues removal. This observation is indicated by the dashed line in Figure 3. However, when the porogen is removed by He/H_2-DSP prior to UV-curing (He/H_2 + UV), the YM of the obtained film exceeds that of the conventionally UV-cured film. The latter observation is indicated by the dotted line in Figure 3. We presume that the porogen removal prior to the UV-curing prevents cross-linking of porogen inside the SiCOH low-*k* skeleton. Therefore, the subsequent creation of the porogen residues inside the low-*k* film skeleton is avoided. Moreover, the cross-linkage of the mechanically strong SiOCH skeleton is not limited by the presence of organic porogen residues. Therefore, much stronger Si-O (as compared to Si-CH_3) links are created in the SiOCH skeleton. The Si-O links significantly improve YM of the film.

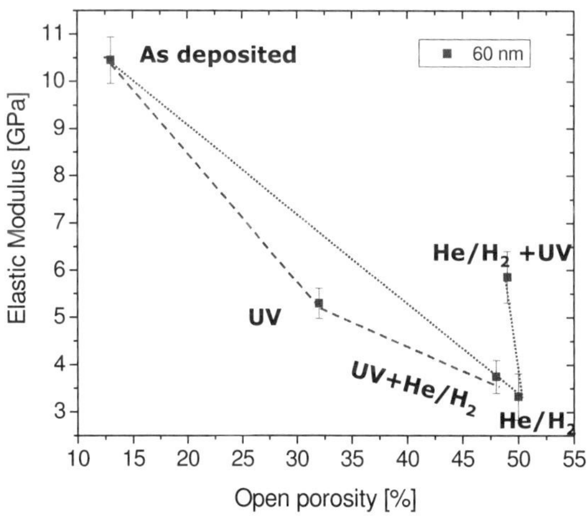

Figure 3. The Young's Modulus as measured by Nano-Indentation versus open porosity as measured by ellipsometric porosimetry for as deposited film and films treated with UV, He/H_2 and combined He/H_2 + UV.

In order to confirm the hypothesis that a porogen residues removal prior to a UV-curing results in a mechanically robust ultra low-*k* film, we cured the films with varied thicknesses. The as deposited films with thicknesses of 65 nm, 120 nm and 190 nm were cured using subsequent He/H_2-DSP and 172 nm UV-cure. The properties of the obtained films are reported in [5]. Two films with thicknesses of 58 nm and 105 nm, uniform from top to bottom were obtained. In the case of the 190 nm film, the porogen was only partly removed due to the limited penetration of H-atoms as described in section II. The subsequent UV-light irradiation of this film results in creation of bi-layer film. Presumably, the transport of decomposed porogen fragments induced by UV-light resulted in significantly higher film shrinkage as compared to the thinner films (presumably caused enhanced cross-linkage of the SiOCH matrix). The bi-layer films show even higher YM than for the 58 nm and 105 nm films, where the porogen was fully removed. The open porosity and the mean pore radii of 58 nm and 105 nm films were measured by EP. The films have 46 % of porosity and mean pore radii of 1.5 nm. The Young modulus were 5.82±0.82 for the 58 nm film and 5.46±0.51 for the 105 nm film. However, it is necessary to mention that values of YM can be influenced by Si substrate effect [11]. The Si substrate effect might vary depending on film thickness. The YM offset for the 60 and 180 nm PECVD films treated with different times of He/H_2-DSP was found to be 0.82 GPa. The 0.82 GPa higher values were measured for 60 nm films.

Table 1. The properties of achieved films.

Initial thickness	After He/H_2 + UV cure	WCA [deg]	k at 100 kHz	YM [GPa]	Open porosity [%]	MPR [nm]
65 nm	58	91.98	1.79	5.82±0.82	46	1.5
120 nm	105	90.78	1.87	5.46±0.51	46	1.5
190 nm	154 nm (Double layer)	87.93	2.24	7.8±0.62	N/A	N/A

CONCLUSIONS

A new curing procedure of enhanced chemical vapor deposited (PECVD) low-*k* film is proposed. The new procedure is performed by conventional UV-curing of PECVD film skeleton when the porogen is already completely removed. This porogen removal is performed by H_2-based plasma afterglow at elevated temperature of an as deposited PECVD film. The effective depth of the porogen removal depends on the penetration depth of the active H radicals into the porous SiOCH matrix and it found to be approximately 160 nm. The proposed method allows us to obtain the porogen residue-free 58 nm and 108 nm thick low-*k* films. The obtained films demonstrate high elastic modulus of ~5 GPa, open porosity of 46 %, mean pore radius of 1.5 nm and *k*-value of approximately 1.8-1.9. The curing of the 190 nm thick film, when porogen was only partly removed, results in mechanically robust bi-layer film with YM of 7.8 GPa and *k*-value of 2.24.

REFERENCES

[1] K. Maex, M. R. Baklanov, D. Shamiryan, F. Iacopi, S. H. Brongersma and Z. S. Yanovitskaya, *Journal of Applied Physics*, **93**, 8793 (2003).
[2] A. Gril and V. Patel, Appl. Phys. Lett., **79**(6), 803 (2001).
[3] F. Iacopi, Y. Travaly, B. Eyckens, C. Waldfried, T. Abell, E. P. Guyer, D. M. Gage, R. H. Dauskardt, T. Sajavaara, K. Houthoofd, P. Grobet, P. Jacobs and K. Maex, *Journal of Applied Physics*, **99,** 053511 (2006).
[4] A. M. Urbanowicz, D. Shamiryan, P. Marsik, Y. Travaly, A. Jonas, P. Verdonck, K. Vanstreels, A. Ferchichi, D. De Roest, H. Sprey, K. Matshushita, S. Kaneko, N. Tsuji, S. Luo, O. Escorcia, I. L. Berry, C. Waldfried, S. De Gendt and M. R. Baklanov, Mat. Res. Soc.: XXIV Advanced Metallization Conference Proceedings, eds. by M. Naik, R. Shaviv, T. Yoda, K. Ueno, p. 594-598 (2009).
[5] A. M. Urbanowicz, K. Vanstreels, D. Shamiryan, S. De Gendt and M. R. Baklanov, *Electrochem. Solid State Lett.*, **12**, H292 (2009).
[6] S. Eslava, G. Eymery, P. Marsik, F. Iacopi, C. E. A. Kirschhock, K. Maex, J. A. Martens and M. R. Baklanov, *Journal of the Electrochemical Society*, **155**, G115 (2008).
[7] M. R. Baklanov and K. P. Mogilnikov, *Microelectronic Engineering*, **64**, 335 (2002).
[8] P. Marsik, A. Urbanowicz, P. Verdonck, K. Ferchichi, D. De Roest, L. Prager and M. R. Baklanov. Proc. of *25th Advanced Metallization Conference*, 2008. Mat. Res. Soc.: XXIV Advanced Metallization Conference Proceedings, eds. by M. Naik, R. Shaviv, T. Yoda, K. Ueno, p. 543-549 (2009).
[9] I. L. Berry, Q. Han, C. Waldfried, O. Escorcia and A. Becknell, in *SEMI® Technical Symposium: Innovations in Semiconductor Manufacturing* (2004).
[10] D. J. Morris and R. F. Cook, *Journal of Materials Research*, **23**, 2429 (2008).
[11] M. Gonzalez, K. Vanstreels and A. M. Urbanowicz, in *10th Int. Conf. Thermal, Mechanical and Multiphysic Simulation and Experiments in Micro-Electronic - EuroSimE*, Delf, Netherlands, p. 486 (2009).

Patterning, CMP and Cleaning

Metal Hard-Mask Based Double Patterning for 22nm and Beyond

Herbert Struyf, Jean-François de Marneffe, Danny Goossens, Dirk Hendrickx, Craig Huffman, Eddy Kunnen, Frédéric Lazzarino, Alexey Milenin, Denis Shamiryan, Adam Urbanowicz, Alain Vandervorst and Werner Boullart
Imec vzw, Kapeldreef 75, 3001 Leuven, Belgium

ABSTRACT

In this paper, metal hard-mask-based patterning schemes are proposed to pattern 30/30-nm line/space structures and 40-nm contact holes at 80 nm pitch. Various litho and etch approaches are compared and discussed.

INTRODUCTION

Looking at the 2008 ITRS roadmap for Logic and DRAM for contacts and metal 1 trenches, it becomes clear that by 2011 contact holes in the order of 40 nm at a pitch of 80 nm are required, while 30-nm trenches at a pitch of 60 nm will need to be standardly patterned by 2013 [1]. To get early access to these dimensions and pitches in order to explore patterning, metallization and other process steps, state-of-the-art single-print 193-nm water-based immersion lithography cannot offer a solution, even at NA values of 1.35. Despite being a very promising technique for the future, EUV lithography needs to further improve through-put, overlay performance and resist systems [2]. In order to bridge the gap between standard 193-nm immersion litho and EUV however, double-patterning and double-exposure techniques can be used.

PATTERNING 30-NM HALF-PITCH TRENCHES

Litho considerations

Several double patterning options have been explored to generate narrow-pitch trenches, such as LELE (litho-etch-litho-etch), LPLE (litho-process-litho-etch) and SDDP (spacer-defined double patterning). In this paper, the focus is on 30-nm half-pitch trenches, patterned with a LELE route, where the design is split over two lithography steps [3]. First, 60-nm trenches are printed at a pitch of 120 nm, which are subsequently shrunk down to 30 nm, using a plasma-assisted polymer deposition technique [4]. After etching trenches in a hard-mask (HM) and ashing the resist, this sequence is repeated, resulting in 30/30 nm trenches in the HM, which are finally etched into the underlying dielectric (Fig. 1). It is obvious that this patterning scheme puts stringent requirements on overlay performance and CD uniformity after the litho steps. As the final CD variation for the 30-nm wide trenches should be less than 10% 3σ after etch (i.e. less than 3 nm 3σ), these strict specifications were already applied to the original 60 nm trenches after litho. Typical 3σ values that were achieved amounted to 2.5 nm after litho and 2.6 nm after the full patterning.

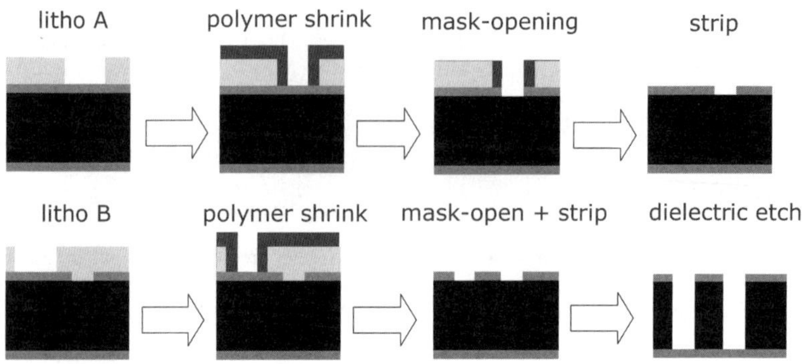

Fig. 1. Schematic overview of a Litho-Etch-Litho-Etch patterning scheme for 30/30 nm L/S trenches

Etch and stack considerations

The etch steps clearly define the final CD and pitch. The shrink process in particular, is very important for the quality of the final result. As the plasma-assisted shrink process was observed to introduce additional proximity effects [4], the shrink process and the subsequent hard-mask opening need to be taken into account when optical proximity corrections are performed. Also, the hard-mask needs to be carefully selected: first of all, it should enable double patterning. This means it should withstand the ashing chemistry used to strip the BARC and resist after the first and second hard-mask openings, and it should be thin enough not to cause too much topography for the second litho step. Secondly, it should offer sufficient etch selectivity to the underlying dielectric layer, minimizing HM faceting, HM corner-rounding, linewidth roughness and CD gain as much as possible. Thirdly, its deposition process should be compatible with the nature of the underlying dielectric, which is likely to be a low-k material with $k < 3.0$ for the technology nodes where 30-nm half-pitch trenches will be used. An additional interesting feature of the masking material could be its limited transparency, which could decouple the litho performance from the nature of the underlying dielectric. It would be equally interesting if it was fully sacrificial, and would be removed in e.g. the CMP step after metallization. Metallic hard-masks (MHM) (e.g. TiN) fit these requirements very well, and even offer additional advantages: they allow shifting the resist strip step from the end of the patterning sequence to before the dielectric etch step. In this way, the low-k sidewalls are not exposed during the strip step, which limits plasma damage to the low-k. Etch and strip damage is widely recognized as one of the key challenges for the integration of advanced low-k materials [5][6].

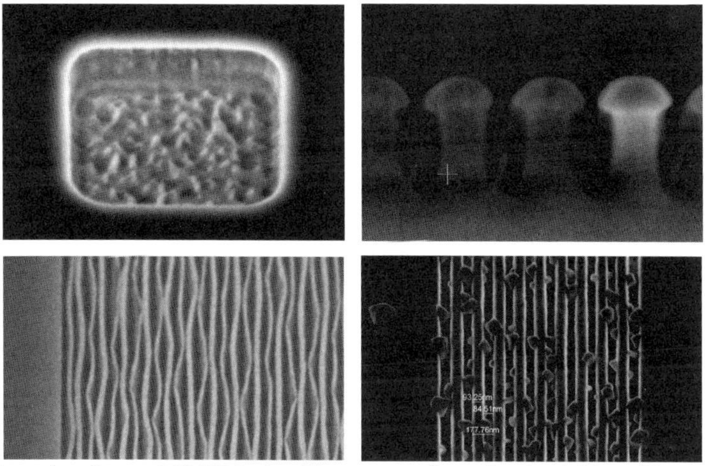

Fig. 2. Examples of potential MHM-related issues: grass formation (top left), profile control (top right), line-wiggling (bottom left) and TiF residue formation (bottom right)

However, next to all the MHM advantages, several disadvantages of the use of MHM have been observed as well (Fig. 2). Grass formation during low-k etch for instance, was found to be linked to the presence of Ti-containing residues on the surface of the low-k material to be etched [7]. Profile control issues may be linked to the redeposition of non-volatile Ti-containing residues on the sidewalls, giving rise to sloped trenches and etch-stop phenomena. This issue can be mediated by increasing the substrate temperature [8]. Alternatively, plasma damage may weaken the low-k material at the sidewalls, causing the damaged layer to be removed during wet clean, when dilute HF is used for instance. This leads to undercut under the MHM. Line-wiggling is typically attributed to the release of residual stress in the HM during etch [9], which results in undulating, bending or even collapsing low-k lines. Fig. 3 shows that the thickness of the MHM and the aspect ratio of the low-k lines both play a role in determining the extent of wiggling. Finally, plasma damage to the dielectric may weaken the structure, which increases the risk of wiggling. Tweaking the properties of the MHM to reduce its stress level without sacrificing too much selectivity to the low-k was found to be the most appropriate way of dealing with line-wiggling. Moreover, the risk of wiggling for a giving structure can be modeled [9]: the model input parameters are basic properties of all the layers in the stack (e.g. elastic modulus, thermal expansion coefficient and residual stress) and the lay-out of the structure. The output parameter – the 'buckling coefficient' - predicts how high the risk of wiggling is. Residue growth on the MHM after low-k etch has often been observed: it is mostly attributed to the formation of TiF outgrowths [10]. Their formation should be carefully avoided as these residues hamper metallization of the trenches. Etch chamber condition issues may arise from sputtering metallic species onto the chamber parts during low-k etch. Appropriate chamber clean, chamber conditioning and chamber monitoring procedures need to be put into place in order to avoid chamber-related issues.

Plasma damage to the low-k dielectric is probably the most important patterning issue to consider for the technology nodes where 30/30 nm trenches will be used for the metal 1 level. As

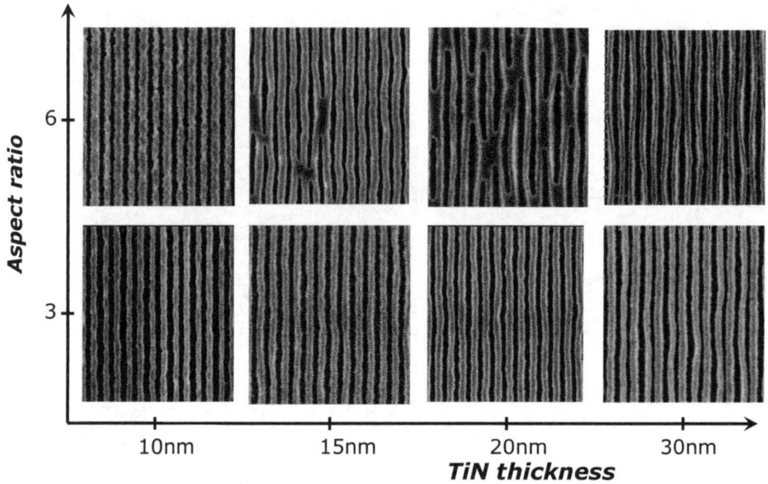

Fig. 3: Influence of aspect-ratio of the low-k lines and MHM thickness on line-wiggling for 30/30 nm L/S trenches

stated above, using a MHM-patterning scheme helps to minimize the damage inflicted by the ashing step. This was proven to be true for materials with k = 3.0 [6], where the sidewall damage depth could be reduced from 25 nm to 5 nm by switching from a resist-based patterning approach to a MHM-based scheme. However, for k = 2.5 materials, the use of a metal hard-mask in itself did not appear to be sufficient [11]: after MHM-opening and strip, EF-TEM carbon intensity maps indicated that carbon depletion extended in an isotropic way underneath the MHM for a large number of strip chemistries. The nature of the ashing plasma was found to be important: as can be seen from Fig. 4, even for 90/90 nm L/S trenches the impact of different strips on time-dependent dielectric breakdown (TDDB) results varies significantly, when all other processing steps are kept constant. For narrower trenches, the effect of plasma damage along the sidewalls will be even more outspoken, as a higher percentage of the low-k material will be damaged than for 90-nm half-pitch lines. Simulations taking into account moisture absorption in 5 nm of damaged material on each side of the line indicated that for a k = 2.5 material, it would be more beneficial to use oxide than low-k for 20/20 nm L/S structures. Probably the simplest way to avoid strip damage altogether would be to introduce a second HM layer underneath the MHM. It should be impenetrable to the ashing plasma, compatible with the underlying low-k dielectric and (partially) removable by CMP. That low-k materials can be integrated successfully in 30-nm half-pitch metal 1 structures, is proven in Fig. 5, where excellent yield for a k= 3.0 low-k is obtained for 1-cm meander-fork structures [12].

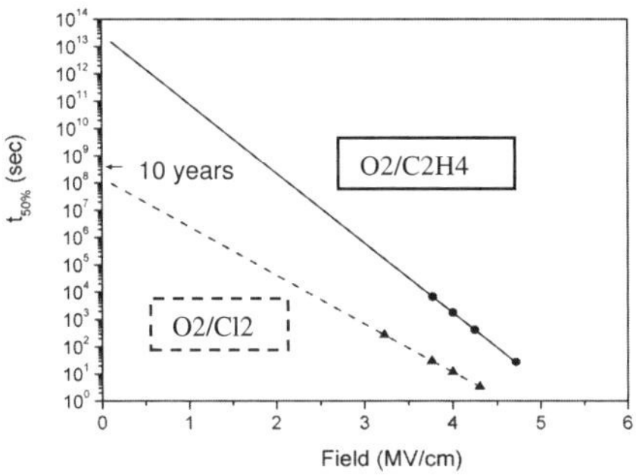

Fig. 4: TDDB results for 90/90 nm meander-forks at 100°C, with measurement points for both O2/Cl2- and O2/C2H4-ashed wafers and extrapolation to 0.11 MV/cm [11] (E-model was used because it gives the most conservative lifetime prediction at low fields).

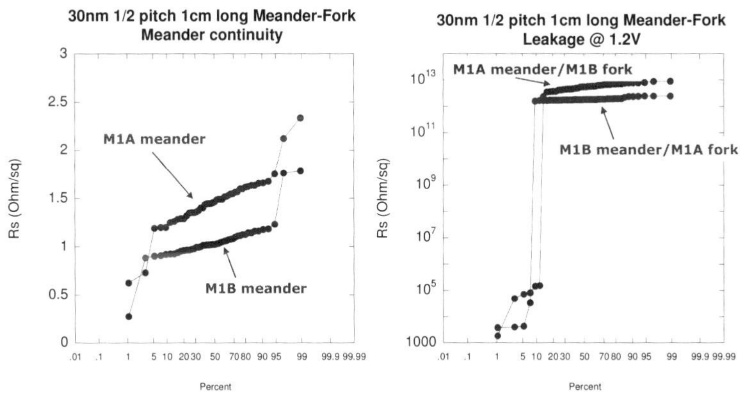

Fig. 5: Electrical yield for 30/30 nm 1-cm meander-forks (continuity and leakage)

Alternative for metal hard-mask

As an alternative for MHM, organic masks have been proposed [13]: these masks appear to show far fewer problems with line-wiggling because the stress in the mask layers is either low or easily controllable. Obviously, none of the specific MHM-related issues that were discussed above, would be present when organic masks are used. On the other hand, stripping these organic layers would have to be done after the full trench etch, which would increase the plasma budget the trench sidewalls would be subjected to. Even more cumbersome would be the fact that double patterning would become very difficult with an organic mask: after the Si-containing layer on top of the organic mask layer would have been opened using the first litho photo, the resist and the BARC would have to be ashed while the carbon mask layer is exposed (Fig. 6). This would be difficult with many strip chemistries. However, it was observed that when a He/H2 chemistry is used in a downstream plasma ash chamber at elevated temperature, ash selectivities of over 150:1 can be reached between 193-nm immersion resist and an amorphous carbon layer deposited at 400 C. This easily allows resist and BARC ash while the carbon layer is exposed. Moreover, the same plasma conditions were proven to inflict no plasma damage to low-k materials with a k down to 2.3 [14]. This would allow stripping the carbon layer (very long strip times would be needed but this chemistry does strip) without damaging the low-k material. All in all, this strip chemistry could enable low-damage low-k patterning using a carbon mask for double patterning of 30-nm half-pitch structures.

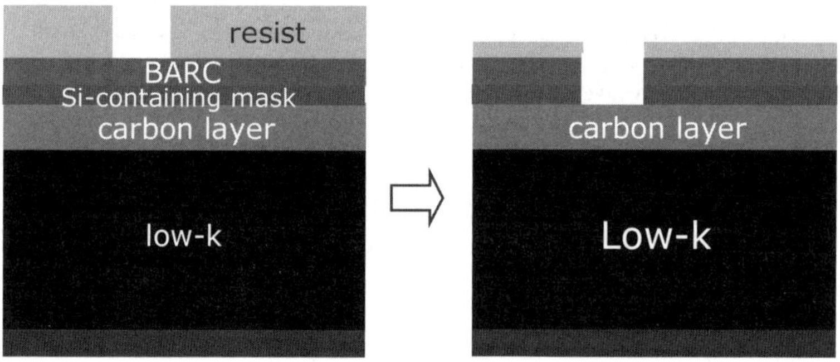

Fig. 6: Stack with carbon mask before and after Si-mask opening (for litho photo A), showing exposed carbon layer material when the resist and BARC need to be stripped.

PATTERNING 40-NM HALF-PITCH CONTACT HOLES

Litho considerations

To generate 40-nm half-pitch contact holes (CH) problems such as limited litho resolution by the low aerial image contrast associated with dark field masks, and the 2-D character of CH as a limiting factor in the use of extreme resolution enhancement techniques, need to be overcome. One way of doing this, is by deconvoluting the 2-D CH into two exposures of lines (essentially 1-D structures). Lines with widths in the order of 40 nm can be printed at a pitch of 80 nm by making use of dipole exposures [15]. When a second exposure is added at an offset angle of 90 degrees, and both positive tone resist and positive tone development are used, the result after development would be dots of 40 by 40 nm. There are several ways to convert these dots into contact holes. Most of them involve complicated and numerous intermediate steps, such as spinning and patterning additional coatings, mask-freezing, etc. Negative tone development, however, was found to constitute a simple method to go straight from exposure to development and end up with holes of the right size and pitch in the resist mask.

Alternatively, EUV lithography offers a straightforward way of obtaining 40-nm half pitch contacts after litho only. However, as stated before, this technique needs optimizations in several areas before it will be ready for production.

Mask considerations

As we move along the ITRS roadmap towards the most advanced technology nodes, the resist thickness for contact hole patterning appears to decrease to values below 100 nm, while the aspect ratio of the contact holes has seen a dramatic increase from less than 3:1 for the 0.35 μm node to over 5:1 for the 22 nm node. The combination of these trends makes clear that an additional mask is needed to pattern the contacts for the most advanced technologies. Both MHM (TiN) and organic masks have been evaluated for contact holes. Selectivities of over 150:1 can be obtained for TiN, but sufficiently good selectivity for carbon masks has been reported as well [16]. The use of TiN does however imply the same etch chamber contamination risk as for the patterning of 30-nm half-pitch trenches (see above). Both TiN and carbon masks were tested on double-exposed wafers and on EUV-litho wafers, and both types of hard-masks were found to yield excellent results. Fig. 7 shows successfully patterned 80-nm pitch contact holes after EUV litho on a carbon mask, while Fig. 8 displays top-down images for 40 nm contacts at a similar pitch after double exposure litho and after TiN HM patterning.

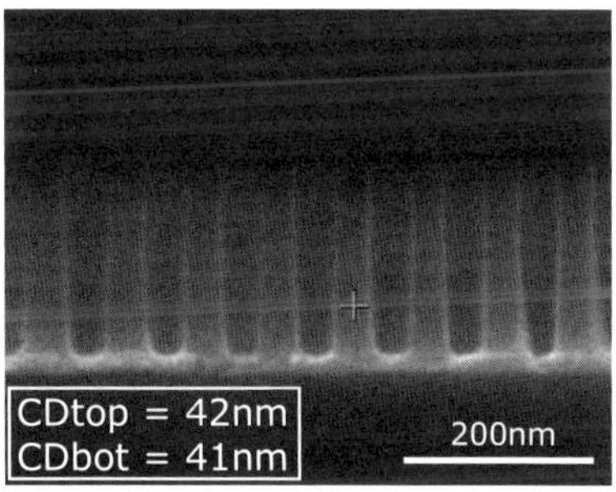

Fig. 7: 80-nm pitch contacts patterned with EUV litho on a carbon mask

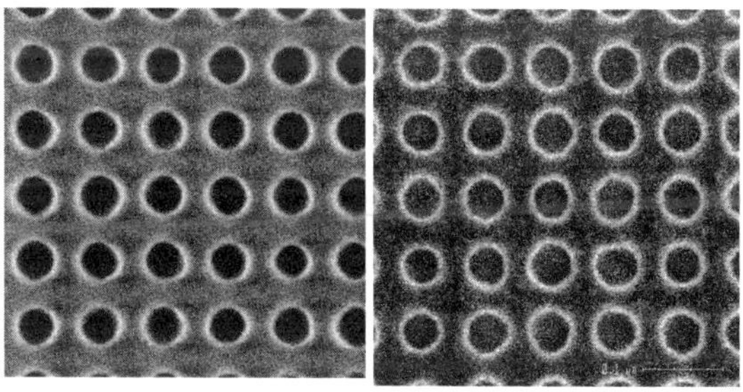

Fig. 8: Top-down pictures of 80-nm pitch contact holes after double exposure litho (left) and after MHM-opening (right) [15]

CONCLUSIONS

For 30-nm half-pitch trenches and 40-nm half-pitch contacts, patterning routes were proposed and discussed, taking into account various litho, etch and stack approaches and considerations. It was proven that for both types of structures, satisfactory results can be obtained after careful litho, etch and stack optimization.

ACKNOWLEDGMENTS

The Imec BEOL litho and integration teams are gratefully acknowledged for valuable input.

REFERENCES

1. ITRS roadmap 2008
2. J.V. Hermans, B. Baudemprez, G. Lorusso, E. Hendrickx, A. van Dijk, R. Jonckheere and A.-M. Goethals, Proc. SPIE Vol. 7271 (2009)
3. J. Versluijs, J.-F. de Marneffe, D. Goossens, M. Op de Beeck, T. Vandeweyer, V. Wiaux, H. Struyf, M. Maenhoudt, M. Brouri, J. Vertommen, J.S. Kim, H. Zhu and R. Sadjadi, Proc. SPIE Vol. 6924 (2008)
4. M. Op De Beeck, J. Versluijs, Zs. Tokei, S. Demuynck, J.-F. de Marneffe, W. Boullart, S. Vanhaelemeersch, H. Zhu, P. Cirigliano, E. Pavel, R. Sadjadi and J. Kim, Proc. SPIE Vol. 6519 (2007)
5. F. Iacopi, M. Stucchi, O. Richard qnd K. Maex., Electrochem. Sol. St. Lett. 7 (4) (2004)
6. H. Struyf, D. Hendrickx, J. Van Olmen, F. Iacopi, O. Richard, Y. Travaly, M. Van Hove, W. Boullart and S. Vanhaelemeersch, Proc. IITC (2005)
7. J.-F. de Marneffe, D. Goossens, T. Conard, D. Hendrickx, H. Struyf and W. Boullart, Proc. PESM (2009)
8. H. Struyf, D. Hendrickx, V. Paraschiv, D. Campos Garcia, G. Mannaert, W. Boullart and S. Vanhaelemeersch, Proc. DPS (2006)
9. M. Darnon, T. Chevolleau, O. Joubert, S. Maitrejean, J.C. Barbe, Appl. Phys. Lett. 91 (2007)
10. M.C. Lin, M.Q. Wang, L. Joe, R. Huang. Ch.M. Weng, J.H. Liao, J. Tang, W. Lu, H.W. Chen, J.T.C. Lee, Solid State Phenomen. 134 (2008)
11. H. Struyf, D. Hendrickx, D. Shamiryan, O. Richard, M. Pantouvaki and W. Boullart., Proc. DPS (2008)
12. S. Demuynck, C. Huffman, M. Claes, S. Suhard, J. Versluijs, H. Volders, N. Heylen, K. Kellens, K. Croes, H. Struyf, G. Vereecke, P. Verdonck, D. De Roest, J. Beynet, H. Sprey and G. Beyer
13. M. Darnon, T. Chevolleau, T. David, J. Ducote, N. Posseme, R. Bouyssou, F. Bailly, D. Perret and O. Joubert, to be published in J. Vac. Sci. Technol. B
14. A. Urbanowicz, K. Vanstreels, D. Shamiryan, S. De Gendt and M. Baklanov, Electrochem. Solid State Lett. (2009)
15. V. Truffert, J. Bekaert, F. Lazzarino, M. Maenhoudt, A. Miller, M. Moelants and T. Wu, Proc. SPIE Vol. 7274 (2009)
16. X.S. Rao, J. Widodo, W. Lu, M.S. Zhou, L.C. Hsia, T. Chu, R.K.H. Lee, A. Jain, H. Yu, B.T. Nguyen and D. Padhi, Proc. AMC (2009)

Conference Proceedings AMC XXV © 2010 Materials Research Society

Investigation of Cu Bimetallic Corrosion in CMP Chemical Environments using Micropattern Corrosion Screening

Praveen R. Nalla, Karthikeyan S.M. Pillai, Kyle K. Yu, Shyam Venkataraman, Oliver Chyan*

University of North Texas, Department of Chemistry, Denton, Texas 76203

ABSTRACT

On-chip corrosion, originating from contact of dissimilar materials, can cause serious reliability issues for integrated circuits and Micro Electro Mechanical Systems (MEMS) devices. The CMP processing exposes Cu interconnects and its adjacent barrier/liner to the continual presence of corrosive chemicals, the resulting bimetallic corrosion could be detrimental to the overall device yields. Furthermore, the minute dimensions (40-60 nm) involved in sub-45nm node can greatly accelerate the corrosion damages that diminish the device's reliability. In this paper, a new rapid screening metrology based on micropattern corrosion testing was developed and applied to study Cu bimetallic corrosion related to CMP processing. Ruthenium (Ru) and other noble metals have been actively investigated for use as alternative diffusion barrier materials to Ta/TaN for Cu interconnects. The increased noble character of diffusion barriers like Ru can further intensify the Cu galvanic corrosion in CMP chemical environments. The micropattern corrosion screening reveals that Cu contacts with Ru leads to enhanced galvanic corrosion of Cu as compared to Cu/Ta. The corrosion mechanisms of Cu/Ru and Cu/Ta were established based on results from Tafel plots, x-ray photoelectron spectroscopy and micropattern corrosion screening.

INTRODUCTION

Since 1997, Copper with high conductivity has replaced aluminum as the metal of choice for interconnects [1, 2]. Due to the lack of volatile byproducts in the reactive ion etching of Cu, the Cu interconnects are fabricated by dual damascene patterning process. As shown in Figure 1, the dual damascene process involves etching a trench/via pattern into a dielectric, the blanket deposition of a diffusion barrier followed by Cu electroplating, and finally removal of Cu overburden by chemical-mechanical planarization (CMP) [3-4]. Since the CMP process exposes the Cu interconnect and metal barrier to the continual presence of corrosive chemicals, the resulting bimetallic corrosion between Cu and diffusion barrier metal can be detrimental to the overall production yields of integrated circuit devices. The same type of microscopic corrosion that arises during fabrication or in operational use can also be a critical reliability issue for MEMS [5].

Previously, we identified that the microscopic on-chip corrosion of a 17 μm x 17 μm Digital Light Processing (Texas Instruments) device is originated from a dissimilar metal contact (TiW and Al-Si-Ti) on the MEMS [6]. With ion chromatographic analysis, our accelerated corrosion experiments further reveal that a specific chemical (adsorbed F^- ions) initiated active bimetallic corrosion. The study suggested that certain energetically favored oxidation and reduction reactions occurring at bimetallic contacts within the micro device structures can initiate catastrophic on-chip corrosion when exposed to appropriate chemical environments. More critically, the minute dimensions (< μm) involved in a miniature MEMS device can greatly accelerate the corrosion damages that shorten operational lifetime and diminish the device's reliability.

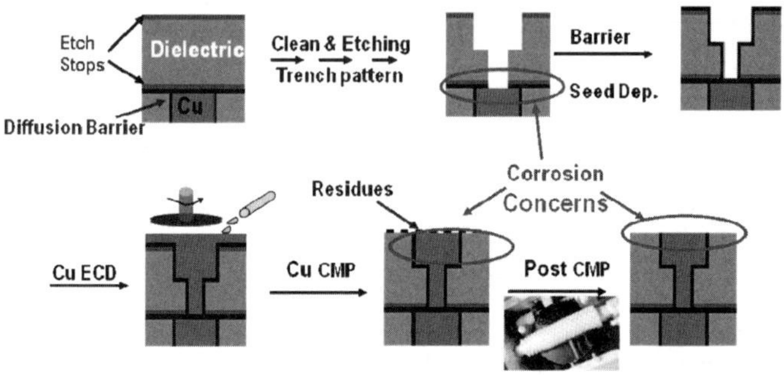

Figure 1. Bimetallic Cu corrosion concerns during CMP and post CMP

In recent years, ruthenium (Ru) and Ru-based barriers have been subjected to intensive investigation as a Cu diffusion barrier [7, 8] and oxygen barriers for high-k, stacked and dynamic random access memory capacitor device applications [9, 10]. With the introduction of a Cu diffusion barrier metal, like Ru having different electrochemical properties compared to Cu, it is anticipated to impose new challenges for CMP processing, despite its excellent Cu plating advantages. As microelectronic circuits move toward sub 45 nm domains, continued shrinkage of device dimensions makes corrosion related reliability threats even more acute. Therefore, there is a need for an effective methodology to evaluate bimetallic corrosion of Cu interconnects when exposed to CMP and post CMP cleaning conditions.

In this paper, a new rapid screening metrology based on corrosion testing micropattern was developed and applied to study Cu bimetallic corrosion related to CMP processing. The Cu micro-dots were deposited, through a micro-dot mask, on selected barrier/liner metal substrates including Ta, Ru, TaN, and carbon. The Cu corrosion on the testing pattern was observed in real time under digital optical microscope to record surface transformations of Cu corrosion and to determine the Cu corrosion rate in different chemical environments, different substrates and ambient controls (Ar, air and oxygen). The micropattern corrosion screening results indicated that Cu corrosion was enhanced when contacts with Ru as compared to Ta. The corrosion mechanisms of Cu/Ru and Cu/Ta were established based on results from Tafel plots, x-ray photoelectron spectroscopy and micropattern corrosion screening. Our data suggest that the enhanced Cu/Ru corrosion is controlled by several inter-dependant factors including galvanic potential differences, chemical speciation, pH values and surrounding ambient.

EXPERIMENT

Micropattern Corrosion Screening

Micro pattern of Cu deposited on Ta, Ru and SiO_2 supported by Si wafer are used to study the bimetallic corrosion. These Cu micro structures of ~700 A° thick with diameters of 130, 200, 400 μm were formed by vapor depositing Cu through a fine mask on pre-cleaned Ru (100 nm) and Ta (100 nm) wafer, as shown in Figure 2. Ru film was prepared on a pre-cleaned HF etched Si wafer by using a standard DC magnetron sputtering tool at a chamber pressure of 20 mTorr and Ar flow rate of 20 sccm. The base pressure prior to Ru sputtering was recorded as 1×10^{-8} Torr. Visual evidence of in-situ corrosion was collected using a metallurgical microscope (Eclipse ME-600, Nikon) by immersing micro-pattern structures in various chemical solutions like 0.1 M ammonium citrate (tribasic) solution and erosion/surface transformation on Cu micro-dots with time is recorded with a computer-controlled digital camera connected to microscope.

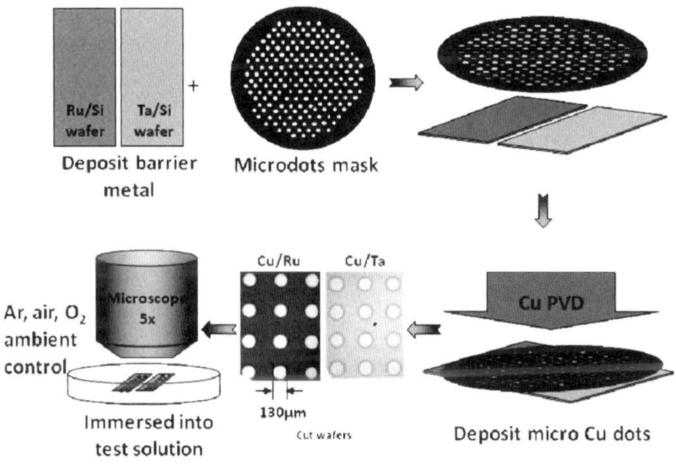

Figure 2. Fabrication process flow for corrosion screening micropattern

Electrochemical and Surface Characterizations

Potentiostat (model 440, CH Instruments) was employed to acquire the potentiodynamic data. The metal shots (Heraeus Inc.) of 6 mm diameter are used as electrodes for all electrochemical data which were pre-cleaned by polishing on 0.5 micron final polishing pad and sonicating in 18.2 MΩcm de-ionized water. The three-electrode system was employed to obtain potentiodynamic data. The Pt sheet and $Hg/HgSO_4$ electrode (satd. K_2SO_4, CH instrument) were used as counter and reference electrodes, respectively. Potassium sulfate is used as a supporting electrolyte in all the solutions. The potential differences and galvanic current measurements were done through a galvanic cell setup. The galvanic current flow was measured using the source meter

(Keithley, 2400 source meter). Electrolytes used were 0.1 M ammonium citrate or sodium citrate or potassium sulfate or as mentioned. The x-ray photoelectron spectroscopy (XPS) (Phi 5000 VersaProbe, with an Al Kα 58.7 eV excitation source) provided surface chemical composition information.

DISCUSSION

In general, the chemicals used for the CMP consist of abrasives, oxidants, corrosion inhibitors, and organic additives, which are added to improve specific surface characteristics during the CMP. As shown in Table I, salt of organic additive containing carboxylic groups and ammonium ion is commonly used as a complexing agent to improve the removal rate of Cu and as a pH buffer. Ammonium citrate is one of the common organic salts used in the CMP slurry and post-CMP solutions. Roy et al. [11] has demonstrated that citric acid acts as an effective complexing agent [12, 13] for Cu in presence of H_2O_2. Citric acid is also known to act as corrosion inhibition in nitric acid medium and achieves surface planarity during CMP. Cu is known to form (+1) and (+2) complexes with ammonia. The mechanism of Cu interaction with ammonia is extensively studied [14]. Though citric acid and ammonia have been evaluated as complexing agents for Cu-CMP slurries, the studies were limited to Cu or barrier material individually. The effect of dissimilar metallic contact between Cu and the barrier metal was not considered. To evaluate the efficacy of micropattern corrosion screening metrology, we started with the 0.10 M ammonium citrate solutions with Cu microdots deposited on Ru and Ta substrates.

Table I. Commercial CMP/Post CMP solutions and their listed chemical components

Commercial CMP	Listed Chemicals	Applications
ECP clean	Citric acid, **Ammonium Citrate** , pH 3.3	Acid Clean
HS T805-1	Silicon compound(SiO2) **+Heterocyclic** + **RCO2H** + **Glycol ether** (**R'OR"OH**) + H2O pH 2.0 - 4	CMP solution
EP C600Y-75	Used for polishing contain Metal Oxide(Alumina particals< 5%) pH 3-5	CMP solution
Cu Clean 811	**Organic Acid**< 3%, aldehyde<3%,(primary) **Amine**< 10% pH 10-11	Base clean
ESC 774	Organic **amine**/ Organic **acid**/ Quartanary **ammonium** Hydroxide compound/H2O; pH 12.5	Concentrated Post CMP cleaner
ESC- T794	(TMAH)4.0% **(CH3)4N(OH)**, **organic amine** 10-30% (may have **ethanolamine**) pH> 11	Concentrated Alkaline Cleaner

Micropattern corrosion screening in ammonium citrate solution

As shown in Figure 3, Cu microdots deposited on Ru (Cu/Ru) were eroded rapidly (~ 9 min) in ammonium citrate solution of pH 9.5. The time (t_{corr}) to completely erode the Cu mircrodots provides a convenient semi-quantitative gauge of bimetallic corrosion rate. In contrast, the corrosion of Cu microdots on Ta (Cu/Ta) was significantly slower (>60 min). In addition to the corrosion rate, the micropattern corrosion screening affords *in-situ* observation of surface transformation occurring in Cu corrosion process. When submerged in alkaline ammonium citrate, Figure 3&4, Cu microdots on Cu/Ru micropatterns remain metallic appearance, but Cu surface on Cu/Ta micropatterns was gradually coated with dark color over coating. Subsequent XPS analyses confirm that the coating is mostly Cu(+1) oxide.

The substrates were found to play a key role in the bimetallic corrosion observed by the micropattern corrosion screening. Figure 4 shows the time dependent corrosion of Cu microdots sputter deposited on Ru, Ta, TaN and C substrates. Ta and TaN are currently used as the Cu diffusion barriers for most of the IC devices. The nobler Ru substrate can result in nearly ten times faster Cu corrosion than TaN and C. The new micropattern corrosion screening metrology provides a flexible platform to rapidly evaluate Cu bimetallic corrosion trend of different barrier/liner materials under different type of CMP chemical formulations.

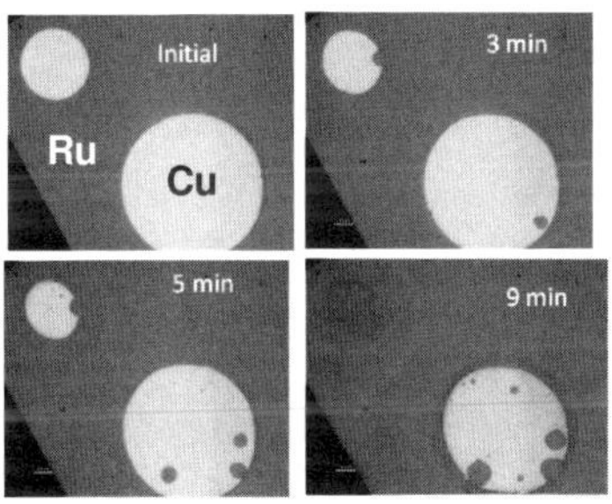

Figure 3. In-situ Microscopy of Cu/Ru Bimetallic Corrosion in 0.1M Ammonium Citrate, pH 9.

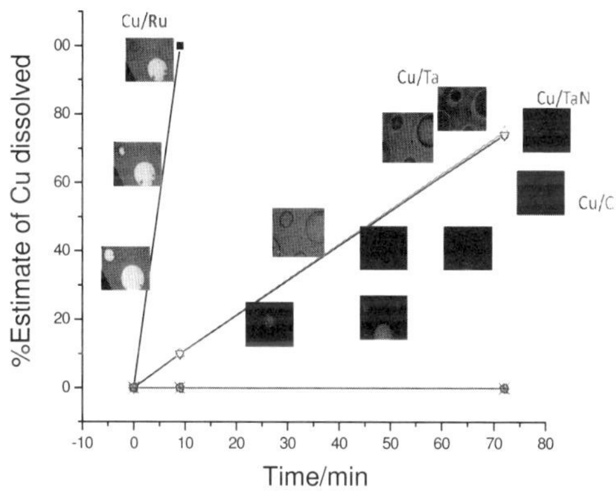

Figure 4. Effect of different substrates (Ru, Ta, TaN, C) on Cu corrosion in 0.1M Ammonium Citrate, pH 9.

Cu Corrosion Mechanism in Alkaline Ammonium Citrate

To establish the Cu corrosion mechanism, we investigated the effects of 1) chemical environments 2) interfacial reactions 3) substrates on Cu corrosion using micropattern corrosion screening combined with electrochemical analysis and x-ray photoelectron spectroscopy (XPS). Tafel plot (Figure 5) and open circuit potential measurements revealed the corrosion potentials of Cu, Ta and Ru follow $E_{Ru} > E_{Cu} > E_{Ta}$. Therefore, Cu was under cathodic protection when contacting with Ta, but could be biased anodically with adjacent contact of Ru. The direct corrosion current measurement by galvanic couple of Cu/Ru and Cu/Ta also confirms the galvanic potential trend. The micropattern corrosion screening metrology allows rapid screening of Cu corrosion under different chemical environments including K_2SO_4, KOH, Na Citrate , NH_4OH and NH_4Cl (all adjusted to pH 9.5). The finding demonstrated that ammonium ion NH_4^+ containing solution, Figure 6, not merely solution alkalinity, resulted in rapid corrosion of Cu microdots deposited on Ru substrate.

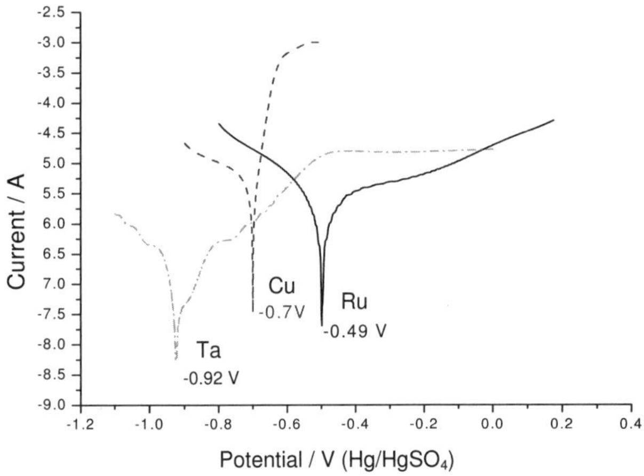

Figure 5. Tafel plot for Ta, Ru and Cu in 0.1M Ammonium Citrate, pH 9.

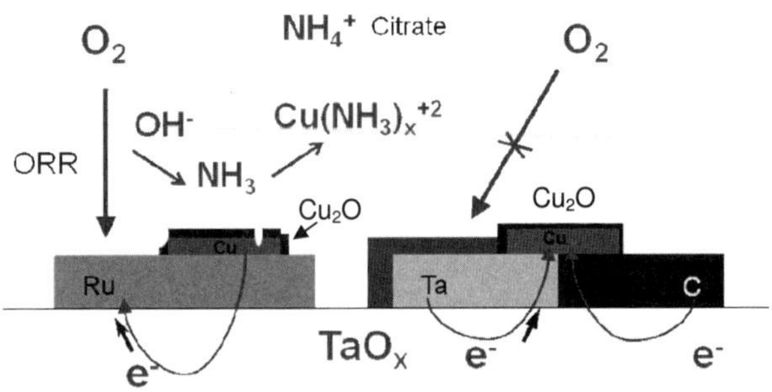

Figure 6. Proposed Cu bimetallic corrosion mechanisms for Cu/Ru, Cu/Ta and Cu/C.

Figure 6 summarizes the proposed Cu bimetallic corrosion mechanisms for Ru and Ta in alkaline ammonium citrate solution. The ambient control corrosion study identified that oxygen can greatly accelerate the corrosion of Cu microdots based on micropattern corrosion screening. Ru and Pt are well known electrocatalysts for the oxygen reduction reaction (ORR). The ORR cathodic reaction is often the rate determination step for the fuel cell operation. With Cu/Ru bimetallic contact, as shown in Figure 6, Ru substrate promotes the ORR reaction to raise local pH from 9 to ~12. The increase alkalinity promotes deprotonation of NH_4^+ to generate more neutral NH_3(aq) that has a strong complexing capacity to Cu ions (formation constant $K_f = 7.24 \times 10^{10}$). The Cu-$NH_3$ complexes are highly soluble and provide efficient pathway to remove oxidized Cu and accelerate the Cu corrosion. In contrast, Ta substrate was oxidized readily based on XPS surface analyses. In contrast, the Ta oxide and carbon substrates are not capable of promoting ORR and consequently a lower Cu corrosion rate was observed.

Micropattern Corrosion Screening in Organic Amine and BTA Solutions

In addition to the inorganic amine like ammonium citrate, other organo-amines like ethanolamine ($NH_2CH_2CH_2OH$) and tetra methyl ammonium hydroxide [$(CH_3)_4NOH$] are commonly employed in the CMP and post CMP formulation. The Cu bimetallic corrosion in three different amine systems was studied using Cu/Ru micropattern. Our micropattern corrosion screening showed that the Cu corrosion trend follows NH_4citrate > ethanolamine > tetra methyl ammonium hydroxide. The finding indicated that Cu corrosion will be slowed down in organo-amines with more alkyl groups attached to N atoms.

Benzotriazole ($C_6H_5N_3$, BTA) is a Cu corrosion inhibitor most commonly used in Cu CMP solution formulations [15]. However, it is highly desirable that BTA addition be kept as low as possible because the strongly surface bonded Cu-BTA layer often becomes the source of organic contamination that can affect the conductivity and interlayer adhesion of subsequent Cu interconnects. The micropattern screening method can be used to identify the minimum BTA inhibitor concentration required for a particular chemical environment.

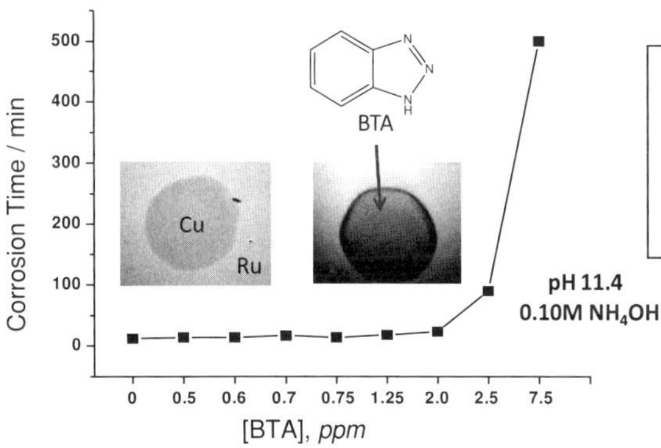

Figure 7: Optimization of BTA inhibition in NH_4OH solution by micropattern corrosion screen metrology. BTA-Cu complex can be seen on Cu with some proper conditions.

For example, Figure 7 shows the effect of BTA concentration on the Cu corrosion time based on Cu/Ru micropattern screening in NH_4OH solution pH = 11.4. With addition of 2.5 ppm BTA, the Cu corrosion rate of Cu/Ru micropattern was decreased by nearly 10 times. The Cu corrosion rate can be slowed down further over 50 times if 7.5 ppm BTA was added into NH_4OH solution (pH = 11.4). As also depicted in the Figure 4 inset, the BTA to Cu (+1) complex formation can be directly observed under microscope with proper condition. The desirable Cu corrosion inhibition was made possible by a strong chemical reaction of BTA to Cu(+1) surface layer on Cu interconnects [16].

CONCLUSIONS

A novel micropattern corrosion screening metrology was developed and demonstrated to be highly effective in rapid bimetallic corrosion screening under CMP relevant chemical environments. The open micropattern screening platform allows easy control of ambient and solution composition that enable us to establish bimetallic corrosion mechanism of Cu/Ru vs. Cu/Ta in ammonium citrate solution. Our experimental data revealed that the balance between oxide formation and complex reaction can affect the extent of Cu bimetallic corrosion. The alkaline NH_4Citrate solutions with strong complexing ammonia ligands will result in faster Cu corrosion. For citric acid and K_2SO_4 solutions, a slower Cu corrosion process that could lead to Cu passivation was observed. The detailed bimetallic corrosion mechanisms, obtained by the new micropattern corrosion screening, help to unveil the real functions of each chemical previously hidden in the complex mixture of CMP solution. It also significantly expands the chemistry tool set that could potentially enable better process designs for the next generation of CMP processes.

ACKNOWLEDGMENTS

The authors acknowledge the Robert A. Welch Foundation (B-1287) and Semiconductor Research Corporation (1292.042) for financial support. Helpful discussion from Trace Hurd toward this work is appreciated.

REFERENCES

1. C. Ryu, K.W. Kwon, A.L.S. Loke, V.M. Dubin, R.A. Kavari, G.W. Ray, and S.S. Wong, in Digest IEEE 1998 Symposium on VLSI Technology (IEEE, 1998) pp. 156.
2. H.C. Chem, M.S. Yang, J.Y. Wu, and W. Lur, in Interconnect Technology 1999 (IEEE International Conference, 1999) pp. 65.
3. J.G. Park, A.A. Busnaina, and Y. Hong, Microelectronic Applications of Chemical Mechanical Planarization, 467, (2008).
4. T.K. Doi, *Manufacturing Engineering and Materials Processing,* **73**, 341, (2007).
5. H Melitta, W.D. Frank, T.W. John, K. Martin, C. Carlo, and M. Roya, *Sen. Actuators* **145**, 323, (2008).
6. J. Prasad, O. Chyan, J.J. Chen, M. Liu, L. Chen, and F. Xu, *J. Mater. Res.* **12**, 3241, (1997).

7. a) O. Chyan, T.N. Arunagiri, and T. Ponnuswamy, *J. Electrochem. Soc.* **150**, C347, (2003). b) O. Chyan, T. Ponnuswamy, *U.S. Patent No. 7247554*, (2007).
8. a) M.W. Lane, C.E. Murray, F.R. McFeely, P.M. Vereecken, and R. Rosenberg, *Appl. Phys. Lett.* **83**, 2330, (2003). b) R. Chan, T.N. Arunagiri, Y. Zhang, O. Chyan, R.M. Wallace, M.J. Kim, and T.Q. Hurd, *Electrochem. Solid-State Lett.* 7, G154, (2004).
9. Woo-Jin. Lee, and Hyung-Soon. Park, *Appl. Surf. Sci.*, **228**, 410, (2004).
10. W.J. Lee, H.S. Park, S.I. Lee, and H.C. Sohn, *J. Appl. Electrochem.* **34**, 119-25, (2004).
11. V.R.K. Gorantla, K. A. Assiongbon, S.V. Babu, and D. Roy, *J. Electrochem. Soc.* **152**, G404, (2005).
12. J. C. Chen, and W.T. Tsai, *Mater. Chem. Phys.* **87**, 387, (2004).
13. J. C. Chen, S.L. Lin, and W.T. Tsai, *Mater. Chem. Phys.* **87**, 387, (2004).
14. Q. Luo, R.A. Mackay, and S.V. Babu, *Chem. Mater.* **9**, 2101, (1997).
15. S. Deshpande, S.C. Kuiry, M. Klimov, and S. Seal, *Electrochem. Solid-State Let.* **8**, 98, (2005).
16. G.W. Poling, *Corrosion Sci.* **10**, 359, (1970).

Conference Proceedings AMC XXV © 2010 Materials Research Society

Modeling and Applications of Copper Interconnect Chemical Mechanical Polishing (CMP) Simulation

Senthil Arthanari, Shia Yu, Marci Liao, Anand Inani, Katsuhiro Shimazu, Brian Stine
PDF Solutions, Inc., 333 W. San Carlos St., Ste. 700, San Jose, CA 95110

ABSTRACT

The shrinking lithography depth of focus budget demands that the CMP topography be controlled. As CMP topography is determined by the pattern density variations within a chip, controlling the density via dummy-fill schemes becomes necessary. By employing a sound model and an efficient simulator, we have shown that the multilayer CMP topography evolution can be successfully simulated. With this predictive ability, we were able to evaluate various dummy-fill schemes on a product with nine metal layers.

INTRODUCTION

As the critical dimensions continue to shrink with each node, the lithography process depth of focus (DOF) budget is shrinking as well. This reduction in DOF budget demands that the within die topography variations incoming into the lithography step be proportionately reduced. In the back end of the line processing, with inherent dishing and erosion issues, copper damascene CMP is the primary contributor to the incoming topography. Any reduction in dishing and erosion from CMP, therefore, directly translates to better control of the incoming topography [1]. In the earlier nodes, much of the copper CMP topography improvements have come from process improvements alone. However, for 130nm and below, as the number of metal layers continue to increase (thereby increasing the cumulative topography), layout density adjustments (dummy-fill or slotting) are also needed. As it is not practical to evaluate the layout adjustments physically, a multilayer simulator that can adequately predict the topography evolution becomes paramount. For such a simulator, we review the methodology used and the results from the simulations run to optimize the dummy-fill schemes for a nine metal layer product.

DISHING-INDUCED LOCAL FORCE REDISTRIBUTION

Dishing is a phenomenon initiated by parity in removal rates that exist in between two materials. High removal rate material (e.g. Cu) moves ahead (dished) and leaves behind the other material (e.g. oxide) as "highlander". However, as dishing progresses, dished material, hiding behind highlander, gradually loses contact with the external pressure. A fair share of the load that the dished material is supposed to carry would have to transfer to the highlander. As a result, the high removal rate of dished material slows down while highlander's low removal rate picks up. Eventually, the parity in removal rates between two materials vanishes when steady state is reached. Undoubtedly, this local force redistribution process depends on the local feature size. Narrow trench will soon reach steady state while wider trench can still carry load into deeper dishing amount. Load carrying capability (LCC) is a phenomenological modeling concept that captures this local force redistribution process.

In a complex layout of a real product where features with different sizes and shapes coexist, a series of negative sizing operations are done to the dished material and their density extracted. Each sized density map is supposed to represent the ability of dished material to carry load (i.,e, it's LCC) at certain dishing amount within the local neighborhood of the density calculation window. The deeper the dishing, the larger the negative sizing performed, the smaller the density of the resized layout which eventually emulates the lower capability to carry load of the external pressure from the CMP process. The conversion factor that relates the dishing to sizing amount is left as a model parameter to be determined globally in the characterization step. This model parameter depends on the rigidity of the polishing pad and the process condition. It is uniformly applied to all windows, but is allowed to change at different stages of the CMP process when condition changes. The aforementioned feature size dependence of the dishing process falls out naturally through this sizing treatment. Small features would vanish during sizing; these are features that have reached steady state and are no longer able to carry any load. With equal initial drawn density, large features would lose LCC slower than a group of smaller features, reflecting the fact that they will dish more. After the sized densities are extracted, discrete local densities at various sizing are joined together to represent the local load carrying capability as a function of local dishing. Linear interpolation is used to find proper LCC value in between those discrete points. Dictated by different local layout environment, different windows have different LCC function. Once the fair share of local load is determined, it is then proportionally distributed between dished material and highlander according to the proper LCC value calculated using local dishing amount so that the updated removal rate can be calculated.

MULTILAYER TOPOGRAPHY SIMULATOR

In this study, a PDF Solutions' proprietary chip level numerical model for Cu CMP is used. To accommodate the complex layout encountered in real product layout, we abandon the conventional width dependent dishing model in this simulator and replace it with the concept of LCC which can be characterized by simple layout sizing operation as described in the previous section. This simulator is based on a generalized contact mechanics and step model consistent with the MIT step model [2]. The interaction between non-uniform chip level topography and uniformly applied external down force is modeled through linear superposition of local height dependent point solution [3]. Periodic boundary condition at die edge falls out naturally from the FFT convolution method used to implement the linear superposition. In addition to large scale force redistribution due to topography, a secondary model is deployed to partition the local force into dished material (Cu) and its environment (oxide) region according to local layout density and dishing amount.

The simulation domain covers the entire chip. To reduce the computational complexity and to account for non-uniform layout styles in typical products, the simulation domain is sub-divided into non-overlapping tiles within which the local oxide height and the Cu height are individually kept. A time iteration loop simulates the evolution of the local topography in small time increments as the CMP process proceeds. Figure 1 illustrates the simulation steps. The first step calculates the local force according to the die level topography. In this step, local topography is represented by the local oxide height as it bears direct interaction with polishing pad. Total force balance is maintained to counter the applied down force. Once the local force within each tile is obtained, it is then partitioned between the Cu and the oxide regions according to the LCC of Cu, which is a function of the average dishing level at the time. Once local pressure is calculated, the Preston model [4] is then applied separately to each of the Cu and oxide regions. Stage 4 simply updates the topography and advances to next time step. With little extra book keeping effort, this simulator supports multi-material processes (e.g. Cu/Barrier/Oxide) where different tiles can operate on different materials according to local situation. At the end of each time step, we carry this topography as the starting topography of the next layer, assuming a conformal deposition has occurred.

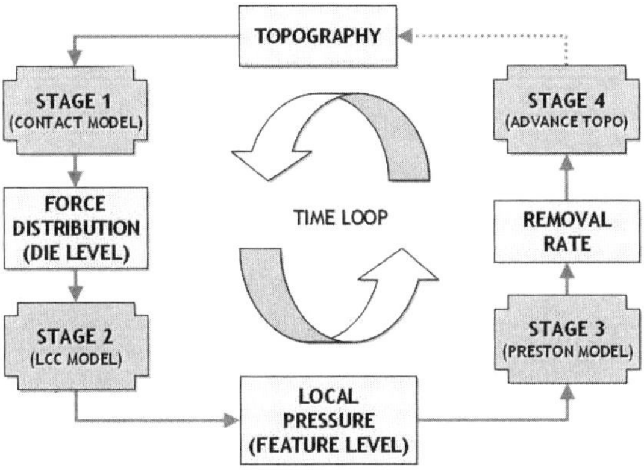

Figure 1: Stages of topography simulation

MODEL CALIBRATION

To effectively simulate the multilayer topography evolution, the first step is to calibrate the input parameters (interaction distance and the polish rates of copper, barrier metal, and oxide) at the main and secondary polish steps. We approached this issue of calibration by mapping the input parameter space centered on the blanket polish rates and an interaction distance of 200 microns. Once this space is mapped, we then ran multiple parallel simulations to determine the multilayer topography at various metal layers. The topography maps thus evolved were then compared with the actual full chip physical atomic force microscopy (AFM) measurements from various process layers. This process is repeated several times, shifting the input parameter space until the desirable match between physical and simulated topographies is achieved (Figure 2). PDF Solutions' in-house pdFAB® parallel simulation framework [5] was used to run hundreds of simulations at a fraction of the time required. Comparison of simulated topography using thus obtained optimum parameters with the actual AFM is shown in Figure 3.

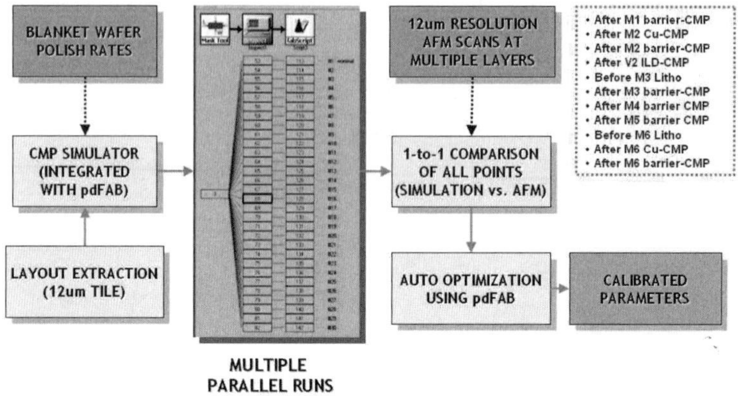

Figure 2: Calibration of input parameters for the simulator.

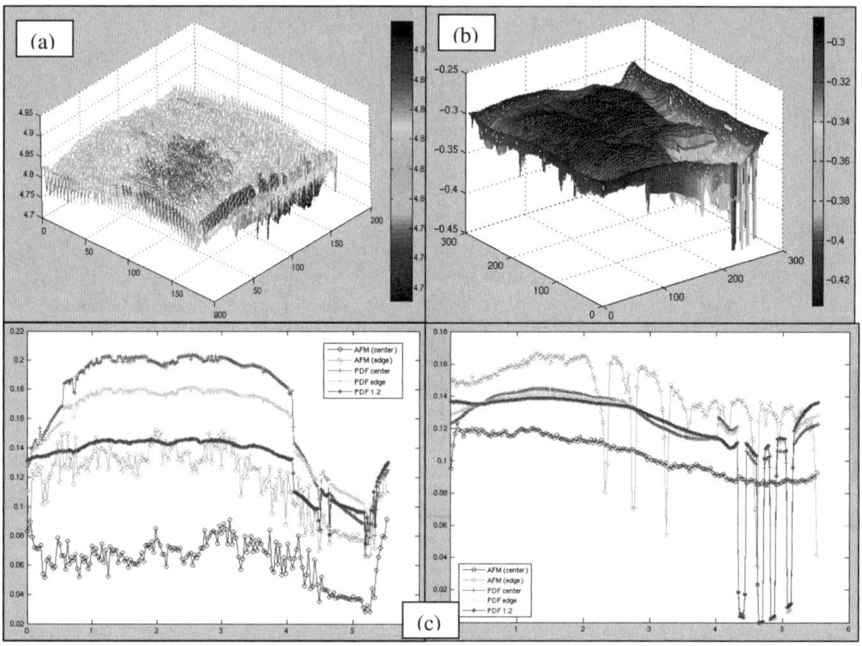

Figure 3: (a) Across chip AFM measurement, (b) Simulated across chip topography, (c) 1-D cross sections comparing AFM and simulation results. Note that an offset has been manually added between AFM and simulated curves.

From these calibration runs, we found that a smaller interaction distance (~150 microns) and the blanket polish rates for most part represent the actual process well. However, we also found that the copper removal rate during the barrier metal CMP needed to be much higher (~5X) than the blanket copper polish rate using the barrier metal process. We believe that this increased removal rate of copper during the barrier metal removal step is because of the electrochemical interaction of the barrier metal and copper. In addition, due to the very large scan area, the AFM measurements were afflicted by large scale non-linear errors. These large scale non-linearities in the AFM measurements were effectively removed by applying a Fourier filter before comparing with the simulation results.

MODEL APPLICATION – DUMMY FILL OPTIMIZATION

Global Step Height (GSH) is defined as the difference in the highest and lowest topography points at any particular layer. We used this as the primary metric to qualify non-uniformity at any given layer and eventually to define layout density adjustment schemes. As GSH is directly proportional to the pattern density ranges within a chip, rule based dummy-fill was used as the primary knob to modify it. Each metal layer was subjected to a pattern density range restriction and the dummy-fill was used to meet that range restriction. Also, all the layers in a given simulation run were subjected to the same range restriction. Figure 4 shows the details of the analysis flow.

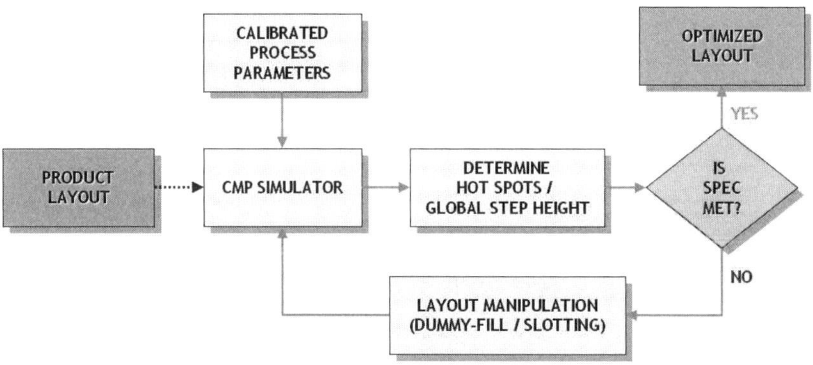

Figure 4: Analysis flow for Global Step Height Reduction

This flow was applied to all metal layers of the product and the evolution of GSH quantified with each subsequent metal layer as shown in Figure 5. As can be expected, for any given dummy-fill scheme the GSH increases steadily with the metal layers. It can also be noted that the GSH with no dummy-fill at all is slightly better than the current dummy-fill scheme used. This apparent discrepancy comes from the fact that the current dummy-fill scheme is not CMP specific, and underlines the need for CMP specific dummy-fill schemes. The GSH was alarmingly high at M9 using the current dummy fill scheme and lithography budget could not permit this. This meant some drastic changes to the dummy fill scheme. CMP simulations were used to find the main factors that will lead to GSH improvement: RR adjustments, range of pattern densities allowed at each layer and pattern density threshold below which DF should be applied. New dummy fill schemes were also simulated [6] and GSH evolution of each quantified as in Figure 5. Finally an optimized scheme with pattern density control was proposed to provide almost 40% reduction in M9 GSH which met the lithography DOF budget (Figure 6).

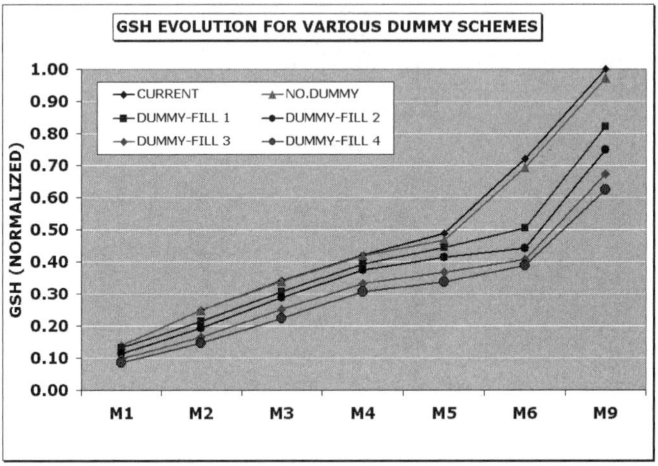

Figure 5: GSH evolution with metal layers for various dummy schemes

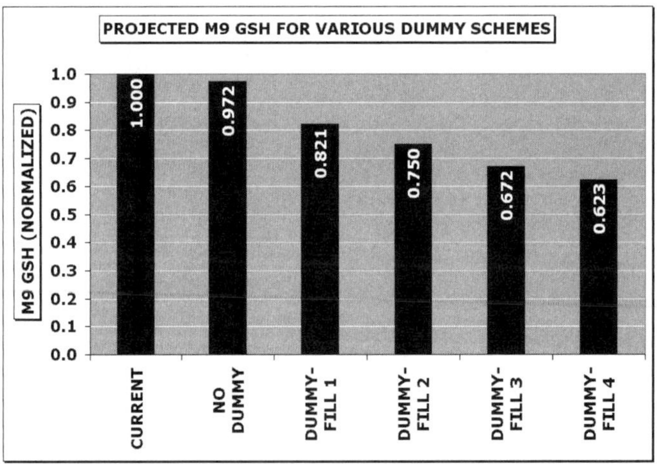

Figure 6: Dummy fill optimization using M9 GSH as the criteria to satisfy lithography DOF budget

This example clearly shows the merits of this method of optimizing dummy fill scheme over not only the current layer but also taking into account the underlying topography. The existing scheme based on single layer dummy fill was not sufficient in controlling topography and in fact did not show any improvement over the "no dummy" fill case. Optimizing the dummy fill scheme across layers did a much better job. There are two possibilities of applying this methodology – either on a product by product basis or a common scheme for a process which could be based upon optimizing the dummy fill scheme across multiple products.

CONCLUSIONS

An advanced CMP simulator based on a new load carrying capability model is presented and a method to calibrate the input parameters across materials is explained. This conceptual model was adapted in a simulator to simulate full chip topography across multiple layers. AFM measurements were performed on a product to confirm the accuracy of the model. Finally, a criterion to determine the within die topography incoming to litho is defined and various dummy-fill schemes to reduce this criterion is analyzed and results presented.

REFERENCES

1. B. E. Stine, V. Mehrotra, D. S. Boning, J. E. Chung, D. J. Ciplickas, "A Simulation Methodology for Assessing the Impact of Spatial/Pattern Dependent Interconnect Parameter Variation on Circuit Performance," *International Electron Devices Meeting*, 1997.
2. T.E. Gbondo-Tugbawa, "Chip-scale modeling of pattern dependencies in copper chemical mechanical polishing processes," *PhD Thesis*, Massachusetts Institute of Technology, 2002.
3. Thye-Lai Tung, "A method for die-scale simulation of CMP planarization," *Proc. SISPAD*, pp. 65-68, 1997
4. F. Preston, " The theory and design of plate glass polishing machines," *Journal of the Society of Glass Technology,* Vol. 11, pp. 214-256, 1927.
5. *pdFab* Users Manual, PDF Solutions Inc., 1998.
6. A.B. Kahng, G. Robins, A. Singh, A. Zelikovsky, "Filling algorithms and analyses for layout density control," *IEEE Trans on CAD of ICS*, Vol. 18, pp. 445-462, 1999.

Conference Proceedings AMC XXV © 2010 Materials Research Society

Improvement of etch processes for SiCOH materials with novel *in situ* diagnostic and evaluation methods

Sven Zimmermann[1], Nicole Ahner[1], Frieder Blaschta[1], Matthias Schaller[2], Henrik Zimmermann[3], Hartmut Rülke[2], Norbert Lang[3], Jürgen Röpcke[3], Stefan E. Schulz[1,4] and Thomas Gessner[1,4]

[1]Department Back-end of Line, Fraunhofer ENAS, Technologie-Campus 3, Chemnitz, D-09126, Germany
[2]Globalfoundries Dresden Module Two GmbH & Co. KG, Wilschdorfer Landstrasse 101, Dresden, D-01109, Germany
[3]INP Greifswald e.V., Felix-Hausdorff-Strasse 2, Greifswald, D-17489, Germany
[4]Center for Microtechnologies, Chemnitz University of Technology, Chemnitz, D-09107, Germany

ABSTRACT

The focus of this paper is the improvement of etch processes for SiCOH dielectrics in CF_4 plasmas with respect to the sidewall damage. Three different additives, Ar, O_2 and C_4F_8, were admixed to the parent gas in different concentrations to provoke a change of the etch process. Argon was added to enhance the sputter effect on the $SiO_{4/2}$ matrix of the SiCOH. Using the additive O_2, a serious damage of the SiCOH in combination with a CH_3 extraction, an enhanced hydrophilic behavior and a surface densification was aspired. Finally the C_4F_8 was used to enhance the polymerization efficiency of the etch process. The real occurrence of the theoretical predicted mechanism was analyzed using different bulk and surface sensitive measurement methods, e.g. contact angle measurements, spectral ellipsometry, Hg-probe investigation and FTIR analysis. A slight reduction of the sidewall damage with C_4F_8 addition and a small increase of this value using argon were measured using SEM cross sections. The plasma dissociation of the process gas and the formation of SiF_x and CO_x compounds as reaction products were clear influenced by the additives, measured with quadrupole mass spectrometry (QMS). Measurements with the quantum cascade laser absorption spectroscopy (QCLAS) show a surprising saturation effect for SiF_4 in the reaction chamber. The maximum SiF_4 concentration was $1.5 \cdot 10^{13}$ molecules/cm^3. It seems that higher SiF_4 concentrations are not possible, when SiCOH will be etched. Finally a new evaluation method on the basis of the correlation coefficient was applied to find dependencies between changes in the surface energy and dielectric constant and the results of the optical emission spectroscopy (OES). Therewith it was possible to identify species in the plasma, which promote and hamper the process induced rise of the dielectric constant and the surface energy. A promoting behavior was found for fluorine and different oxidizing species. Compounds of carbon, nitrogen and silicon as well as argon hamper the process induced material degradation.

INTRODUCTION

The low-k material SiCOH replaces the long time used SiO_2 in the field of interlevel dielectrics. Conventional SiCOH materials consist of a $SiO_{4/2}$ matrix with embedded hydrophobic CH_3 groups in an $O_3{\equiv}Si\text{-}CH_X$ structure. These groups avoid moisture absorption, which increases the effective dielectric constant [1].

The integration of these new materials results in special challenges concerning the further scaling of interconnects dimensions. Usually such materials will be structured using reactive ion etching (RIE) with fluorocarbon plasmas. During such processes the SiCOH material will be damaged with the result of an undesirable increase in its *k*-value. Typical damage causes are ion bombardment, damaging radicals and UV radiation [2]. The damage mechanisms are the removal of hydrophobic CH_3 groups, the surface densification by ion bombardment and photochemical modifications initiated by UV radiation [1].
Aside from etch and damage mechanisms, accelerated ions and hot radicals are able to sputter the SiCOH material. This results in the formation of $Si_XO_YH_Z$ clusters in the etch chamber. The stoichiometry of this $Si_XO_YH^+_Z$ clusters (x=1-22, y=1-44, z=0-7) depends basically on the oxygen to silicon ratio and the hydrogen content in the plasma [3].
Finally fluorocarbon plasmas show a polymerization inside the plasma on both, the reactor walls and the sample surfaces. The CF_2 fragment of the parent gas is the real precursor for this effect. The reaction proceeds like $C_XF_Y(CF_2)_n+CF_2\rightarrow C_XF_Y(CF_2)_{n+1}$ [4]. CF_2 macromolecules called oligomers are formed in the gas phase. It was reported that gas phase grown CF_2 chains formed polymer particles with molecular weights of more than 100.000 [5]. The etch and polymerization efficiency of the process depends on the carbon to fluorine ratio inside the plasma. Higher fluorine concentrations promote the etch reaction. Higher carbon contents increase the polymerization effects [6]. Different additives like H or O influence the C/F ratio. The combined etch and polymerization mechanism results in the formation of a polymer film on the sample surfaces during the etch process. This polymer film, which is the main inhibitor for transport of species and delivery of activation energy to the SiCOH films, will be consumed by fluorine etching and sputtering effects [7]. In other works the formation of a two layer stack on the sample surfaces, which consists of a reactive layer and a polymer layer above, was reported [8].
Currently, *in situ* diagnostic methods replace the empirical process optimization due to the exact insight into complex mechanisms in the reactor, e.g. etch, damage and polymerization. Common *in situ* methods are the optical emission spectroscopy (OES), quadrupole mass spectrometry (QMS) and Langmuir probe analysis (LP) [9,10]. Results of such measurements are generally complex data sets with frequently a lack of comparability to process results. Often, very small changes in the plasma conditions lead to large and irreproducible impacts on etch profiles and damages.

EXPERIMENTAL DETAILS

All etch processes were performed in an Oxford Plasmalab system 100. The ICP power was adjusted to 1000 W and bias power to 60 W. The pressure in the etch chamber was 7 mTorr and the CF_4 flow was predefined with 12 sccm. The flow rates of the additives Ar, O_2 and C_4F_8 were varied from 2 to 8 sccm.
Process induced bulk and surface modifications of SiCOH blankets were analyzed with Hg-probe measurements, spectroscopic ellipsometry, contact angle measurements and FTIR analysis. A trench etch process on structured samples was performed to analyze the influence of the process and the additives on the sidewall damage. A special SiC hard mask in combination with a dip in diluted HF solution allows the analysis of the sidewall damage quantitatively. To measure the dissociation of the parent gas and the formation of reaction products a VISION 1000E quadrupole mass spectrometer was used. It was connected to the plasma chamber using a standard KF-40 flange. Atoms and molecules in the reactor residual gas were ionized with 70 eV

and measured in the range from 1 amu to 200 amu. The resolution of the QMS is 0.1 amu. The total concentration of SiF_4 in the process chamber was measured with a QMACS Process Fibre infrared quantum cascade laser optical absorption spectroscope. The measurement principle and characteristics of this *in situ* diagnostic tool were published in [11]. Finally OES spectra were collected by means of a USB 4000 fiber optic spectrometer in combination with a 74-UV silica collimating lens. They were measured from 180 nm to 890 nm in steps of 0.2 nm. The OES spectra as well as the measurements of the *k*-value and the polar parts of the surface energies were used to apply a method for spectra interpretation. The latter is developed further at the Fraunhofer ENAS and called correlations analysis. It bases on the calculation of the statistical quantity correlation coefficient ($Korr_{SDS,PR}$), see equation 1 (SDS denotes spectral data sets, PR denotes process results).

$$Korr_{SDS,PR} = \frac{\frac{1}{n}\cdot\sum_{i=1}^{n}\left(SDS_i - \frac{1}{n}\cdot\sum_{i=1}^{n}SDS_i\right)\cdot\left(PR_i - \frac{1}{n}\cdot\sum_{i=1}^{n}PR_i\right)}{\sqrt{\frac{1}{n}\cdot\sum_{i=1}^{n}\left(SDS_i - \frac{1}{n}\cdot\sum_{i=1}^{n}SDS_i\right)^2}\cdot\sqrt{\frac{1}{n}\cdot\sum_{i=1}^{n}\left(PR_i - \frac{1}{n}\cdot\sum_{i=1}^{n}PR_i\right)^2}} \qquad (1)$$

The result of this analysis is the statistical correlation based on comparing each two variables separately. To perform the correlation analysis a special computer program was established, which consists of a correlation coefficient calculation routine and different sorting algorithms. The calculated coefficient is a measure of the coherency between species in the plasma, identified with OES measurements, and the process induced modification in the bulk and at the surface of the SiCOH material, identified with Hg probe and contact angle measurements.

DISCUSSION

Etch-induced material modifications

Figure 1 shows the influence of different process conditions on the SiCOH surface. The etch processes generally increase the surface energy and its polar part. This indicates a higher wetability and an increased hydrophilic behavior. The additive O_2 strengthens this effect. For C_4F_8, the flow dependent decrease in the surface energy in combination with an increased polar part results in the formation of a surface polymer film. Argon fails as candidate to alter the etch process with respect to a surface change of the SiCOH material. The changes of the dielectric constant and the refractive index, see figure 2, indicate the etch induced modification of the SiCOH bulk. Generally, all etch processes provoke an enhanced *k*-value. The addition of small amounts of O_2 increases this effect. However, C_4F_8 reduces the etch induced bulk damage due the formation of a surface polymer. Using no additives, the refractive index shows the highest value. A decrease was observed due the dillution effect of argon and the enhanced polymerization with C_4F_8. The reduced values for O_2 do not correspond with the measured *k*-values and the surface energies and are not yet completely understood in the moment.

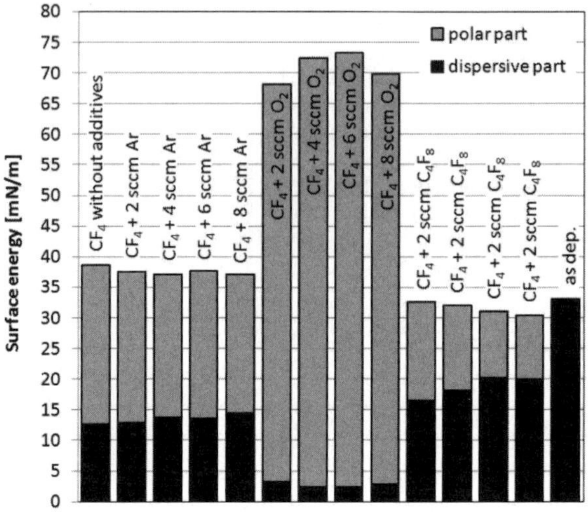

Figure 1. The surface energy and its polar and dispersive parts in dependence of etch chemistry.

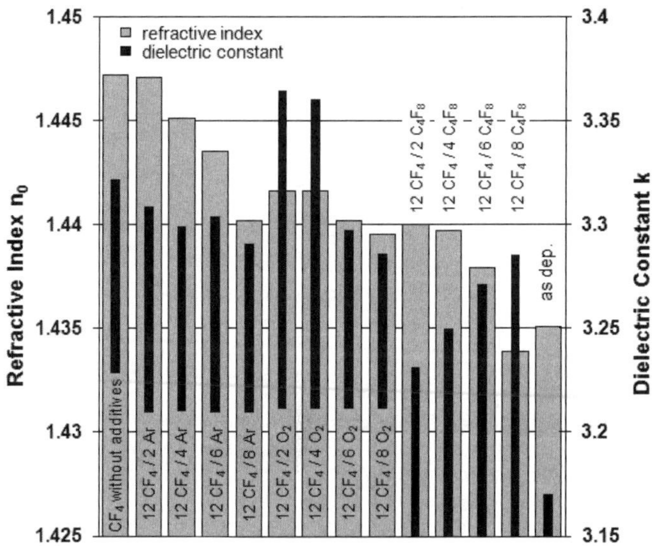

Figure 2. Behavior of the refractive index and the dielectric constant in dependence of etch chemistry.

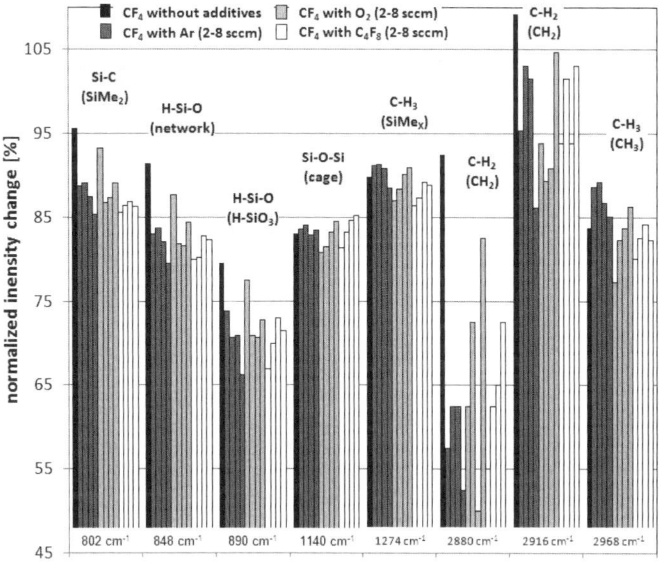

Figure 3. FTIR shows the influence of the etch chemistry on material properties *(the interpretation of the spectra was performed with respect to [12]).*

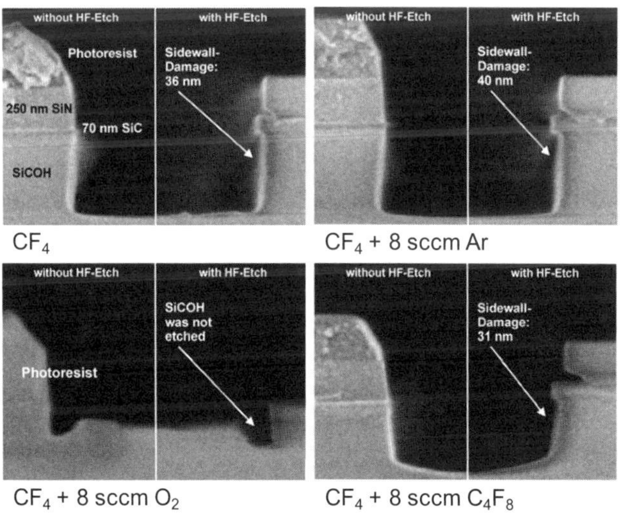

Figure 4. Sidewall damage in dependence of the etch chemistry using an SiC hard mask and HF decoration etch *(The graphics were generated by Fraunhofer ENAS).*

The FTIR measurements in figure 3 show a clear influence of the etch processes on the material properties. In comparison to the pristine state (100 %) a modified $SiO_{4/2}$ network and a depletion of methyl-groups, which are bonded and not bonded to the network, can be observed. Surprisingly the additives have a comparatively small influence on the etch induced material modifications. Only an decrease of the concentration of CH_2 groups which are not network bonded was observed. This seems to be the result of a parent gas dillution. Finally, the measured sidewall damages are shown in figure 4. Only the additive C_4F_8 produces a small reduction. However, argon provokes a small increase and O_2 should be avoided due a significant photoresist-attack in combination with a hampered SiCOH etch. A disadvantage of C_4F_8 is a rounding of the etch profiles, see figure 4. It was reported, that the sidewall damage fluctuates along the trench geometries, e.g. due the formation of sidewall striations [13].

Plasma conditions during the different etch processes

In the following section of the paper some selected results of the applied *in situ* diagnostics will be shown. These measurements were done to understand the physical and chemical mechanisms in the plasma, which are responsible for the SiCOH material damage.

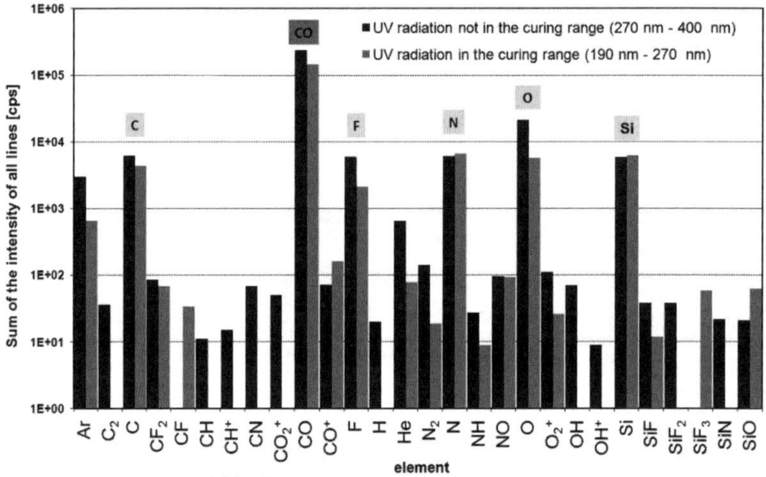

Figure 5. Calculated emissions of UV radiation of different elements, the values are sums of intensities of all spectra lines in the considered wavelength range.

The emission of UV and vacuum UV (VUV) radiation from the plasma initiate photochemical modifications of the SiCOH material [1]. Figure 5 shows spectra line intensity sums of selected species in the wavelength range of 190 -400 nm, whereas the curing range between 190-270 nm were presented separately. In order to avoid UV induced material damage, the concentration of the species C, CO, F, N, O and Si should be reduced. The most destructive molecule is CO, which has the ability to emit UV intensities more than two orders of magnitude higher than all other plasma species.

The plasma dissociation of the process gas shows an increased concentration of all fragments using C_4F_8 as additive, see figure 6. Especially the concentrations of C_2F_X and C_3F_X were increased by C_4F_8 addition. However the addition of O_2 reduces the concentration of these species. Furthermore a CF-depletion was observed for O_2 addition. Using argon the flow dependent decrease of all parent gas fragments indicates only a dilution effect. This rebuts the thesis, that argon enhances the process gas fragmentation due ion bombardment [10]. CF_3 is the dominant plasma species with the highest partial pressure, agreement with [10]. In figure 7 can be observed, that O_2 as well as C_4F_8 enhances the formation SiF_X fragments, which indicates a higher etch rate. C_4F_8 increases the fluorine incorporation into the etch chamber. O_2 stimulates the break of C-F bonds in the parent gas, which also become visible in the clearly higher CO_2 concentration. Argon nearly do not shows an influence on the reaction product formation.

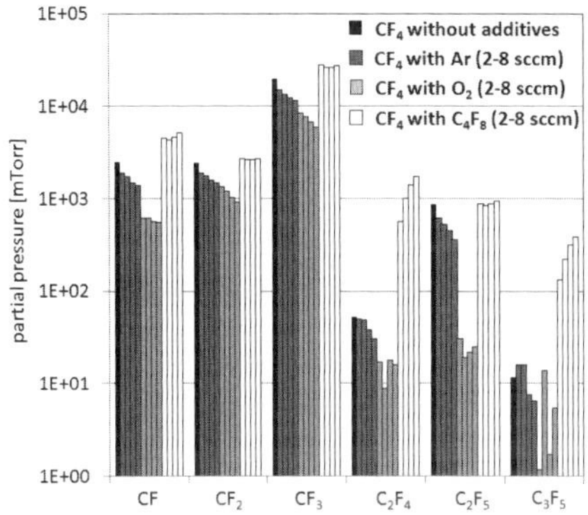

Figure 6. Influence of the additives on the dissociation of the process gas CF_4, measured with QMS.

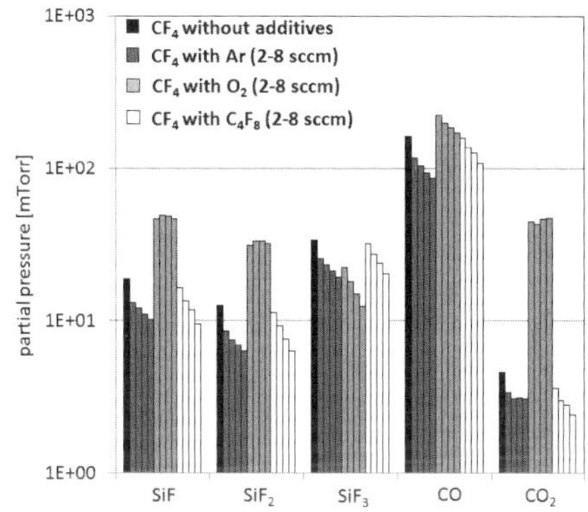

Figure 7. Formation of important reaction products in dependence on the additive concentration.

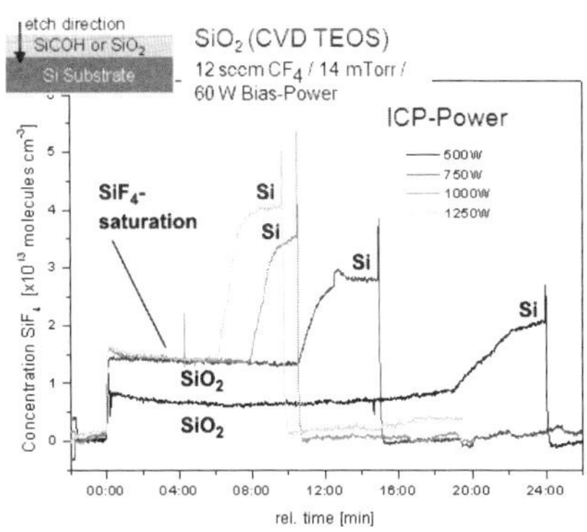

Figure 8. Behavior of the SiF_4 concentration during the etch of SiO_2 in CF_4, measured with QCLAS.

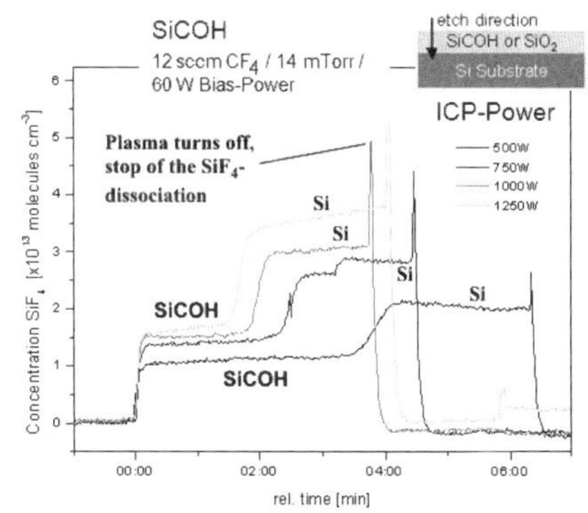

Figure 9. The SiF_4 concentration during the etch of SiCOH in CF_4.

QCLAS measurements were used to monitor the time dependent SiF_4 concentration during etch processes of SiO_2 and SiCOH in CF_4. To the surprise the SiF_4 concentration shows a sample material independent saturation point at nearly $1.5 \cdot 10^{13}$ molecules/cm^3. This point can be observed for SiCOH as well as SiO_2. Not until the films were etched down and the silicon substrates were reached, the SiF_4 concentration shows the expected power dependent increase. The reasons for this saturation are not yet known. Another effect is an abrupt SiF_4 increase when the plasma stops. In this moment stops also the dissociation of the SiF_4, what results in an increase of its concentration. After a short time the SiF_4 will be pumped down and the concentration becomes virtually zero.

Correlations between process results and plasma conditions

The last section of the paper deals with the complex correlation between the results of the *in situ* analysis and the processes. Figure 10 and 11 show a correlation analysis between OES spectra and the *k*-value as well as the polar part of the surface energy. The change of the *k*-value represents a measure for bulk modifications and the polar part indicates surface conditions of the SiCOH material. It is obviously, that the surface is significantly more sensitive on the different plasma species. As destroying species fluorine, different oxidizing compounds, CN for the bulk and SiF for the surface were identified. Species, which hamper the SiCOH degradation, are Ar, silicon-, nitrogen- and carbon-compounds. The increasing influence of the carbon-compound C_2 on the polymerization efficiency was reported in another publication [14]. It is interesting, that the main hampering element for the bulk degradation is a Si-compound and for the surface degradation an N-compound. Using these results a special plasma with a minimum material damage on the surface and in the bulk can be formed.

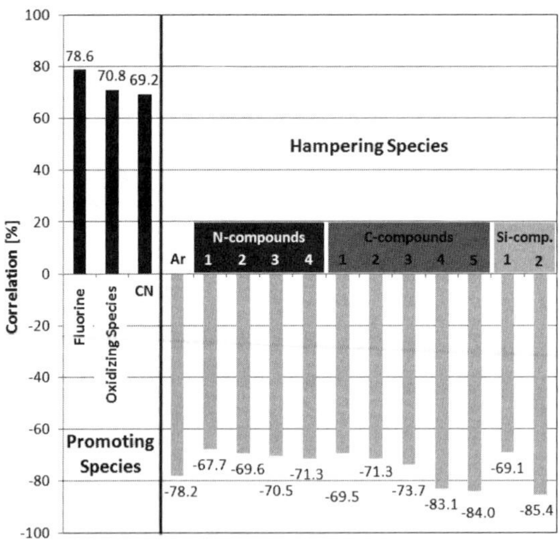

Figure 10. Species in the plasma, which promote or hamper the process induced rise of the dielectric constant *(the calculation method is declared in the experimental).*

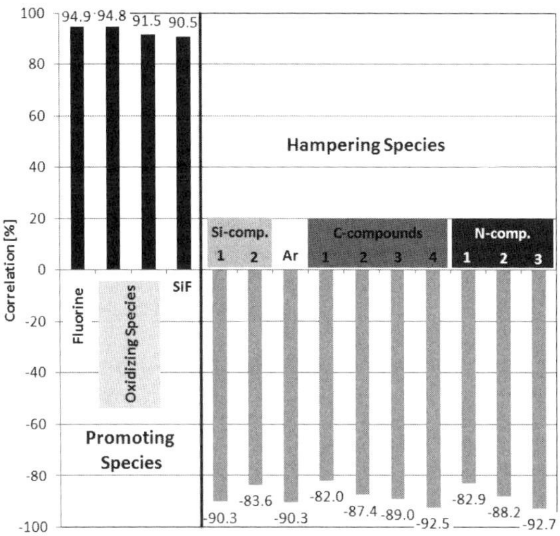

Figure 11. Species in the plasma, which promote or hamper the process induced rise of the polar part of the surface energy *(the calculation method is declared in the experimental).*

CONCLUSIONS

This paper underlines the high importance of *in situ* diagnostic methods for development and optimization of plasma etch processes for low-*k* materials. With respect to the performed measurements, the conclusion is admissible, that not only the *in situ* measurements with complex apparatuses but also the interpretation of the data and the identification of hidden correlations to the process results are big challenges. Another fact, which becomes clear in this paper, is the necessary replacement of the empirical process optimization. Only with the change of process parameters in the absence of fundamental knowledge about the species in the plasma and their interactions, among each other and with the low-*k* material, no improved plasma condition can be found. The correlation analysis, described in this paper, could be powerful tool in this topic. In the near future, the identification of species in the plasma, which are responsible for the sidewall damage, should be identified with this method. Finally the conclusion can be given, that only with the additive C_4F_8 an improvement of the sidewall damage was reached in comparison to an etch process without additives. But a significant etch profile deviation limits the C_4F_8 addition. Therewith, a final etch recipe represents always a compromise between sidewall damage and etch geometry.

ACKNOWLEDGEMENTS

The project described in this publication is funded by the German Federal Ministry of Education and Research, registered under funding number 01M3173. The author named in the publication bears responsibility for all published contents.

REFERENCES

1. M. R. Baklanov, A. Urbanowicz, G. Mannaert and S. Vanhaelemeersch, in International Conference on Solid-State and Integrated Circuit Technology 2006 (ICSICT 2006), edited by T.-A. Tang, G.-P. Ru and J.-L. Jiang, (Wiley-IEEE Press), pp. 291-294.
2. S. Uchida, S. Takashima, M. Hori, M. Fukasawa, K. Ohshima, K. Nagahata and T. Tatsumi, J. Appl. Phys. **103**, 073303-1, (2008).
3. T. Schenkel, T. Schlathölter, M. W. Newman, G. A. Machicoane, J. W. McDonald and A. V. Hamza, J. Chem. Phys. **113 (6)**, 2419, (2000).
4. G. Cunge and J. P. Booth, J. Appl. Physics, **85 (8)**, 3952, (1999).
5. K. Takahashi and K. Tachibana, J. Appl. Phys. **89 (2)**, 893, (2001).
6. J. W. Coburn and E. Kay, IBM Journal of Research and Development **23 (1)**, 33, (1979).
7. A. Sankaran and M. J. Kushner, Appl. Phys. Lett. **82 (12)**, 1824, (2003).
8. T. Tatsumi: Appl. Surf. Sci. **253**, 6716, (2007).
9. M. V. V. S. Rao, S. P. Sharma, B. A. Cruden and M. Meyyappan, Plasma Sources Sci. Technol. **11**, 62, (2002).
10 F. Gaboriau, G. Cartry, M.-C. Peignon and Ch. Cardinaud, J. Phys. D: Appl. Phys. **39**, 1830, (2006).
11. G. D. Stancu, N. Lang, J. Röpcke, M. Reinicke, A. Steinebach and S. Wege, Chem. Vap. Deposition **13**, 351, (2007).
12. A. Grill and D.A. Neumayer, J. Appl. Phys. **94 (10)**, 6697, (2003).
13. R. M. Badam, M. Mukherjee-Roy, W. Shaoyu and A. Naman, Jap. J. Appl. Phys. **43 (11A)**, 7381, (2004).
14. S. Zimmermann, F. Blaschta, M. Schaller, H. Rülke, S. E. Schulz and T. Gessner, in Advanced Metallization Conference 2008 (AMC 2008), edited by M. Naik, R. Shaviv, T. Yoda, and K. Ueno, (Mater. Res. Soc. Conf. Proc.AMCXXIV, Warrendale, PA, 2009), pp. 387-392.

Conference Proceedings AMC XXV © 2010 Materials Research Society

The Effect of Plating and Anneal Conditions on Post-CMP Defects

Marcelle Marshall, Roey Shaviv, Tom Mountsier, Jon Reid, Andrew McKerrow
Novellus Systems Inc., 4000 North First Street, San Jose, CA 95134

ABSTRACT

Low post-CMP void count in copper metal lines is a key requirement for robust metallization processes. In this study we evaluated the effect of process queue time, anneal conditions, and plating bath chemistry on post-CMP defects. All three parameters affect defect levels. Short queue-time between post-plating anneal and CMP significantly reduces post-CMP voids density. A similar trend is observed with lower anneal temperatures, regardless of anneal time, and with low concentration of trace elements in the plated Cu. We conclude that minimum defect levels can be achieved by plating with bath chemistry that results in low concentration of trace elements, by using low temperature anneal, and by reducing and controlling post-plating anneal to CMP queue time.

INTRODUCTION

As the dimensions of metal lines shrink with advancing technology nodes, eliminating post-CMP defects is required to maintain good yield and reliability performance.[1] That demand is enhanced by the smaller cavity sizes and the higher aspect ratios of the conductor lines, which make void-free electroplating more challenging. Recent advances in plating technology show promise in void-free Cu fill for high aspect ratio small features.[2] In an earlier study we explored the effect of anneal conditions on plated Cu.[3] In this study, the effect of plating bath chemistry, anneal conditions, and process queue times on post-CMP defects are examined. We find all three factors to have an impact on defect density.

EXPERIMENTAL

Cu lines were formed in low k dielectric (k = 3) using single level, damascene processing. Minimum nominal dimensions were 90 nm for the line width and 110 nm for the space. After processing through plating and anneal splits and through CMP, wafers were inspected for post-CMP defects using a KLA-eV300 top view SEM. Top-down scanning electron micrographs of the metal lines arrays were used to determine defect density. Cu voids and pits were counted from 360 SEM pictures with 3 μm Field of View, collected from 36 dies across each wafer, constituting 10 pictures per die. Total inspected area is 3240 μm^2/wafer. Each split contained 3 wafers. Results were consistent within each split in all cases so only an average is reported in this paper. In a previous experiment a CMP process with two clean steps, acidic and basic, was used.[2] It was found that the acidic clean step promotes corrosion of the Cu lines and distorts the counting of Cu voids caused by plating chemistry, anneal condition or queue time. In this study, the acidic clean step was excluded. The first set of experiments tested the effect of room temperature anneal, in situ anneal and the queue time between plating and CMP (3 days or 3 weeks) on post-CMP Cu voids. The second set of experiments tested five different plating bath formulations, with similar plating conditions, (chemistries A, B, C, D and E as detailed in Table 1) and four post-plate anneal conditions: 180° C/ 90 s, 225° C/ 90 s, 300° C/ 90 s, and oven anneal of 180° C/ 1800 s. The concentrations of in-film trace elements such as C, S and Cl were measured using Secondary Ion Mass Spectrometry (SIMS) at two locations on each wafer, center and edge. There was no significant difference between center and edge results so the concentration average for each element is reported in Table 1.

	Concentration (ppm)		
	C	S	Cl
Chemistry A	13.5	8.3	28.3
Chemistry B	35.4	8.6	44.7
Chemistry C	0.5	0.1	0.9
Chemistry D	5.8	0.5	4.7
Chemistry E	3.6	0.4	1.7

Table 1: The concentrations of trace elements for the plating chemistries used in this experiment, as measured by SIMS.

RESULTS AND DISCUSSION

Longer queue times between post-plating anneal and CMP significantly increases the post-CMP defect density (primarily Cu voids) for chemistry A (Table 1), whereas room temperature anneal reduces the defect density (Figure 1). Long anneal to CMP queue time allows post-anneal stress relief which promotes void formation at the weakest points; the interface between the Cu overburden and the narrow metal lines.

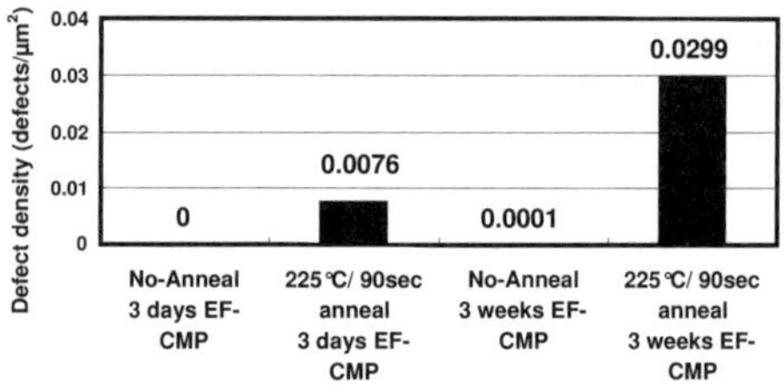

Figure 1: The effect of Anneal and queue time on post-CMP defects: Measured Cu voids density for chemistry A with and without the 225° C/90 s anneal and with two plating to CMP queue times (3 days and 3 weeks).

For all chemistries, lower anneal temperature leads to lower defect density (Figures 2 and 3). This is likely due to lower thermal stress and/or less plastic deformation of the Cu. The high anneal temperature of 300° C, used in this study, is known to result in high tensile stress in the annealed Cu.[4,5] Stress relief leads to void formation at grain boundaries, in particular where three grains come together. Those voids are subsequently exposed when the CMP process removes the overburden and the top portion of the dielectric layer. Since grain growth kinetics differs between the overburden and inside the lines, in particular narrow lines,[6] the transition area at the top of the line is particularly sensitive to void formation. This is evident from Figure 4 where STEM cross sections demonstrate that the voids appear to be confined to the top of the Cu lines and to coincide with grain boundaries, while no voids are apparent deeper in the line.

Grain orientation inside the lines is also dependent on anneal conditions. Bamboo-like, vertical grain-boundaries are achieved by either long anneal time or high temperature anneal, regardless of chemistry. Since bamboo-like grain boundaries are beneficial to reliability, in particular electromigration lifetime,[7] we expect a benefit in long anneal times or high anneal temperatures.

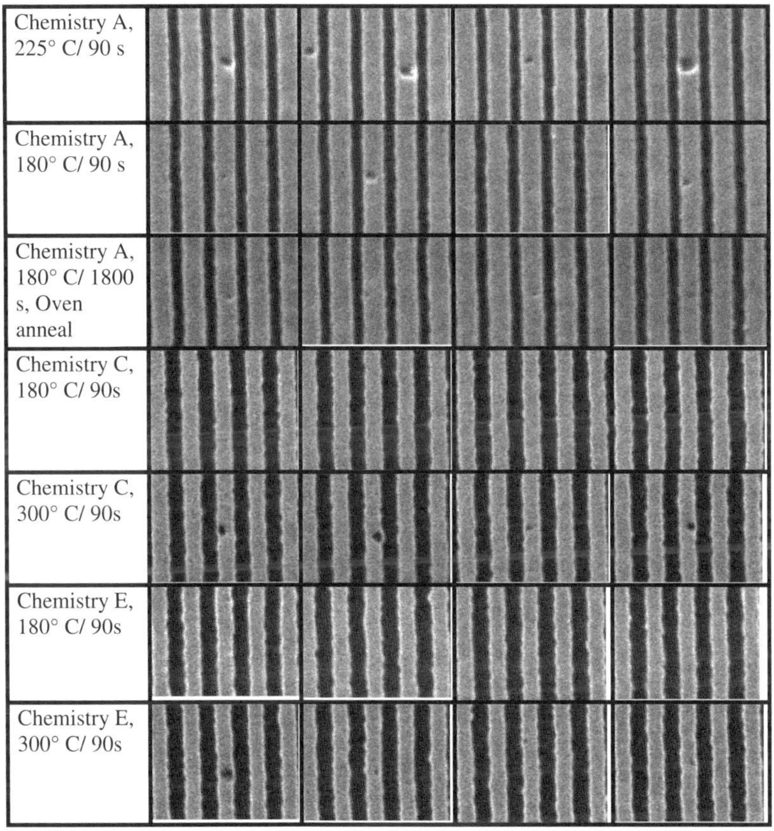

Figure 2: SEM top view pictures show post-CMP Cu voids, when present, for each split.

Cu with lower concentration of trace elements showed lower counts of post-CMP defects, as illustrated in Figure 3. The dependence of the defect density on anneal temperature is stronger in chemistries with high concentration of trace elements. High purity chemistries resulted in lower Cu void counts overall and hence had a weak dependence on the anneal temperature. While trace elements appear detrimental to void density, they are believed to be beneficial for preventing via stress migration.[8] Thus, a trade off between defect performance and stress migration reliability exists. This conflict can be resolved by design rules and integration.

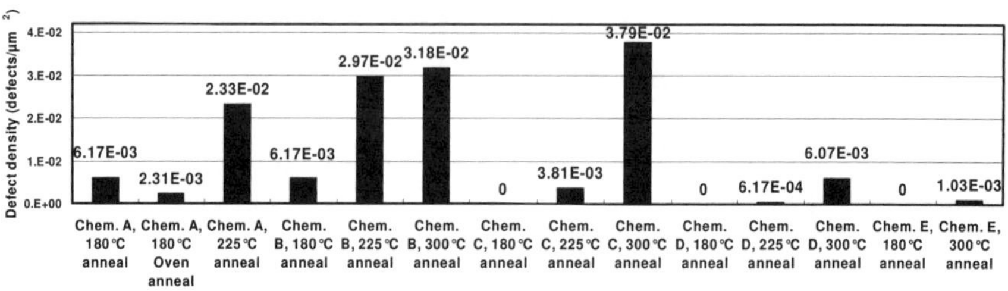

Figure 3: Measured Cu voids density for each chemistry with each anneal condition (anneal time is 90 s for all conditions except the "oven anneal" where anneal time is 1800 s).

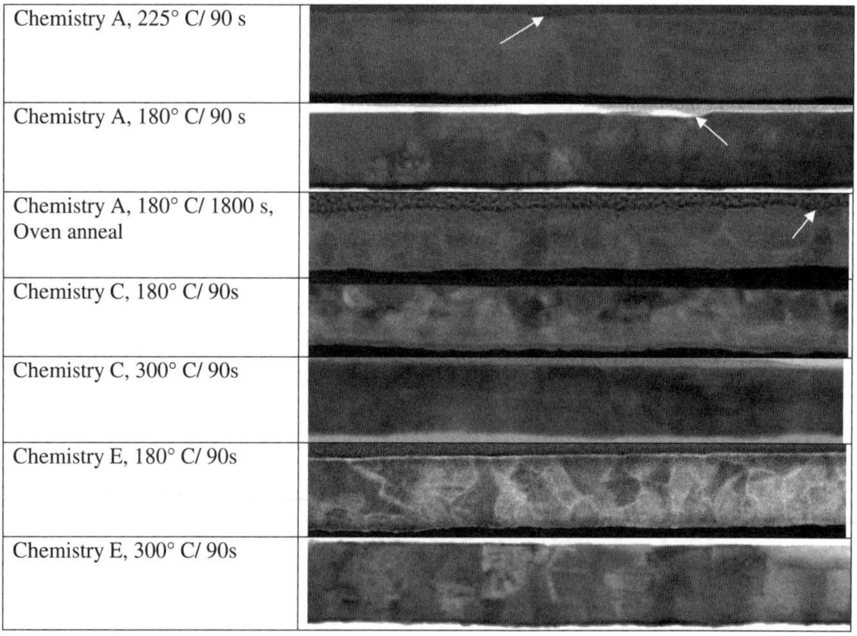

Figure 4: FIB STEM sidewall images of Cu lines plated with three chemistries and annealed with 4 different conditions. Depressions in the Cu at the top of the lines are highlighted with arrows for clarity. These depressions may be identified as post-CMP defects in top view SEM inspection.

Anneal time did not have a significant effect on Cu voids, as illustrated in Figure 5, but it had some effect on Cu grain morphology inside the lines, as illustrated in Figure 4. This observation is consistent with earlier reports.[9] The longer the anneal time, the larger the Cu grain size.

Higher anneal temperature results in lower resistance of the metal lines, as illustrated in Figure 6, a known effect attributed to lower scattering of charge carriers by grain boundaries, associated with larger grains and vertically oriented grain boundaries. The effect of anneal temperature on line resistance is more significant when using chemistries with low concentration of trace elements. This is more a result of grain size and grain boundaries orientation than a direct effect of the trace elements on Cu resistivity. Thus, the larger grains and bamboo-like orientation that result with low concentration of trace elements help in reducing grain boundary scattering, leading to lower line resistance with narrower distribution.

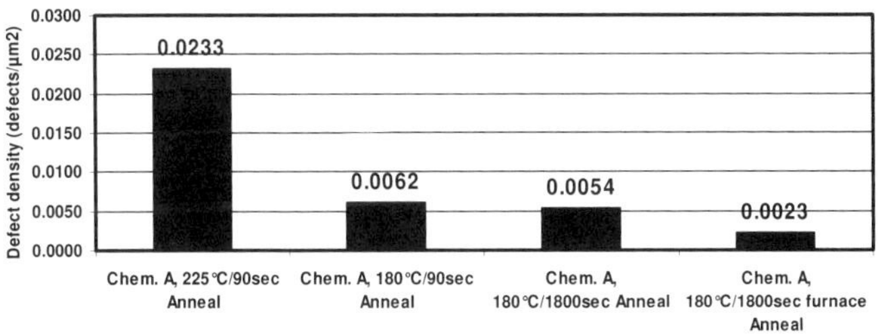

Figure 5: Measured Cu voids density for chemistry A with 4 different anneal conditions.

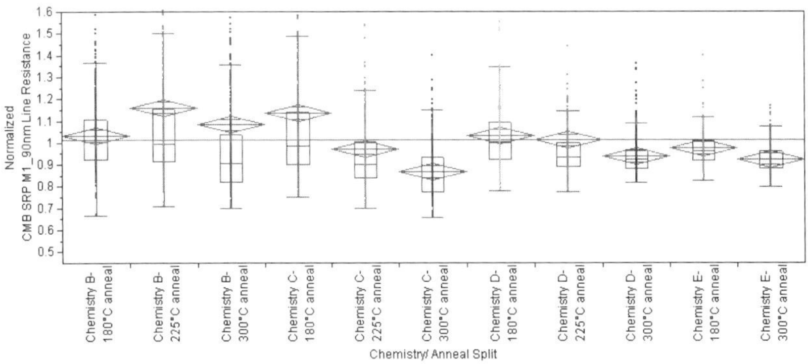

Figure 6: Normalized serpentine line resistance of plating chemistry and anneal splits.

The impact that trace element concentration, anneal conditions, and queue times have on post-CMP Cu voids and grain boundaries is believed to be related, in part, to segregation of trace elements to grain boundaries during anneal.[6] In general, grain boundaries are susceptible to chemical attack, which explains the low level of defects seen even with low levels of trace elements and high temperature anneals. As concentrations of trace elements increase, the bonding between adjacent Cu grains weakens. In places where three Cu grains meet in proximity to the top of the line, CMP may dislodge the top laying grain leaving a void behind. We conclude that optimized anneal conditions and tight queue time control are required in cases where concentration of trace elements is elevated. Plating chemistries leading to low concentration of trace elements can be beneficial to achieving low defect density copper metallization.

CONCLUSION

We report that short queue times between post-plating anneal and CMP is beneficial in reducing post-CMP defect counts. A similar trend is observed with lower anneal temperatures, regardless of anneal time. Plating chemistries leading to low concentration of trace elements can be beneficial to achieving low defect density copper metallization. We conclude that minimum defect levels can be achieved by lowering the concentration of trace elements in the plated Cu, reducing anneal temperatures, and tightening the queue time between post-plating anneal and CMP.

ACKNOWLEDGEMENTS

The authors would like to thank Natalie Tran from the Novellus Customer Integration Center (CIC) for her help on KLA-Tencor eV300 and to Terri Shofner and Daniel Pullman from the Novellus SEM lab for the STEM work.

REFERENCES

1. Alers, G.B. Dornisch, D. Siri, J. Kattige, K. Tam, L. Broadbent, E. Ray, G.W. Reliability Physics Symposium (IRPS), 2001. Proceedings. 39th Annual. 2001 IEEE International p.350.
2. J. Reid, Electrofill Challenges and Directions for Future Device Generations ADMETA 2007.
3. J. Sukamto, J. Reid, Electrochemical Society Proceedings Spring 2004: Electrochemical Processing in ULSI and MEMS: edited by L. Deligianni et al, p82-96, 2005.
4. Ryung-l Yul Park, Sang-Rok Hah, Chan-Geun Park, Dong-Kwon Jeong, long-Seong Son, Hyeok-Sang Oh, Ju-Hyuk Chung, Jeong-Lim Nam, Kwang-Mycon Park and Jae-Dong Byun Proceedings of the IEEE IITC p. 130 (2002).
5. R.P. Vinci, E.M. Zielinski, J.C. Bravman; Thin Sol. Film 262 (1995).
6. J. Reid and J. Zhou, ECS Transactions, Volume 2, No. 6, (2007).
7. P.S. Ho, et al AMC 2007 (unpublished).
8. A. Uedono, T. Suzuki, T. Nakamura, T. Ohdaira, and R. Suzuki, Jpn, J. Appl. Phys. 46, 1938 (2007).
9. L. M. Gignac, C.-K. Hu, B. W. Herbst, B. C. Baker-O'Neal, Advanced Metallization Conference (AMC 2007), p. 82 (2007).

Conference Proceedings AMC XXV © 2010 Materials Research Society

Aluminum pad cleaning with diluted organic acids for CMOS image sensors

S. Bilouk[1, 2*], L. Broussous[1], R. P. Nogueira[2], C. Jayet[3], C. Pernel[3]
[1] STMicroelectronics, Centre de Recherche Crolles, 850 rue Jean Monnet, 38926 Crolles Cedex, France
[2] UMR 5266 et 5631 INPG-CNRS-UJF SIMAP/ LEPMI, 1130 rue de la piscine, BP 75, 38402 St. Martin d'Hères, France
[3] CEA-LETI MINATEC, F38054 Grenoble, France

ABSTRACT

The aluminum-copper alloy reactivity in different acid solutions was studied by Scaning Electron Microscopy (SEM) observations and electrochemical methods using open-circuit monitoring and polarization curve measurements. In the mineral acid, the dissolution rate is significant due to the Al oxide and substrate dissolution. On contrary, in diluted oxalic acid, the dissolution rate is very low, and almost negligible in diluted glycolic acid. While in the oxalic solution, the dissolution of aluminum occurs, a passive behavior of the Al-glycolic acid interface is observed. This is confirmed by the OCP evolution and by the shape of the potentiodynamic curve. Indeed, an anodic plateau appears, which reflects the metal passivation . In addition, the cathodic shift of the corrosion potential reflects a nobler behaviour of Al in glycolic acid. Thus, according to thermodynamic considerations, the reaction of proton reduction cannot occur at these conditions of pH and potential. This can explain the very low dissolution of aluminum in glycolic acid solution.

INTRODUCTION

CMOS-based image sensors are more and more used in the digital imaging industry; these sensors are common in digital cameras, cellular phones, webcams, toys and security cameras. Image sensors integration with standard CMOS logic process allows realizing integrated single-chip device, however it also brings new requirements and integration challenges. In the Back End of Line interconnect levels, a number of process steps such as dry etch and subsequent wet clean must be optimized in order to improve overall sensor performance. Regarding wet clean process, the integration of resist micro-lens in the final aluminum interconnect levels also yields new issues. Indeed, organic solvent blend, initially designed for resist stripping are widely used for aluminum cleaning. However, this cleaning solution may also strongly degrade micro-lens array. As an alternative cleaning solution, aqueous acidic solutions were studied as they cannot dissolve resists. This paper deals with the understanding of aluminum behavior during various acidic clean processes. The action of two organic acids (oxalic and glycolic acids) and a mineral one (sulfuric acid) will be investigated by electrochemical and physical characterizations.

EXPERIMENT

A three-electrode cell (volume 500mL) was used for Open Circuit Potential (OCP) monitoring and Linear Sweep Voltammetry experiments (LSV) with a large-area Pt foil as counter electrode and a Saturated Calomel reference Electrode (SCE). The working electrode was made of a thin film of $AlCu_{0.5\%}$ deposited on Si <100> mounted on a Radiometer Analytical CTV 101 rotating disk electrode. The active area was 2 cm^2. All experiments were performed at room temperature. The potentiostat used was a GAMERY PCL 4750 potentiostat galvanostat ZRA from Gamery Instruments.inc. LSV tests were conducted by scanning the applied potential from -1 V /OCP to +1 V /OCP at a scan rate of 1 mV s^{-1}. All measurements were carried out at 200 rpm.

Aqueous solutions were prepared with deionised water. A 1%wt glycolic and oxalic acid based solutions and a 10%wt diluted sulfuric acid solution are used in the present study.

RESULTS AND DISCUSSION

Open Circuit Potential monitoring

The aluminum electrode was immersed in the different solutions and the variation of OCP with time was recorded until a steady state value was reached, as reported in figure 1.

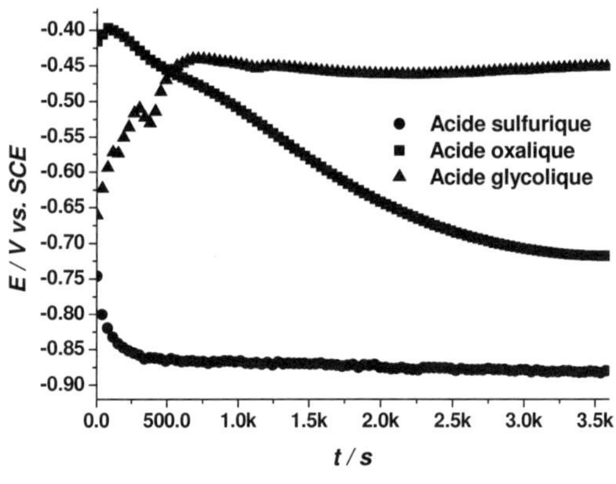

Figure 1. Open circuit potential evolution of Al in the different acidic solutions

In sulfuric acid, the OCP quickly decreases from -0.75 V_{ECS} to -0.890 V_{ECS}. However, the variation of the potential is not significant and its stabilization is fast. This trend can be attributed to a pronounced dissolution of the oxide or to the Al metal in the acidic media.

In presence of organic acids, the OCP variations are more pronounced (several hundreds millivolts), which could be attributed to significant modifications of the surface.

In oxalic acid, the OCP of Al slowly decreases from -400mV$_{ECS}$ to -730mV$_{ECS}$ after 300s of exposure. According to the literature, in oxalate ions containing solution, a similar initial negative shift has been reported [1]. It was suggested that the shift was due to the alteration of the native oxide previously formed in contact with air. In the present work, alterations could possibly take place with the participation of the oxalic ions [2,3]. The evolution of OCP towards more negative values can be attributed to the specific adsorption of the oxalate anions on the metal surface or on the native oxide layer. This adsorption may induce oxide dissolution.

For glycolic acid, the time evolution of the OCP is significantly different. The increase of OCP during the immersion time can be due to a protective action of the organic acid on aluminum. Unlike oxalic acid, the action of the glycolate anion seems to induce aluminum passivation. Its specific adsorption can thus enhance protective properties of the native oxide film.

Similar results for dissolution kinetics of an oxide covered aluminum electrode were observed in solutions containing complexing agents [4-6]. The detachment of Al-complexes in solution is generally a very slow process [5]. Therefore, the fact that OCP steady state is slowly reached means that surface means that surface coordination mechanisms play an important role besides the electrochemical reaction.

Polarization measurements

After the OCP reached a stable value, the polarization curves of aluminum were performed from the corrosion potential ($E_{corr} = E_{OCP}$) to +1.6 V vs. E_{corr} with a scan rate of 1 mV.s^{-1} (figure 2). The corrosion parameters obtained from these measurements are given in table 1.

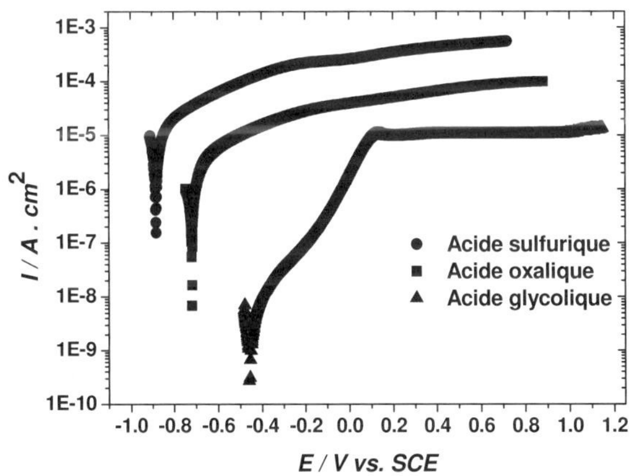

Figure 2. Potentiodynamic polarization curves of aluminum after 1h of immersion in the different studied solutions. Scan rate 1 mV s^{-1} and Ω = 200 rpm.

Table 1. Relevant Al corrosion parameters issued from the polarization curves.

Solution	E_{corr} (V vs ECS)	J_{corr} (µA.cm^{-2})	Etch Rate (Å.min^{-1})
Mineral acid	-0.890	11.6	160
OA_1: Oxalic Acid	-0.720	1.2	14
OA_2: Glycolic Acid	-0.450	0.011	0.12

In sulfuric acid, the curve shape reveals a relatively strong dissolution of the oxide or / and the metal. Indeed, the etch rate deduced from the corrosion current J_{corr} is 160 Å.min^{-1}, i.e 10 to 1000 times higher than in oxalic and glycolic acid solutions respectively. In addition, it is worth to notice that in presence of glycolic acid the aluminum dissolution is negligible (0.12 A.min^{-1}). This can be explained by a protective action of the glyclic acid. In agreement, the shape of the polarization curve is typical of a passivation behavior: a plateau appears with a constant anodic current between +0.2 V vs. ECS and +1.6 V vs. ECS. The obtained current of the limiting plateau (nearly 10 µA.cm^{-2}) is characteristic of metal passivation in the organic acid. This confirms the assumption deduced from the OCP measurements: the probable stability of the glycolate anions on the surface can form a protective organic-mineral mixed film. For the Al/oxalic solution interface the mechanism is likely similar but in this case the detachment of the Al-complex is more significant, which induces a slight dissolution of the alumina or of the Al metal (14 A.min^{-1}). This assumption is in agreement with the shift of the open circuit potential to more negative value commonly attributed to an active behavior of the metal/ electrolyte interface.

Morphology of Al surface after the different acidic treatments

In order to verify the results obtained by electrochemical methods, SEM analysis were carried out after one hour of immersion in the different acids (Fig. 3).

Before the cleaning step, aluminum surface is not homogeneous (fig.3a). Indeed, it is composed by smooth and rough grains. The mineral acid strongly affects the surface morphology (fig.3b). , pitting corrosion occurs on the smooth grains while the roughness of the other grains tends to increase. A similar result is observed for the oxalic treatment (fig.3c), even if, in this case the attack seems less significant.

In glycolic solution, no specific morphology change is observed (fig.3d) compared to the untreated substrate.

All these results are in agreement with the potentiodynamic study.

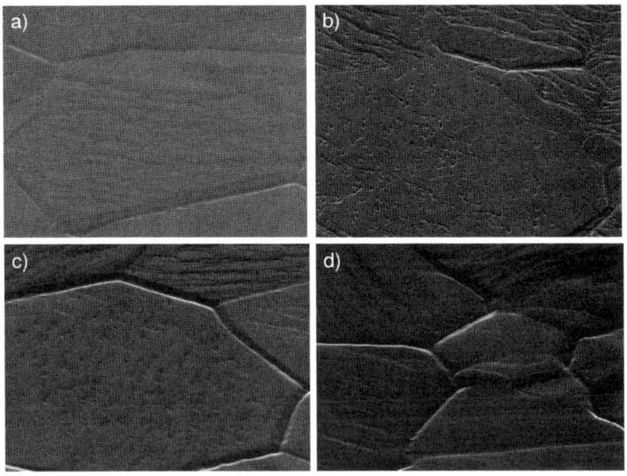

Figure 3. SEM pictures of Aluminum samples immersed in the different acidic solutions: a) reference, b) sulfuric acid, c) oxalic acid, d) glycolic acid X20000

Thermodynamic study

According to thermodynamic aspects, Al^{3+} ions are the predominant form of Al for the corrosion potential previously determined in acidic media [7].

The E-pH diagrams presented in figure 4 are plotted for the Al-organic systems using the JChess software [8]. In presence of organic acids, aluminum ions can form different complexes noted $AlOA_1$ and $AlOA_2$ in oxalic and glycolic solutions respectively.

These complexes could form an adsorbat on Al surface and thus modify the mechanism and the kinetic of Al dissolution. In oxalic solution, the $AlOA_1$ complex formed with Al is probably unstable on aluminum surface and tends to desorb dissolving the substrate. On the contrary, the $AlOA_2$ complex formed with glycolate anions can have a stronger protective effect. Moreover, the thermodynamic study highlights that in oxalic solution, the corrosion potential is under the oxygen and proton reduction equilibrium. Thus, the cathodic process is governed both by the reduction of oxygen and protons. In glycolic solution, the potential of aluminum is higher and slightly exceeds the potential beyond which proton reduction no more occurs. Consequently, only the reduction of oxygen can thermodynamically occur at the considered corrosion potential, the dissolution of aluminum becomes negligible and is limited by oxygen content.

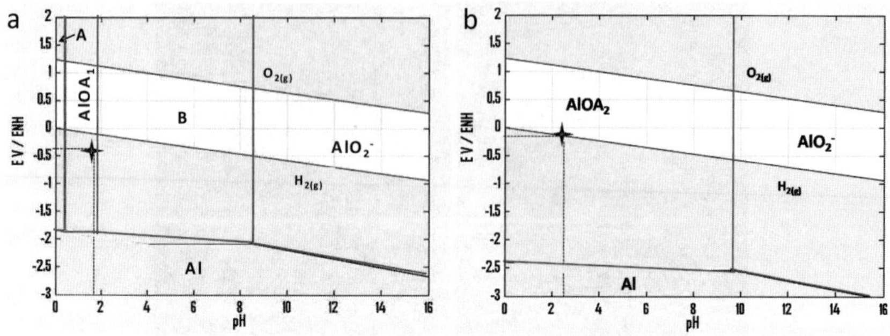

Figure 4. E-pH diagrams of the Aluminum –organic acid system at 25°C:

a) Oxalic Acid, b) Glycolic Acid (cross is set for Aluminum E_{corr} in the studied solutions calculated versus ENH).

Discussion

In aqueous solutions, the aluminum-oxygen bonds of the native oxide are significantly hydrated which induces the formation of surface-OH groups with amphoteric properties. This leads to the establishment of acido-basic equilibrium on the aluminum surface with a characteristic iso-electric point at pH comprised between 7 and 9 [9]. This parameter also defines the pH at which the zeta potential equals to zero. According to the literature, below pH 9, the surface is positively charged due to the formation of $-OH_2^+$ from the adsorption of protons.
In organic acid solutions, the possible surface complexing agents are OA_1^- and OA_2^- for oxalic and glycolic acid respectively. They can be attracted by the positive surface and react to form aluminum hydroxy-ligand surface species according to the following mechanism (Eq. 1 and Eq. 2):

$$(Al-OH_2^{\ +})_{ads}+OA^- \rightarrow (Al-OH_2-OA)_{ads} \quad \text{Eq. 1}$$

$$(Al-OH_2-OA)_{ads}+OA^x \rightarrow (Al(OA)_2^{\ x})_{(aq)}+H_2O+2H^+ +2e^- \quad \text{Eq. 2}$$

It can be supposed that in organic solutions, the second reaction is the rate determining step which can explain the evolution of the OCP which takes a long time to stabilize. In oxalic solution, the mixed film is probably not stable while in presence of glycolic acid, the oxide layer is protective allowing a weak dissolution of the metallic substrate. This assumption is confirmed by both the OCP evolution and the shape of the polarization curve where an anodic plateau reflecting a passivation of the metal was observed.
Moreover, because in the pH range 1.8-2.4, the concentration of H^+ ions is sufficiently high, it can be supposed that the dissolution of aluminum is under the control of electrochemical processes. This means that the main cathodic reactions are the proton reduction and the reduction

of oxygen while the anodic reaction is the dissolution of aluminum in the form of Al^{3+} aqueous complexes.
However, in presence of glycolic acid, a singular passivation behavior is noticed. The E-pH diagram shows that in presence of glycolic acid, the shift of the corrosion potential to less negative values implies that the proton reduction cannot occur thermodynamically. Only the oxygen reduction can be involved in cathodic processes and thus controls the corrosion mechanism. As dissolved oxygen concentration is very low (~9ppm), the kinetic of aluminum dissolution is significantly slowed down.

CONCLUSION

The electrochemical and surface properties of aluminum in sulfuric, oxalic and glycolic acid solutions were investigated by means of open circuit monitoring, polarization curve measurements and Scanning Electron Microscopy observations.
The decrease of the OCP and its very fast stabilization in the mineral medium is typical of a pronounced dissolution of aluminum. In both organic solutions, the OCP reaches a steady state value very slowly likely due to the low detachment rate of surface compounds in the solution. Indeed, the thermodynamic study demonstrates that in these medium, aluminum surface can be complexed by both organic acids. However, their complexing action is different. An interesting behavior of the Al/ glycolic acid interface is observed. In oxalic solution, Al dissolution and pitting corrosion occur whereas in glycolic solution the surface seems not to be attacked. Besides, the aluminum/ glycolic acid interface has been shown to exhibit a passive behavior at anodic potentials. Finally, in this solution, aluminum is nobler and consequently, oxydation by protons cannot occur: the corrosion is limited only by the reaction of oxygen reduction. All these assumptions deduced from electrochemical and morphological analysis confirm the protective action of glycolic acid on aluminum.

REFERENCES

[1] A.M. Shams El Din, N.J. Paul, Thin Solid Films, **189**, 205, (1990).
[2] L. Kobotiatis, N. Pébère and P.G. Koutsoukos, Corros. Sci., **41**, 941 (1999).
[3] W. Wilhelmsen, A.P. Grande, Electrochimi. Acta, **33**, 269 (1988).
[4] G. Furrer and W. Stumm, Chimia, **37**, 338 (1983).
[5] M. Segura and D. Hasenay, J. App. Electrochem., **31**, 961 (2001).
[6] W. Stumm, G. Furrer and B. Kunz, Croat. Chem. Acta, **56**, 593 (1983).
[7] C.M.A Bett, Corros. Sci., **33** (2), 203 (1992).
[8] J. Van der Lee and L. de Windt, CHESS Tutorial and Cookbook/Version 3.0, users manual LHM/RD/02/13, Ecole des mines de Paris, Fontainebleau (2002).
[9] G.A. Parks, Chem. Rev., **65**, 177 (1965).

Materials and Characterization

Conference Proceedings AMC XXV © 2010 Materials Research Society

Electronic Transport Properties of Copper Atomic Wires

Saeideh Mohammadzadeh[1,*], Reinhard Streiter[1,2], and Thomas Gessner[1,2]
[1] Center for Microtechnologies, Chemnitz University of Technology, 09126 Chemnitz, Germany
[2] Fraunhofer Research Institute for Electronic Nano Systems, ENAS, 09126 Chemnitz, Germany

ABSTRACT

Based on the non-equilibrium Green's function density functional tight binding method, electronic transport properties of copper atomic wire through (001) direction have been analyzed. Numerical calculations suggest that transport properties behave relatively similar and independently from the wire length and depend strongly on the geometry defects if the wire length approaches to atomic dimensions.

INTRODUCTION

Atomic sized contacts fabricated by scanning tunneling microscope and the mechanically controllable break junction [1, 2], have been of great interests to both experimental and theoretical studies for the two past decades. In present work, the electronic transport properties of copper atomic wires are analyzed theoretically in the framework of the non-equilibrium Green's function technique on the density functional tight binding method [3-6]. The copper atomic wires are sandwiched between (001) electrodes, and the electronic current is deduced according to the Landauer formulation [7, 8]. Performing this approach, transmission coefficients, current-voltage characteristic, and potential difference are computed to study the effect of geometry structure on the electronic transport properties.

THEORY

Copper atomic wire attached to the (001) electrodes is assumed with experimental atomic distance. To study geometry effects on electronic transport, atomic wire is considered to be defect free in three lengths and for the longest wire an extra copper atom is placed at the interface of electrode and scattering region as geometry defect shown in figure1.

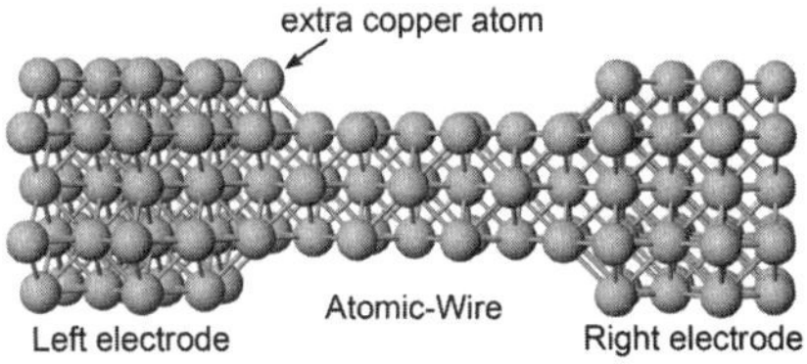

Figure 1. Schematic view of geometry structure of copper atomic wire in (001) direction. An extra copper atom is added to the interface of left electrode and wire as geometry defect.

Calculations are performed using the non-equilibrium Green's function within the density functional tight binding method [3-6] and the Landauer-Büttiker theory [7, 8] to obtain the differential conductance of the system. The Green's functions are calculated for ideal atomic positions of the wire and electrodes under circumstances of zero bias voltage and external bias voltage up to 1.0 V self consistently. Interaction between the valence electrons is treated in the local density approximation (LDA) [9]. Since the applied bias changes the transport properties from those at the zero-bias limit, it is important to consider the effect of the applied bias in the theoretical studies. Within this method the current can be written with an expression formally equivalent to the scattering theory [10-12].

$$I(V) = \frac{2e}{h}\int_{\mu_1}^{\mu_2} Tr[\Sigma_L^< G^> - \Sigma_L^> G^<]dE \qquad (1)$$

where $\Sigma_L^{<,>}$ are the self energy functions for the left contact and the Keldysh Green's functions, $G^<(E)$ and $G^>(E)$, are defined in terms of the retarded Green's function G^r, the advanced Green's function G^a, and the non-equilibrium self energies $\Sigma^<$ and $\Sigma^>$. The limits of the integration, μ_1 and μ_2, are the electrodes chemical potentials. e is the electron charge, and h is Planck's constant. Based on the Landauer's scattering approach to electronic transport and assuming that transport is ballistic for finite bias voltages, the current can be established by noting that the transmission matrix $t(\varepsilon)$ is written as

$$t(\varepsilon) = [\Gamma_R(\varepsilon)]^{1/2} G^r(\varepsilon)[\Gamma_L(\varepsilon)]^{1/2} \qquad (2)$$

where the $\Gamma_{L(R)}(\varepsilon)$ is given by

$$\Gamma_{L(R)}(\varepsilon) = i(\Sigma_{L(R)}^r(\varepsilon) - [\Sigma_{L(R)}^r(\varepsilon)]^\dagger) \qquad (3)$$

By substituting the transmission matrix into 1, the current is obtained as follows

$$I(V) = \frac{2e}{h}\int Tr[\Gamma_R G^r \Gamma_L G^a](f_R(E) - f_L(E))dE \qquad (4)$$

where $Tr[\Gamma_R G^r \Gamma_L G^a]$ gives the transmission coefficients and $f_{R(L)}(E)$ describe the Fermi distribution functions for right and left electrodes.
The differential conductance can be obtained from the current flowing through the system by taking the derivative of current with respect to bias voltage [13-14].

DISCUSSION

Total transmission amplitude with respect to the electron incident energy shown in figure 2 reduces by applying external bias voltage of 0.0-1.0 V. Furthermore, total transmission spectra for different wire lengths behave similar, but differently after placing geometry defect illustrated in figure 2(c). Current-voltage characteristic of each atomic wire, depicted in figure 3, shows the dependency of current on wire length and geometry defect. The conductance of 0.7, 1.1, and 1.4 nm wire lengths considered at room temperature and in the limit of zero bias voltage is computed as $3.4\,G_0$, $3.5\,G_0$, and $0.7\,G_0$, respectively.

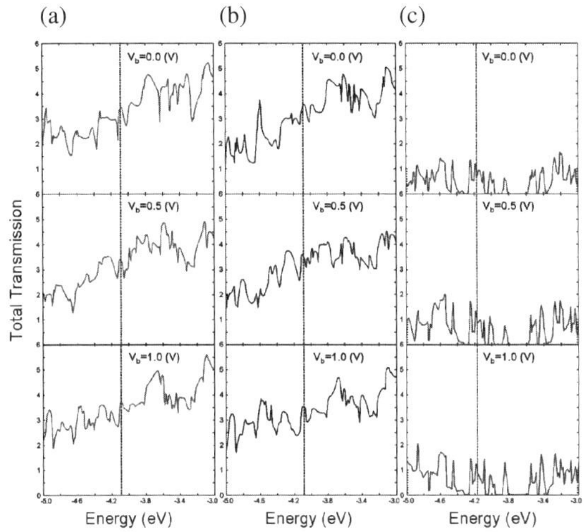

Figure 2. Total transmission spectra as a function of incident electron energy at bias voltage of 0.0, 0.5, and 1.0 V for three wire lengths (a) 0.7 nm, (b) 1.1 nm, and (c) 1.4 nm. Wires of 0.7 nm and 1.1 nm lengths are defect free, while an extra copper atom is considered as geometry defect in 1.4 nm wire. Dotted line shows the Fermi energy of system.

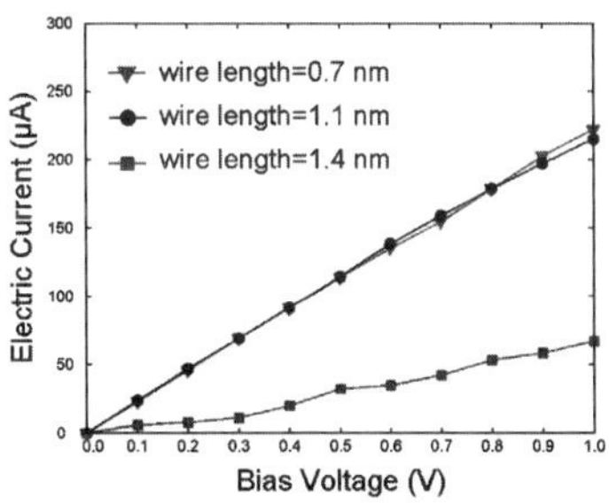

Figure 3. Current-voltage characteristic of copper atomic wire through (001) direction. The curve specified by square signs is plotted in presence of geometry defect. Current gives the same results for different wire lengths, and differs obviously for the structure including defect.

The differences in electrostatic potential with 0.5 V and 1.0 V from zero bias voltage in plane along the copper atomic wires is plotted in figure 4. Unlike the potential difference of two short and defect free atomic wires, the potential difference of the structure with defect does not behave smoothly and drops additionally near the interface.

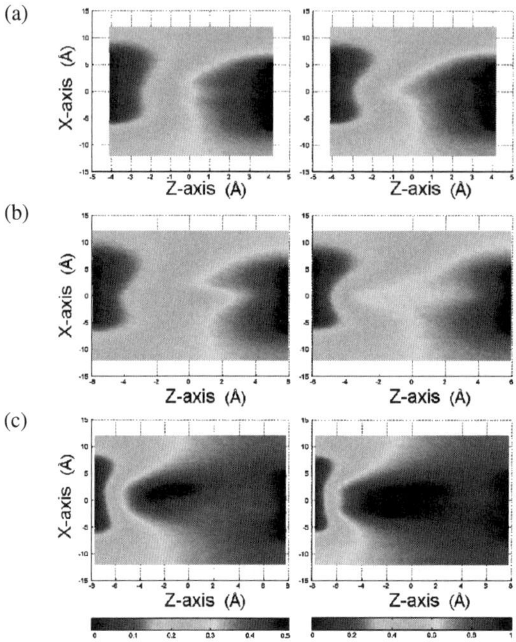

Figure 4. Potential difference in plane along the copper atomic wire, (a) wire length=0.7 nm, (b) wire length=1.1 nm, and (c) wire length=1.4 nm in two potential differences of 0.5 V (left) and 1.0 V (right). Figure 4(a) and (b) indicate potential difference of defect free structures, and figure 4(c) shows voltage drop in presence of extra copper atom at the interface.

CONCLUSIONS

In this paper, performing non-equilibrium Green's function density functional tight binding method in copper atomic wires bridged between (001) electrodes, transmission spectra, current-voltage characteristic, conductance, and potential difference of copper atomic wires have been computed as a function of atomic wire geometry structure. Since these structures form the situation of ballistic conduction, Landauer-Büttiker theory has been applied. Copper atomic wires are supposed to be defect free at room temperature in three lengths and for the longest wire an extra copper atom at the interface of wire and electrode is considered to study the role of geometry defect. Results have shown that as the wire length approaches to more atomic-sized; the current-voltage characteristic behaves relatively the same and independently from the wire length, while current-voltage characteristic, conductance, and potential difference strongly depend on the geometry defect presented in the longest contact. Based on numerical calculations, geometrical changes thus suggest being an important factor for varying the electronic transport properties.

ACKNOWLEDGMENTS

We gratefully acknowledge useful discussions with Dr. Alessandro Pecchia concerning non-equilibrium Green's function density functional tight binding method and the Deutsche Forschungsgemeinschaft and International Research Training Group "Materials and Concepts for Advanced Interconnects" for the financial supports.

REFERENCES

1. J. M. Krans, et al, "One-atom point contact", Phys. Rev. B48, pp. 4721-4724 (1993).
2. H. Yasuda and A. Sakai, "Conductance of atomic-scale gold contacts under high-bias voltages", Phys. Rev. B56, pp. 1069-1072 (1997).
3. T. Frauenheim, et al, "Atomistic simulations of complex materials: ground-state and excited-state properties", J. Phys. Condens. Matter, 14, pp. 3015-3074 (2002).
4. A. Pecchia and A. Di Carlo, "Atomistic theory of transport in organic and inorganic nanostructures", Rep. Prog. Phys. 67, pp. 1497-1561 (2004).
5. G. C. Solomon, et al, "Understanding the inelastic electron-tunneling spectra of alkanedithiols on gold", J. Chem. Phys. 124, pp. (094704) 1-10 (2006).
6. A. Pecchia, L. Latessa, A. Gagliardi, Th. Frauenheim, A. Di Carlo, "Molecular Nano Electronics", p. 205, (2007).
7. S. Datta, "Electronic Transport in Mesoscopic Systems", Cambridge University Press, NewYork, p. 48 (1995).
8. M. Büttiker, Y. Imry, R. Landauer, and S. Pinhas, "Generalized many-channel conductance formula with application to small rings", Phys. Rev. B31, pp. 6207-6215 (1985).
9. G. Seifert, D. Porezag, T. Frauenheim, "Calculations of molecules, clusters, and solids with a simplified LCAO-DFT-LDA scheme", Int. J. Q. Chem. 58, pp. 185-192 (1996).
10. Y. Xue, S. Datta, and M. A. Ratner, "First-principles based matrix Green's function approach to molecular electronic devices: general formalism", Chem. Phys. 281, pp. 151-170 (2002).
11. J. C. Cuevas, A. L. Yeyati, and A. Martin-Rodero, "Microscopic Origin of Conducting Channels in Metallic Atomic-Size Contacts", Phys. Rev. Lett. 80, pp. 1066-1069 (1998).
12. L. I. Glazman and A. V. Khaetskii, "Nonlinear Quantum Conductance of a Lateral Microconstraint in a Heterostructure", Europhys. Lett. 9, pp. 263-267 (1989).
13. E. Castano and G. Kirczenow, "Theory of nonlinear transport in narrow ballistic constrictions", Phys. Rev. B41, pp. 3874-3877 (1990).
14. L. Martin-Moreno, J. T. Nicholls, N. K. Patel, and M. Pepper, "Non-linear conductance of a saddle-point constriction", J. Phys. Condens. Matter 4, pp. 1323-1333 (1992).

Low temperature silicidation of Pd layers on crystalline silicon monitored via *in situ* resistance measurements

Erik J. Faber[1], Rob A. M. Wolters[1,2], Bijoy Rajasekharan[1], Cora Salm[1] and Jurriaan Schmitz[1]
[1]MESA+ Institute for Nanotechnology, University of Twente, Semiconductor Components Group, P.O. Box 217, 7500AE Enschede, The Netherlands
[2]NXP Semiconductors, Eindhoven, The Netherlands

ABSTRACT

Metal-silicon contacts may degrade during (SOI) wafer processing at unknown, high temperature conditions such as caused by sputtering or etching. We observed this phenomenon for Pd contacts on SOI wafers. Silicide formation of Pd contacts on crystalline silicon has therefore been investigated for low temperatures (57 – 135° C). We present an elegant measurement technique for monitoring the silicide formation process using *in situ* four point resistance measurements. From this the Pd_2Si thickness was determined via a resistance model. The results obtained via this simple measurement technique showed that silicide formation of Pd contacts on Si occurs already at 57° C, which is a temperature easily reached during metal sputtering onto thin SOI wafers. Furthermore, below approximately 100° C our results indicated that apart from Pd_2Si formation also mixed layers of Pd in Si and vice versa exist whereas at higher temperatures these mixed layers are absent or not significantly present.

In addition, we consider this measurement technique as an elegant, universally applicable method for monitoring silicide formation of metal-silicon contacts.

INTRODUCTION

Palladium is of interest as a high work function (5.12eV [1]) metal to form good contacts to p-type silicon. Pd on Si, however, is known to form Pd_2Si at temperatures even below 200° C [2,3]. Silicide formation at such low temperatures should be taken into account when devices are processed under conditions with unknown temperature, such as sputtering. As long as good thermal conducting bulk materials are used (such as silicon wafers of several hundred micrometer thickness), the heat transport to the bulk is sufficient to prevent excessive heating at the interface of Si and the deposited Pd. Silicon On Insulator (SOI) technology [4], has opened the way to new experimental silicon devices with nanoscale dimensions such as Multi-gate transistors and FinFETs [5]. During the sputter deposition of Pd layers on SOI wafers, however, the actual temperature at the Si-Pd interface can easily lead to silicide formation due to poor heat transport through the thin Si and the underlying poor thermal conducting SiO_2 layer. This effect is illustrated in Figure 1.

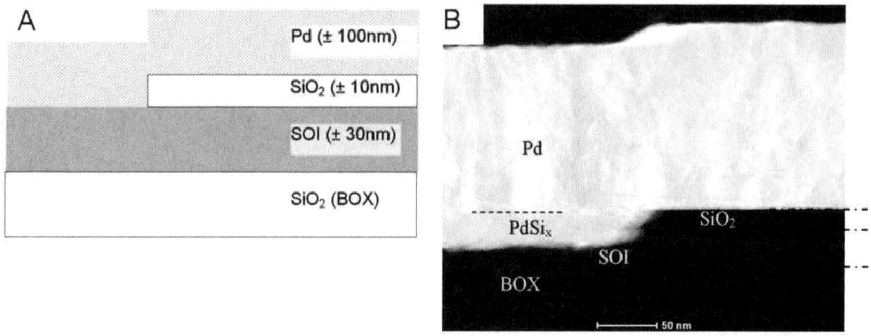

Figure 1. Schematic drawing of a SOI device with Pd contact (A) and High Angle Annular Dark Field (HAADF) Transmission Electron Microscopy (TEM) picture of the resulting device (B). The device consisted of a 100nm of Pd layer sputtered on a SOI wafer with 30nm of silicon.

In Figure 1B it can be seen that palladium has reacted with a large part of the silicon (SOI) into either a mixture of Pd and Si or Pd_2Si as was determined using an Energy Dispersive X-ray (EDX) detector.

Above findings resulted in this study in which we have focused on the low temperature (< 150° C) reactions of Pd layers on Si. This was done via *in situ* four point resistance measurements. From this the Pd_2Si thickness was determined via a simple resistance model as outlined below.

THEORY

The reaction of metallic Pd layers with bulk Si has been extensively studied. Below approximately 800° C only one stable phase is formed, namely Pd_2Si. The formation of this silicide follows diffusion limited kinetics [6]. The formed layer thickness of the silicide d_{Pd2Si} grows according to the Arrhenius equation (1):

$$d_{Pd2Si}^2 = D_0 t e^{-E_A/kT} \tag{1}$$

where D_0 [m^2s^{-1}] is the diffusivity pre-exponential term, t [s] is the time, E_A [eV] is the activation energy, k [$eV \cdot K^{-1}$] is Boltzmann constant and T [K] the temperature. Values for the activation energy vary between 0.9 – 1.5 eV [3,7]. Most studies in literature calculated the Pd_2Si thickness via backscattering techniques [7-9] or X-ray Diffraction [2,3]. In this study we determined the Pd_2Si thickness via in-situ resistance measurements. The basis for the calculation of d_{Pd2Si} is a three layer resistive model of the samples in the form of a stack of Pd, Pd_2Si and Si. The resistances of these individual layers are in parallel and this gives for the total measured sheet resistance:

$$\frac{1}{R_{sheet,meas}} = \frac{d_{Pd}}{\rho_{Pd}} + \frac{d_{Pd2Si}}{\rho_{Pd2Si}} + \frac{d_{Si}}{\rho_{Si}} \quad (2)$$

Where d is the thickness of the layer and ρ the resistivity of the layer. d_{Si} and ρ_{Si} are considered constant and insensitive to temperature in the investigated temperature range (50 – 225° C). The initial Pd layer thickness $d_{Pd,0}$ and the resulting Pd_2Si thickness $d_{Pd2Si,\infty}$ after all Pd has been consumed have been determined using Rutherford Backscattering Spectroscopy (RBS). Subsequently, we have applied a simple linear relation between d_{Pd} and d_{Pd2Si} according to:

$$d_{Pd} = d_{Pd,0} - \left. d_{Pd,0} d_{Pd2Si} \right/ d_{Pd2Si,\infty} \quad (3)$$

Finally, substituting equation (3) in (2) gives for d_{Pd2Si}:

$$d_{Pd2Si} = \frac{\frac{1}{R_{sheet,meas}} - \frac{d_{Pd,0}}{\rho_{Pd}(T)} - \frac{d_{Si}}{\rho_{Si}(T)}}{\frac{1}{\rho_{Pd2Si}(T)} - \frac{d_{Pd,0}}{d_{Pd2Si,\infty}\rho_{Pd}(T)}} \quad (4)$$

Here the resistivities have been compensated for temperature. The procedure used for temperature compensation is illustrated in Figure 2.

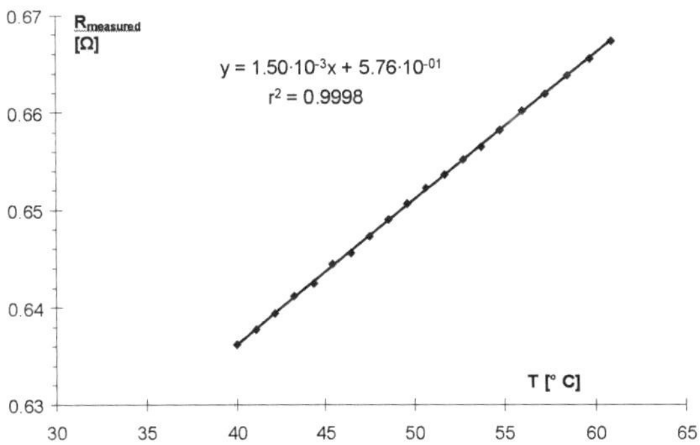

Figure 2. Measured resistance versus temperature for the heating up phase in the temperature range 40° - 60° C. The solid line shows the best linear fit ($r^2 = 0.9998$).

During the heating up phase the measured resistance increased linearly with temperature. It is assumed that during the relatively fast heating up no intermixing of Pd in Si or vice versa or the formation of silicide occurs. The resistance change is then solely due to the heating up of the (unaffected) Pd layer and thus reflects the Temperature Coefficient of Resistance (TCR). The r^2 values for the linear fits were near unity for all samples and this supports our assumption. The parameters of the best linear fit have been used for calculating the resistivity of ρ_{Pd} as a function of temperature. A similar procedure was used for obtaining ρ_{Pd2Si} as a function of temperature. This was derived using the resistance vs temperature data during the cooling down phase of completely silicidized samples.

EXPERIMENT

Silicon substrates (100 mm, p-type, <100>, resistivity 5-10 Ωcm) were cleaned using standard cleaning. After a 1% HF dip, a layer of nominal 45 nm of Pd was sputter deposited at 500W at room temperature. Hereafter, samples were put in a vacuum oven and annealed at constant temperatures in the range of 57° - 135° C. In situ 4-point probe resistance measurements were performed during these thermal treatments. Additionally, samples were analyzed using X-ray Diffraction (XRD) and RBS.

DISCUSSION

The normalized resistance measurements at constant temperature are shown in Figure 3.

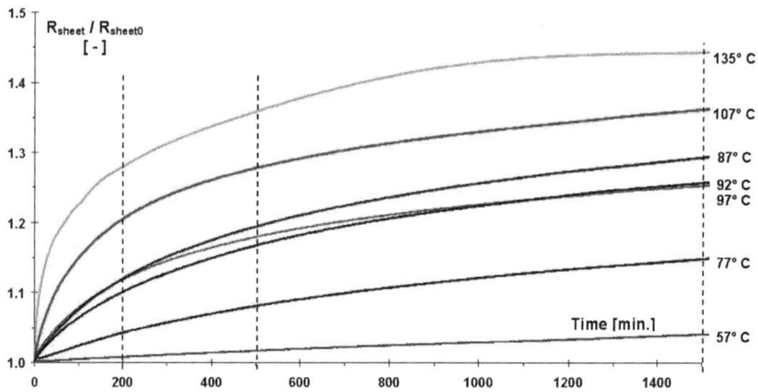

Figure 3. Relative resistance change (Rsheet / Rsheet at t = 0) vs time for temperatures in the range 57° - 135° C.

The measured resistance increases in time, even at a temperature as low as 57° C. This is attributed to the formation of a higher resistance Pd_2Si layer at the cost of the lower resistance Pd layer. We also evaluated the room temperature resistance change during a measurement of 20,000 minutes (app. 14 days). Although we observed a change over time it was hardly visible on this timescale. After 20,000 minutes the change was 2.3% and after 1500 minutes it was less than 0.2% as compared to R_{sheet0}, respectively. The latter would not be visible in Figure 3. Therefore the error due to a room temperature resistance change was neglected and the samples were regarded to be stable during storage prior to measurements.

The trend in Figure 3 shows a discrepancy between 80 – 100° C. In general an increase in resistance change is observed with increasing temperature. For the measurements at 87, 92 and 97° C, however, we see a decrease in resistance change with increasing temperature. This effect is clearly observed in Figure 4, where the resistance increase as a function of temperature is shown after 200, 500, and 1500 minutes. This deviation at 87° and 92° C is still under investigation in our group.

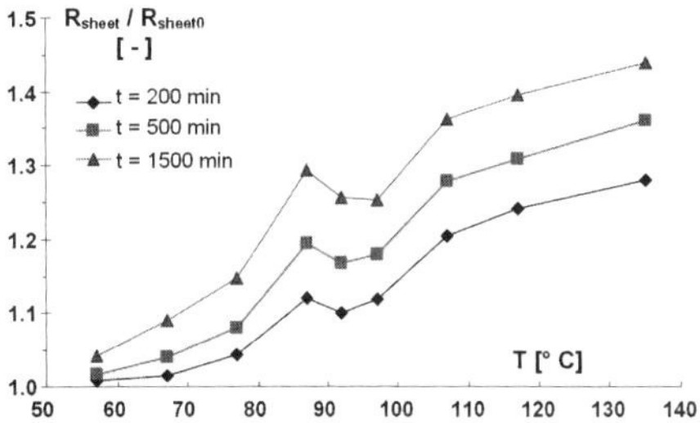

Figure 4. Relative resistance change (Rsheet / Rsheet at t = 0) vs isothermal temperature for three different times.

Apart from the anomaly mentioned above, a trend towards higher resistance for higher temperature/ longer time is observed: At first approximation there is a transition towards a smaller derivative for temperatures higher than 100° C: The change in resistance seems to level off or increases not as severe as a function of temperature as compared to the lower temperatures.

Both Pd and Si are assumed to be diffusion species during the Pd_2Si formation [10]. An explanation for the observed phenomenon is that for lower temperatures mixed layers of Pd and Si may exist via diffusion of Pd into Si and vice versa without the formation (or next to partial formation) of Pd_2Si. This process would also result in a resistance change just like the sole formation of Pd_2Si from Pd and Si. Our simple model so far is not capable of distinguishing between the two processes (mixing and silicide formation). The data is therefore further investigated and complemented with XRD and RBS.

XRD and RBS analysis

Figure 5 shows the XRD diagram from samples isothermally heated at temperatures from 57° - 107° C. Also shown is a reference sample with an as deposited Pd layer.

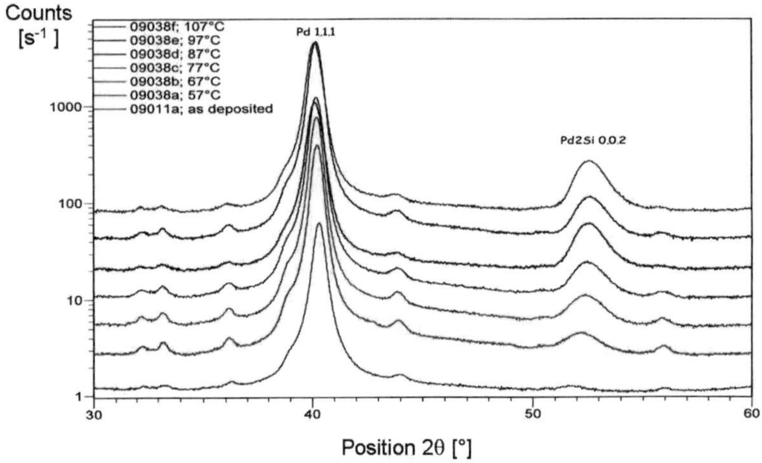

Figure 5. XRD diagram of a reference "as deposited" sample and six samples with isothermal anneal at temperatures between 57 – 107° C. Scans have been shifted vertically to make comparison easier. The other peaks next to Pd(111) and Pd_2Si(002) are all caused by spectral impurities because no monochromator was used.

For the reference samples the Pd peak was clearly visible but the Pd_2Si peak was hardly visible as compared to the annealed samples. For the other six annealed samples both Pd and Pd_2Si are present. Therefore, we conclude that the appearance of the Pd_2Si peak was due to the annealing step and was not present prior to measurement. In general the Pd peak becomes smaller and the Pd_2Si peak becomes larger at higher temperatures. In addition, the Pd_2Si peak shifts to the right because the unit cell becomes smaller. This is probably caused by a slight decrease in the Si content of this phase. XRD indicates that even at 57° C Pd_2Si is formed. However, XRD analysis does not provide information on the presence of mixed layers of Pd in Si.

RBS analyses of the samples (not shown here) reveal a gradient in the Pd and the Si profile over several nanometers, ranging from $PdSi_{0.3}$ (at the Pd side) up to $PdSi_4$ (at the Si side). This indicates that for these samples besides Pd_2Si also mixed layers are present.

Arrhenius plots

From the resistance measurements and using equation 4 the formed Pd_2Si layer thickness was calculated. The activation energy was determined using equation 1. Figure 6 shows the Arrhenius plot for the temperature range of 57-135° C.

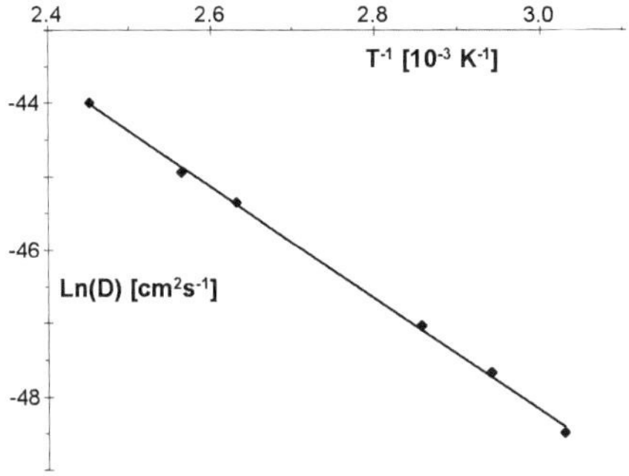

Figure 6. Arrhenius plot of the data for 45 nm Pd samples for temperatures in the range 57 – 135° C. Data points for 87, 92, and 97° C have been omitted. The solid line shows the best linear fit.

A good linear fit was obtained using our resistive model. The resulting activation energy E_A of the data of figure 6 is 0.7 ± 0.1 eV. This is somewhat lower than the data in literature that vary between 0.9 – 1.5 eV [7]. Those data, however, were in most cases determined at temperatures above 150° C. Therefore, in order to compare our model with literature data, we also determined the activation energy from samples with approximately 150nm Pd on Si in the temperature range of 150-225° C. The Arrhenius plot hereof is depicted in figure 7.

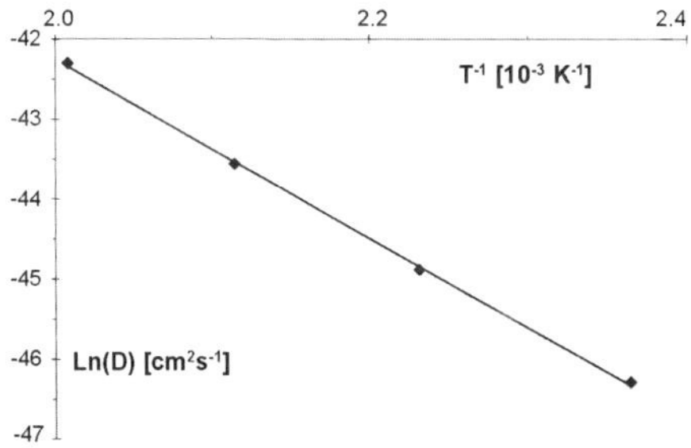

Figure 7. Arrhenius plot of the data from 150 nm Pd on Si for temperatures in the range 150 – 225° C. The solid line shows the best linear fit.

Again a good, linear fit was obtained. From this data an activation energy was derived of E_A = 1.0 ± 0.1 eV, which is in good agreement with literature [7]. To explain the observed discrepancy in found values for E_A for the high and the low temperature range we propose a difference in layer structures of the samples for high (> 150° C) and low (<150° C) temperatures. At high temperatures the diffusing Pd and Si species react instantaneously to Pd_2Si without substantial formation of intermixed layers of Pd and Si. For low temperatures besides Pd_2Si (as shown via XRD) a substantial amount of intermixed (but no silicide) layers may be formed which also result in an increase in measured resistance. In our resistive model these mixed layers are not taken into account. They are assumed to behave like Pd_2Si. This can eventually lead to an error in the derived activation energy. Currently, we try to extend our electrical model to fit the resistance measurements not only to Pd_2Si formation, but also to include the Pd rich and Si rich regions at the Si and Pd side.

CONCLUSIONS

Sputtered Pd layers on bare Si wafers have been examined on isothermal behavior in the low temperature range of 57° - 135° C. We used simple *in situ* four point probe resistance measurements during the thermal treatment (and silicide formation). The resistance data was analyzed using a simple three layer resistive model for deriving the silicide thickness. The silicide growth rate and the activation energy were determined. This method is relatively simple compared to commonly used XRD and RBS measurements.

So far this study has shown that Pd layers on Si are not stable even at temperatures as low as 57° C. XRD analysis revealed the formation of a Pd_2Si phase, whereas RBS indicated that intermixed layers of Pd and Si exist in the temperature range below 100° C. The above findings are of crucial importance for metallic Pd contacts to Si on SOI wafers. During sputter deposition of Pd the heat transport is limited through the thin Si/SiO_2 stack. Significant heating and substantial Pd_2Si formation was observed, resulting in Pd-Si interfaces of ill-defined composition.

REFERENCES

1. E.H. Rhoderick and R.H. Williams, (Clarendon Press, Oxford, 1988), pp. 48.
2. B. Coulman and H. Chen, *Journal of Applied Physics* **59**, 3467 (1986).
3. G. E. White and H. Chen, *Journal of Applied Physics* **67**, 3689 (1990).
4. J.-P. Colinge, *Silicon-on-Insulator Technology*, 3rd ed. (Springer, New York, 2004).
5. J.-P. Colinge, *FinFETs and Other Multi-Gate Transistors*. (Springer, New York, 2008).
6. F. M. d'Heurle and P. Gas, *Journal of Materials Research* **1**, 205 (1986).
7. D. Levy, A. Grob, J. J. Grob et al., *Applied Physics A: Materials Science and Processing* **35**, 141 (1984).
8. N. W. Cheung, M. A. Nicolet, M. Wittmer et al., *Thin Solid Films* **79**, 51 (1981).
9. M. Wittmer and K. N. Tu, *Physical Review B* **27**, 1173 (1983).
10. W. K. Chu, S. S. Lau, J. W. Mayer et al., *Thin Solid Films* **25**, 393 (1975).

Conference Proceedings AMC XXV © 2010 Materials Research Society

Determinant of Electrical Leakage Current for porous SiOC film

Kiyohiro Matsushita, Akinori Nakano, Ippei Yanagisawa, Yuya Nonaka, and Nobuyoshi Kobayashi

ASM Japan K.K., Nagayama 6-23-1, Tama-shi, Tokyo, Japan
Corresponding Author: Kiyohiro Matsushita: tel.: +81 42 337 6314, fax: +81 42 389 7555
e-mail: k.matsushita@asm.com

ABSTRACT

This paper describes determinant for leakage current of PECVD SiOC film, where dielectric constant (k) ranged from 2.4 to 3.0 with a wide variation of processes. The SiOC films include porogen type and non-porogen type porous SiOC films and UV cure was applied for the SiOC films in order to enhance mechanical strength and reduce dielectric constant by removing porogen.[1, 2] Summarized data show that the leakage current increases with an increase in porosity (i.e. lowering k-value) and with an increase in carbon-containing species (i.e. Si-CH3, C-H chemical bonds) in the film. It is shown that UV cure consists of two steps. One is the reduction of k-value with increasing the leakage current during porogen (C-H bond) removal, and the other is the reduction of leakage current due to an annealing restoration effect with decreasing the carbon-containing species of Si-CH3 and C-H.

INTRODUCTION

With scaling the line space at 32 nm node and beyond, leakage current characteristics of porous SiOC film becomes more dominant in the interconnect reliability. Moreover, line space narrowing is enhanced due to line edge roughness and line-via misalignment.[3-6] On the other hand, it has been reported that the leakage current density of low-k film increased with a decrease in k value since defect density in the porous film increased with an increase in porosity. [5, 7]

In this work, we evaluate porous low-k films with various carbon content and k-value in order to investigate determinant for leakage current. We found that the leakage current increased with an increase in porosity and in carbon-containing species in the film.

EXPERIMENT

Porogen type and non-porogen type SiOC films were deposited by PECVD using 4 different siloxane precursors (A, B, C, D). The thickness ranged from 450 to 500 nm. UV cure was executed to reduce the dielectric constant (k-value) by removing porogen and increasing porosity, and to enhance the mechanical strength of SiOC films with k=2.4, as shown in Table 1. A UV lamp having a broadband wavelength (>200nm) was used.[8] The cure time was tuned for these films to minimize k-value. The k-value and leakage current density were measured by Hg-probe and the elastic modulus and hardness by nanoindentation. The film thickness and refractive index were determined by spectroscopic ellipsometry and the microscopic structures by Fourier Transform Infrared Spectroscopy (FT-IR).

Table 1. Film properties of different porous SiOC films

	k2.4				k2.6				k2.8	k3.0
Process	porogen + UV cure				non-porogen + UV cure				no UV	
Chemical	A	B	C	D	A	B	C	D	A	
k	2.41	2.43	2.40	2.39	2.64	2.64	2.64	2.70	2.8	3
EM (GPa)	3.2	4.9	6.4	5.9	6.5	7.7	11.4	4.3	8.5	12
H (GPa)	0.33	0.72	0.93	0.86	0.94	1.30	1.80	0.70	1.5	2.1
Leakage current (A/cm2@2MV/cm)	1.4E-07	2.1E-08	4.6E-09	2.4E-08	1.1E-08	4.0E-09	2.2E-09	2.1E-08	5.5E-09	4.3E-10
C-H/Si-O (%)	3.45	1.82	1.33	1.18	1.49	2.92	1.07	3.03	1.50	2.76
Si-CH3/Si-O (%)	3.30	3.12	1.63	2.78	2.97	3.91	1.77	4.37	3.00	3.4

RESULTS and DISCUSSION

Fig. 1 shows the relation of leakage current density with electric field curves to different k-value of SiOC films all made of precursor A. In all the samples, the leakage current slightly increases with electrical field and shows the breakdown at certain field. It is noticeable that the leakage current level increases and the breakdown voltage decreases with reducing the k-value. It is shown that the leakage current increases with increasing the porosity if the precursor is same.

In order to investigate the behavior of the leakage current among different processes, the leakage current density at 2MV/cm is plotted as a function of dielectric constant (Fig. 2). These data include a variety of processes, which consists of porogen and non-porogen chemistry with and without UV cure, as is shown in Table 1. The data are scattered among processes, but it generally shows the leakage current increases with decreasing k-value. Consequently, it is concluded that the leakage current increases with the increase in film porosity. which can be explained by the percolation model where pores are associated with defect centers for electron transport [5, 7].

Moreover, it should be noted that leakage current density varies among different precursors when k-value is fixed at 2.4 (porogen type) and 2.6 (non-porogen type), respectively. It indicates the leakage current correlates with the chemical structure as well as the porosity. In order to study the correlation with chemical structure, leakage current density is plotted as a function of FT-IR peak area ratios of Si-CH3/Si-O and C-H/Si-O (Figs. 3 and 4). It shows that the leakage current density tends to increase with the increase in carbon-containing peak area ratios of Si-CH3 and C-H. By comparing the slopes of data plots in Figs 3 and 4, the correlation factor between the leakage current and carbon-containing chemical bonds is found to be larger for porogen-type SiOC films as compared with non-porogen SiOC films.

The correlation of leakage current density with k-value and carbon-containing chemical bonds was investigated as a function of UV cure time for both porogen type and non-porogen type SiOC films (Figs. 5 and 6). For porogen type SiOC films in Fig. 5, k-value initially strongly decrease with porogen removal (i.e., reduction of CH bonds). But the leakage current slightly increases and then gradually decreases with a decrease in carbon-containing species during UV cure. On the other hand, for non-porogen SiOC films in Fig. 6, k-value also decreases. This reduction is limited as compared to the porogen-type because the reduction of CH bonds is smaller. However reduction in leakage current is fairly correlated with decreasing carbon-containing species during UV cure after initial reduction of C-H bonds. It implies UV cure process consists of two steps for improving the low-k film properties. In an initial step, it removes C-H bonds thus reducing the k-value in porogen and non-porogen type films. However it does not reduce the leakage current because it increases the defects associated with C-H bonds removal. In the following second step, the defects are continuously restored by an annealing effect of the UV cure so that the leakage current decreases with decreasing the carbon-containing species. We should optimize the UV cure process considering these steps.

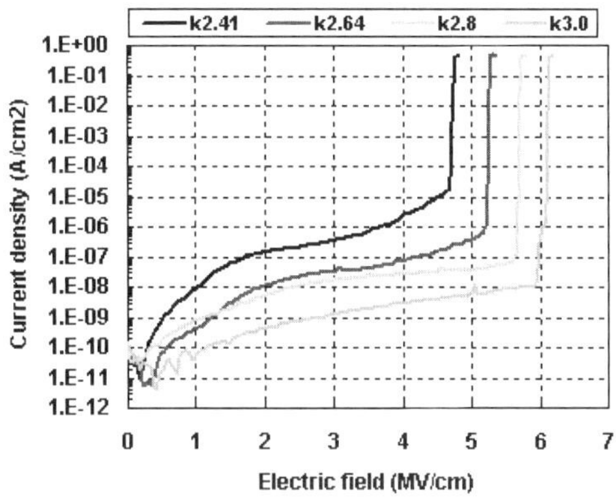

Fig.1 Current density vs. Electric field for different low-k films made of Chemical A

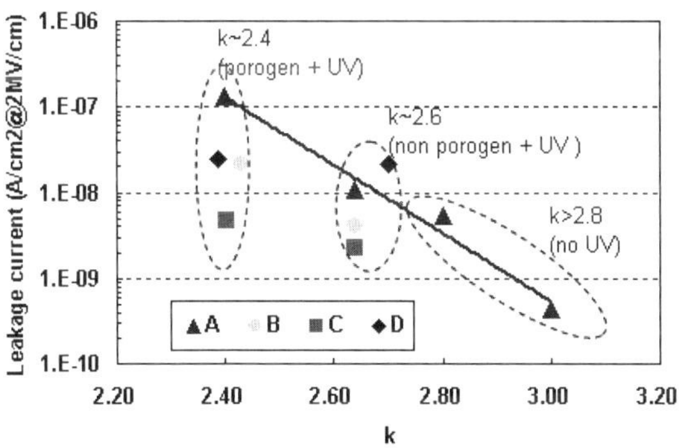

Fig.2 Leakage current density (@2MV/cm) as a function of dielectric constant of different low-k films

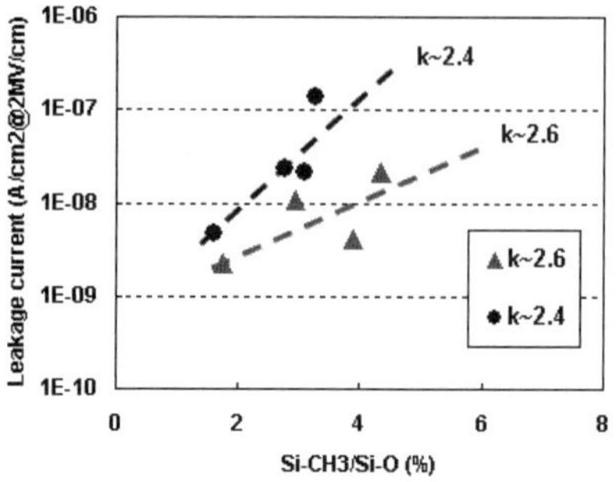

Fig.3 Leakage current as a function of Si-CH3/Si-O area ratio of different porous SiOC films

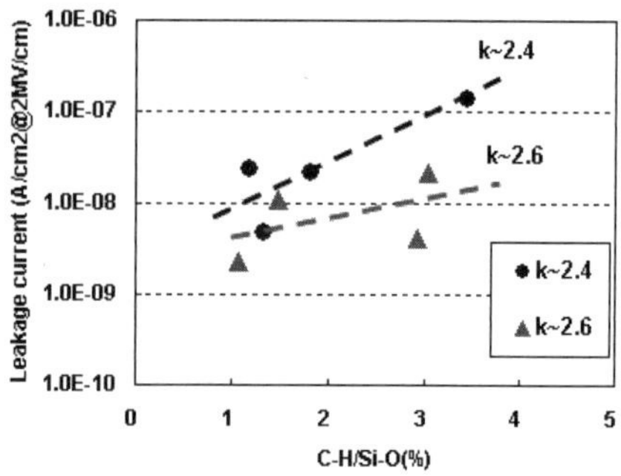

Fig.4 Leakage current as a function of C-H/Si-O area ratio of different porous SiOC films

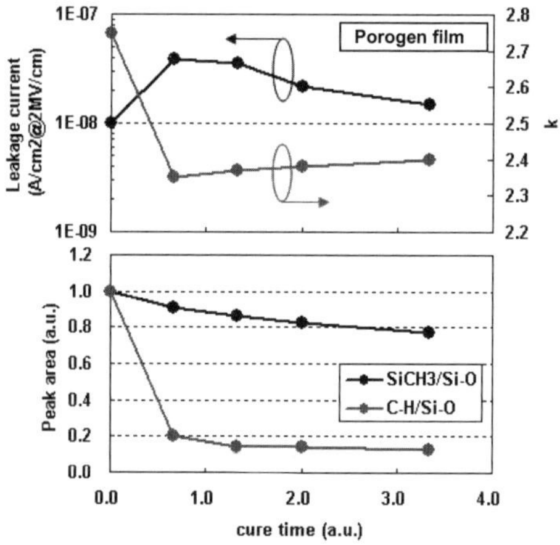

Fig.5 UV cure time dependency of film properties of k2.4 (Chemical D) film

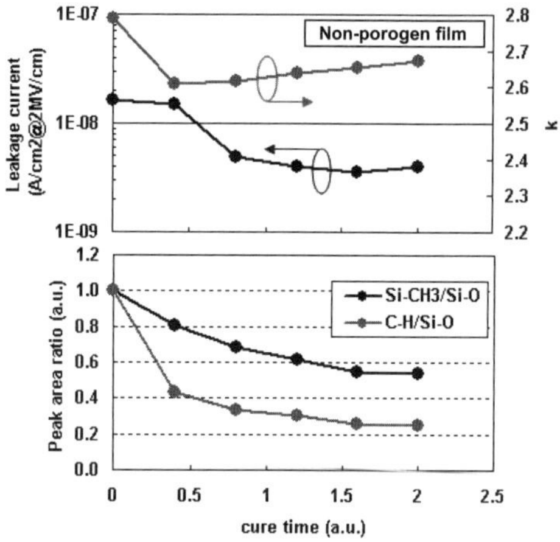

Fig.6 UV cure time dependency of film properties of k2.6 (Chemical C) film

CONCLUSION

Determinant of leakage current of porous SiOC films (k=2.4-3.0) was investigated with a wide variation of processes. In the summarized results, it is concluded that the leakage current increases with increasing the porosity, and moreover increases with carbon-containing species in the film. Effects of UV cure on leakage current were different between porogen-type and non-porogen SiOC films. The leakage current of non-porogen films is reduced by UV cure because an effect of reducing carbon-containing species is dominant, compared with that of increasing porosity. On the other hand, an effect of porosity is more dominant on the leakage current for porogen film because thc defect formation is initially caused during UV cure.

REFERENCES

[1] S. I. Nakao, J. Ushio, T. Ohno, T. Hamada, Y. Kamigaki, M. Kato, K. Yoneda, S. Kondo and N. Kobayashi in Proc. International Interconnect Technology Conference 2006 (IEEE, Burlingame, CA, 2006), p.66-68

[2] M. Kato, K. Yoneda, S. Nakano, S. Kondo, N. Kobayashi, N. Matsuki, K. Matsushita, N. Ohara, A. Fukazawa, T. Ikegawa and T. Kimura in Proc. Advanced Metallization Conference 2005 (AMC2005), edited by S.H. Brongersma, T.C. Taylor, M. Tsujimura,and K. Masu, (Mater. Res. Soc. Conf. Proc. AMCXXI, Warrendale, PA, 2006) p.271-276.

[3] Fen Chen and Mike Shinosky, IEEE Transactions on electron devices, vol. 56, No. 1, January 2009, p.2-12

[4] Michele Stucchi, Zsolt Tokei in Proc. International Interconnect Technology Conference 2008 (IEEE, Burlingame, CA, 2008), p.174-176

[5] E.T. Ogawa, J. Kim, G.S. Haase, H. C. Mogul, and J.W. McPherson in Proc. 41st Annual International Reliability Physics Symposium 2003 (IEEE, Piscataway, 2003), p.166-172

[6] Y. Ono, M. Matsutani, T. Harada, M. Tada, and Y. Ando, in Proc. ADMETA 2008 p.58-59

[7] Robin Degraeve, et al., IEEE Transactions on electron devices, vol.45, No. 4, April 1998, p.904-911

[8] K. Matsushita, N. Tsuji, K. Kagami, M. Kato, S. Kaneko and N. Kobayashi, in Proc. Advanced Metallization Conference 2007 (AMC2007), edited by A.J. McKerrow, Y. Shacham-Diamand, S. Shingubara, and Y. Shimogaki, (Mater. Res. Soc. Conf. Proc. AMCXXIII, Warrendale, PA, 2008) p. 533-537.

Conference Proceedings AMC XXV © 2010 Materials Research Society

Evolution of Cu/SiO2 and Cu/Ta Interface Roughness With Annealing

Andrew P. Warren[1], Tik Sun[1], Bo Yao[1], Katayun Barmak[2], Michael F. Toney[3], and Kevin R. Coffey[1]
[1]Advanced Materials Processing and Analysis Center, University of Central Florida, 4000 Central Florida Boulevard, Orlando, FL 32816
[2]Department of Materials Science and Engineering, Carnegie Mellon University, 5000 Forbes Avenue, Pittsburgh, PA 15213
[3]Stanford Synchrotron Radiation Lightsource, 2575 Sand Hill Rd., Menlo Park, CA 94025
[4]Department of Physics, University of Central Florida, 4000 Central Florida Boulevard, Orlando, FL 32816

ABSTRACT

The scattering of conduction electrons by surface roughness has been considered a key contribution to increased resistivity of nanoscale Cu conductors. To study the evolution of surface (interface) roughness with annealing, a series of SiO_2 and Ta/SiO_2 encapsulated Cu thin films were prepared. Samples were subsequently annealed at 150° and 600° for 30 minutes in a reducing gas to assess the thermal stability of the Cu interfaces and the tendency for void formation. Specular X-ray reflectivity was used to separately determine the root mean square roughness for both the upper and the lower Cu/SiO_2 (or Cu/Ta) interfaces. Notable differences in the roughness evolution between upper and lower interfaces were observed for SiO_2 encapsulated thin films. The Ta/SiO_2 encapsulated films are found to evolve little as compared to the SiO_2 encapsulated films.

INTRODUCTION

The morphological evolution of chemically stable metal/air or metal/vacuum interfaces (free surfaces) with annealing temperature and/or time have been explored in the past[1,2,3] Relatively little work has been done, however, to extend that understanding to chemically stable buried interfaces.

X-ray reflectivity is a powerful technique for the study of surfaces and buried interfaces in thin film systems[4,5,6,7,8.] There are two primary variations of this technique; specular and diffuse; herein we discuss only the former. In specular X-ray reflectivity, a narrow beam of X-rays is incident upon a specimen and the intensity of the beam reflected at an angle equal to the incident angle is measured. The variation of reflected beam intensity as a function of the common incidence and reflection angle will contain information regarding:

a) Sample density - by determining the critical angle for total external reflection and by the amplitude of the Fresnel oscillations,
b) Sample thickness - by measuring the period of the Fresnel oscillations, and
c) Root-mean-square (RMS) roughness - by measuring the decay of the reflected intensity.

It is known that the intensity of a specular reflection from a rough surface is attenuated exponentially, and can be described by[9, 10, 11, 12.]

$$I = I_0 \exp[-q_z^2 \sigma^2] \quad (1).$$

Where I_0 is the intensity of a perfect sample, q_z is the normal component of the momentum transfer vector and σ is the root-mean-square (RMS) roughness. The intensity that is lost from the specular reflection due to roughness is scattered diffusely in a manner described theoretically by Sinha et. al[12].

This reflectivity model is applicable to scattering from a single rough surface. In the case of multiple non-ideal interfaces, the calculation of the specularly reflected intensity must be further modified by Fresnel coefficients that account for multiple interfacial reflections. To model the experimentally obtained scattering profiles, commercially available software from Bede was used which employs a genetic algorithm referred to as "Differential Evolution"[13]. The above approach is used in this work to obtain the Cu film thicknesses and the RMS roughnesses (σ_{upper}, σ_{lower}) of the upper and the lower Cu surfaces of the sample films

This work describes a series of experiments using X-ray scattering to study the interfacial evolution of Cu thin films encapsulated by either SiO_2 or Ta/SiO_2. Both specular and diffuse X-ray reflectivities are highly sensitive to surface and interfacial roughness and existing theories of X-ray scattering, described briefly above, also make these techniques highly quantitative.

EXPERIMENT

SiO_2 and Ta/SiO_2 encapsulated Cu thin films were prepared on Si(100) substrates with 150 nm of thermally grown SiO_2 (1100 °C in dry O_2) to provide a homogenous surface for later depositions. Prior to film deposition, the substrates were RF sputter cleaned and cooled to −40 °C by contact with a liquid nitrogen cooled Cu plate. An underlayer of 20 nm of SiO_2 was sputter deposited prior to the Cu film deposition and a 20 nm SiO_2 overlayer was subsequently similarly deposited to form a SiO_2/Cu/SiO_2 encapsulation structure. The Ta/SiO_2 encapsulated samples differ only by the addition of a 2nm Ta layer, DC sputter deposited immediately prior to and again after, the Cu layer deposition, and therefore have a SiO_2/Ta/Cu/Ta/SiO_2 structure. The Cu layer was deposited by DC magnetron sputtering from high purity (99.9999%) Cu targets at a rate of 0.6 nm/s onto the electrically grounded substrate. All depositions were performed in 4 mTorr of Ar. A deposition chamber pressure in the 10^{-9} Torr range was obtained prior to film deposition and the nominally 99.999% purity Ar was passed through a hot reactive metal getter purifier (SAES Pure Gas) prior to introduction to the chamber. Samples were subsequently annealed at 150° and 600° to asses the evolution of the Cu interfaces.

The annealing of thin films may also introduce voiding, which is a concern for many metallization applications. In this work voiding was assessed by transmission electron microscopy (TEM). A back-etching technique[14] was used for sample thinning and high angle annular dark field (HAADF) imaging in scanning TEM mode at relatively low magnifications was used to assess the void fraction present in the film sample. The void area fraction for these samples has been previously reported.[15] The void area fraction was found to increase with decreasing film thickness and increase upon annealing for the SiO_2 encapsulated films, and the greatest void area fraction observed was 2.4% for a 33.5 nm thick film annealed at 600°C. In contrast, no voiding was found for any of our Ta/SiO_2 encapsulated films samples.

All X-ray scattering experiments were performed at the Stanford Synchrotron Radiation Lightsource (SSRL). Data were collected on thin-film diffraction beam line 2–1. This beam line is equipped with a Huber 2-circle goniometer, a pair of 1 mm slits as the analyzer[16], and a He filled sample stage used to decrease the air scattering background. X-rays with a wavelength of 1.549 Å were monochromated with a double bounce Si(111) crystal. To determine film thickness and RMS roughness (of top and bottom Cu interfaces), both specular and off-specular scans were collected: for specular scans $\omega=2\theta/2=\theta$ whereas for off-specular scans $\omega=2\theta/2 \pm 0.15°$. The off-specular scans were used to subtract the contribution of diffusely scattered X-rays to the specular reflection, resulting in a purely specular reflectivity pattern. Data were collected for 2θ values ranging between 0.2° to 12°, with a step size of 0.02°, 0.01°, or 0.005°, depending on film thickness. The X-ray data was analyzed using Bede REFS software[19]:

DISCUSSION

X-ray scattering techniques reveal significant differences in the evolution of the interfacial morphology of Cu thin films encapsulated by SiO_2 as compared to films encapsulated by Ta/SiO_2. Tabulated results can be found in table 1 for SiO_2 and Ta/SiO_2 encapsulated films. This difference in evolution will be attributed to differences in Cu/SiO_2 and Cu/Ta interface diffusion. Figure 1 shows specular reflectivity of (a) SiO_2 encapsulated and (b) Ta/SiO_2 encapsulated Cu films annealed at 150°C and 600°C. A qualitative conclusion, regarding the effects of annealing on RMS roughness can be readily determined by inspection of the figures without the need for numerical modeling of the data. SiO_2 encapsulated films are seen to undergo a decrease in roughness upon annealing. This is evidenced by the more gradual decay in scattering intensity as the scattering angle is decreased.

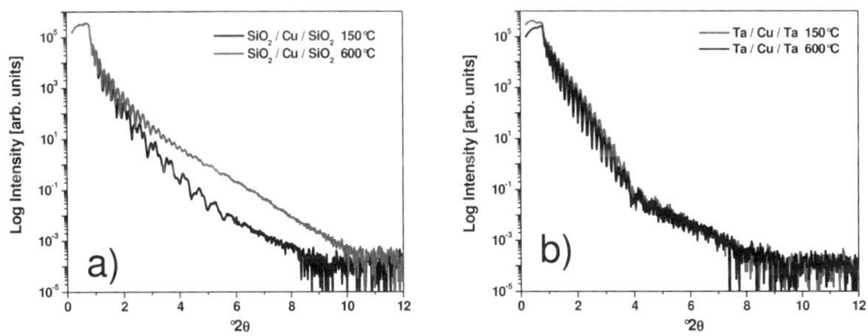

Figure 1. (a) Specular X-ray reflectivity patterns of SiO2 encapsulated Cu thin films (samples D and J). Annealing at 600°C results in a smoother Cu/SiO2 interface, as can be readily seen from the slower decay of specular intensity as compared to its 150°C counterpart. (b) Specular X-ray reflectivity patterns of Ta/SiO2 encapsulated Cu thin films (samples P and V). Contrary to SiO2 encapsulated films, annealing at 600°C does little to smooth the Cu film, as can be seen from the essentially identical reflectivity patters. The above patterns were modeled to quantify the film thickness and interfacial RMS roughness.

In the case of the Ta/SiO_2 encapsulated film, however, little evolution in roughness is observed. Furthermore, the Fresnel oscillations are notably different between 150°C and 600°C annealed SiO_2 encapsulated films; indicating significant interfacial evolution of the Cu thin films. This is not the case for Ta/SiO_2 encapsulated films whose reflectivity patterns are nearly unchanged upon annealing at 600°C, indicating little interfacial evolution.

The specular intensity profiles were modeled to allow for quantitative values of RMS roughness of the upper and lower Cu interfaces to be determined in addition to determining the film thickness. Quantifying the RMS roughness of top and bottom interfaces reveals significant changes upon annealing at 600°C. Samples annealed at 150°C exhibit no correlation between top and bottom RMS interface roughness: there is no relation between these. The relatively rough bottom interface evolves little upon annealing while upon annealing the top Cu/SiO_2 interface roughness is notably reduced.

It is also noted that the upper Cu interface evolves in such a way that its RMS roughness is now correlated to that of the bottom interface (Fig. 2.) Samples annealed at 150°C exhibit no correlation between upper and lower RMS roughness (Fig. 2a). Annealing at 600° results, however, in a notable trend (Fig. 2b). Furthermore, the relationship between upper and lower Cu interface roughness does not seem to depend on the encapsulant, since both SiO_2 and Ta/SiO_2 encapsulated films exhibit this tend with equivalent slopes.

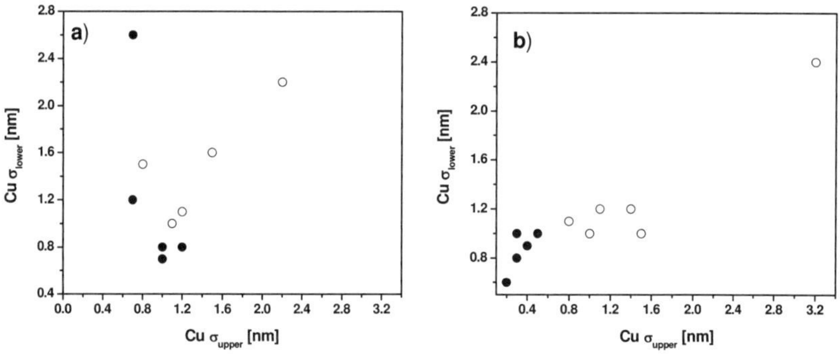

Figure 2. RMS roughness of top and bottom Cu thin films, encapsulated by SiO_2 (closed circles) or Ta/SiO_2 (open circles), as determined by specular X-ray reflectivity. (a) Samples annealed at 150°C, regardless of encapsulant, exhibit no correlation between top and bottom interface roughness. (b) Samples annealed at 600°C however, show a strong correlation between top and bottom Cu interface roughness.

In the thickness range of 20 to 80 nm, the roughness of the SiO_2 encapsulated Cu thin films (Fig. 3a) inversely tracks thickness for samples annealed at 150°C. Upon annealing at 600°C, a constant 0.3±0.1nm RMS roughness is obtained for all thicknesses. In the case of Ta/SiO_2 encapsulated films (fig. 3b) a linear relation between thickness and RMS roughness is observed. This trend for Ta/SiO_2 encapsulated films changes little upon annealing. As was qualitatively discussed above, Ta/SiO_2 encapsulated thin films evolve little as compared to SiO_2 encapsulated Cu thin films.

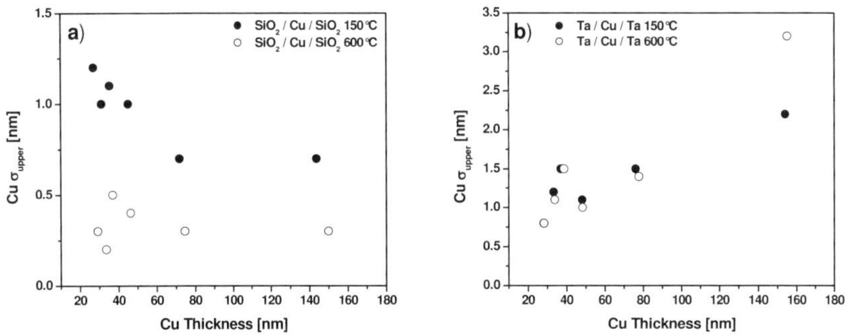

Figure 3. RMS roughness and thickness of SiO_2 (a) and Ta/SiO_2 (b) encapsulated Cu thin films as determined by secular X-ray reflectivity. In the thickness range of 20 to 80 nm, annealing at 600°C reduces the average RMS roughness considerably and also serves to remove any thickness dependence. For Ta/SiO_2 encapsulated Cu, it is apparent that the roughness evolves little as compared to SiO_2 encapsulated Cu thin films.

The effect of interfacial roughness on the resistivity of our sample set has been discussed in previously published work[15] and is being further studied using both theoretical and empirical models[17]. The inclusion of interfacial roughness terms in modeling the classical size effect reduced the modeled summed squared error but not sufficiently to establish a high level of confidence.

CONCLUSION

Specular X-ray reflectivity was used to study the interfacial evolution of Cu thin films encapsulated by SiO_2 or by Ta/SiO_2. The roughness of the lower interface was consistent amongst all SiO_2/Cu samples, with an average roughness of 1.1±0.4 nm, independent of annealing temperature. By contrast, the roughness of the upper Cu/SiO_2 interfaces was very different for the two annealing conditions, decreasing from 1.0±0.3 nm to 0.3±0.1 nm upon annealing at 600°C. An increase in void area fraction was also observed upon annealing the Cu/SiO_2 samples. For the samples with Cu/Ta interfaces, a reduction in roughness with annealing was not observed for either upper or lower interfaces, nor was void formation observed. The lack of interface evolution observed for the Ta/SiO_2 encapsulated Cu films may be associated with a lower surface diffusivity of Cu on Ta than that on SiO_2. However, the role of microstructure in these polycrystalline films is also of further interest.

ACKNOWLEDGMENTS

We gratefully acknowledge the financial support of the Semiconductor Research Corporation, Task No. 1292.008, and NSF MRSEC DMR-0520425. Portions of this research were carried out at the SSRL, a national user facility operated by Stanford University on behalf of the U.S. Department of Energy, Office of Basic Energy Sciences.

Table 1a

Sample	Anneal T [°C]	Thickness [nm]	σ_{upper} [nm]	σ_{lowerr} [nm]
A	150	27.0 ± 0.2	1.2 ± 0.1	0.8 ± 0.1
B	150	31.2 ± 0.1	1.0 ± 0.2	0.7 ± 0.1
C	150	35.3 ± 0.3	1.1 ± 0.5	1.0 ± 0.1
D	150	45.1 ± 0.2	1.0 ± 0.1	0.8 ± 0.1
E	150	71.8 ± 0.2	0.7 ± 0.1	1.2 ± 0.2
F	150	143.8 ± 0.3	0.7 ± 0.1	2.6 ± 0.1
G	600	29.2 ± 0.1	0.3 ± 0.1	0.8 ± 0.1
H	600	33.5 ± 0.1	0.2 ± 0.1	0.6 ± 0.2
I	600	36.9 ± 0.3	0.5 ± 0.1	1.0 ± 0.1
J	600	46.4 ± 0.1	0.4 ± 0.1	0.9 ± 0.1
K	600	74.5 ± 0.1	0.3 ± 0.1	1.0 ± 0.1
L	600	149.9 ± 0.1	0.3 ± 0.1	1.0 ± 0.1

Table 1b

Sample	Anneal T [°C]	Thickness [nm]	σ_{upper} [nm]	σ_{lowerr} [nm]
M	150	28.1 ± 0.7	0.8 ± 0.1	1.5 ± 0.3
N	150	33.3 ± 0.5	1.2 ± 0.2	1.1 ± 0.1
O	150	37.0 ± 0.9	1.5 ± 0.5	1.0 ± 0.2
P	150	48.2 ± 0.4	1.1 ± 0.1	1.0 ± 0.1
Q	150	76.2 ± 1.2	1.5 ± 0.1	1.6 ± 0.5
R	150	154.1 ± 0.5	2.2 ± 0.1	2.2 ± 0.3
T	600	28.3 ± 0.2	0.8 ± 0.1	1.1 ± 0.1
V	600	33.8 ± 0.4	1.1 ± 0.1	1.2 ± 0.2
U	600	38.7 ± 1.5	1.5 ± 0.4	1.0 ± 0.1
V	600	48.4 ± 0.3	1.0 ± 0.1	1.0 ± 0.1
W	600	77.9 ± 1.0	1.4 ± 0.2	1.2 ± 0.3
X	600	155.2 ± 0.3	3.2 ± 0.1	2.4 ± 0.2

Table 1. Annealing temperature, Cu film thickness, Cu grain size, and root mean square roughness (upper, σ_{upper}, and lower, σ_{lower}, of the Cu/encapsulant interfaces) for: SiO_2-encapsulated Cu thin films (table 1a), and the Ta/SiO_2-encapsulated Cu thin films (table 1b).

REFERENCES

[1] F. A. Nichols and W. W. Mullins, Trans. AIME 233, 1840 (1965)

[2] R. W. Balluffi, S. M. Allen, W. C. Carter, Kinetics of Materials, John Wiley and Sons, Hoboken, New Jersey (2005)

[3] E. Jiran, C. V. Thompson, Journal of Electronic Materials, Vol 19, No 11 (1990).

[4] A. Gupta, A. Paul, S. Mukhopadhyay, J. Appl. Phys., Vol. 90, No. 3, 1 August 2001

[5] R. Schad, P. Belien, G. Verbanck, V. V. Moshchalkov, and Y. Bruynseraede, H. E. Fischer, S. Lefebvre, M. Bessiere. Phys. Rev. B 59, 1242 (1999)

[6] A. Ulyanenkov, R. Matsuo, K. Omote, K. Inaba, and J. Harada , M. Ishino, M. Nishii, and O. Yoda. J. Appl. Phys., Vol. 87, No. 10, 15 May 2000

[7] J. M. Freitag, B. M. Clemens . J. Appl. Phys., Vol. 89, No. 2, 15 January 2001

[8] R. Schlatmann, J. D. Shindler, and J. Verhoeven. Phys Rev. B 54, 10 880 (1996)

[9] V. Holy, U. Pietsch, T. Baumbach. Higher resolution X-ray scattering from thin films and multilayers. (Springer, Berlin, 1998)

[10] M. Schmidbauer. X-ray diffuse scattering from self-organized Mesoscopic semiconductor structures. (Springer, Berlin, 2004)

[11] M Tolan, X-ray scattering from soft-mater thin films: Materials science and basic research. (Springer, Berlin, 1999)

[12] S. K. Sinha, E. B. Sirota, S. Garoff, and H. B. Stanley, Phys. Rev. B 38, 2297 (1988).

[13] M. Wormington , C. Panaccione, K. M. Matney, D. K. Bowen. Phil. Trans. R. Soc. Lond. A (1999) 357, 2827–2848

[14] B. Yao, R. V. Petrova, R. R. Vanfleet, K. R. Coffey Journal of Electron Microscopy 55 (4) 209 2006

[15] T. Sun, B. Yao, A. P. Warren, K. Barmak, M. F. Toney, R. E. Peale, and K. R. Coffey, Phys. Rev. B 79, 041402 (2009)

[16] http://www-ssrl.slac.stanford.edu/beamlines/bl2-1/

[17] T. Sun, Bo Yao, A. P. Warren[1], K. Barmak, M. F. Toney, R. E. Peale, K. R. Coffey "Quantitative analysis of surface and grain boundary scattering in the resistivity of nanometric Cu films" SUBMITTED FOR PUBLICATION TO PHYSICAL REVIEW B

Integration

Conference Proceedings AMC XXV © 2010 Materials Research Society

Key Aspects of CoWP and ULK Implementation Into High-Volume Manufacturing for Future Technologies

M. Nopper*, J. Bömmels, O. Aubel, H. Hartz, A. Preusse
GLOBALFOUNDRIES Dresden Module One LLC & Co. KG
GLOBALFOUNDRIES Dresden Module Two GmbH & Co. KG
Wilschdorfer Landstraße 101, 01109 Dresden, Germany
*Corresponding author: markus.nopper@globalfoundries.com

ABSTRACT

Electromigration (EM) is one of the most limiting factors for back-end reliability in recent and future technologies. EM enhancement is necessary to pass reliability requirements [1]. The introduction of ultra-low-K (ULK) materials with k-values below 2.5 reduces the RC-delay significantly. Unfortunately, these materials are very challenging for unit process and reliability performance. Most of the ULK materials are porous and therefore mechanically weaker than conventional dense low-k films [2]. This mechanical weakness further reduces reliability performance and therefore increases the need for EM enhancements techniques.

In this paper, results of the CoWP deposition investigations and the successful implementation of this process into 45nm and 32nm high-volume production environments are presented. Figure 1 shows a smooth and continuous CoWP film on top of 32nm structures.

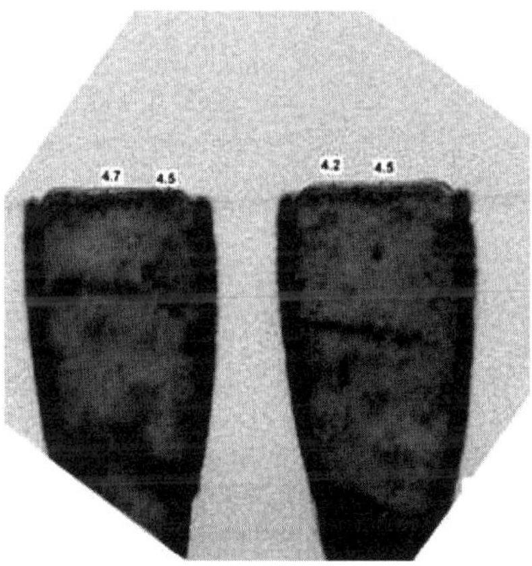

Figure 1. TEM image of a 32nm structure which has been covered with CoWP

The focus of this paper is the interaction between CoWP and ULK processes and how they can be combined to meet aggressive reliability targets, including a discussion of the process demands for introduction into high-volume (HVM) manufacturing.

ELECTROLESS METAL CAP PROCESS BASICS

Typically, CoWP metal cap layers are deposited using an electroless plating process. Unlike electrochemical plating, where the electrons are donated by an external power supply, electroless plating uses the electrons donated from a reducing agent. During this reaction, the metal ions get reduced and the reducing agent is oxidized:

Reduction of metal ions:

$$Me^{n+} + ne^- \rightarrow Me$$

$$Co^{2+} + 2e^- \rightarrow Co$$

Oxidation of reducing agent:

$$Red - ne^- \rightarrow Ox$$

$$H_2PO_2^- + OH^- \rightarrow H_2PO_3^- + H + e^-$$

$$Co^{2+} + 2H_2PO_2^- + 2OH^- \rightarrow Co + 2H_2PO_3^- + H_2$$

The deposition solution (electrolyte) consists of metal ions, reducing agents, stabilizers, complexing agents, buffers, and wetting agents.

CoWP plating requires pre- and post-treatment of the wafer. The pre-treatment is necessary to remove the copper oxide and passivation layers from the interconnect (copper) surface; the electroless CoWP process needs a clean copper surface for proper plating initiation. The post-treatment is necessary to clean and dry the wafer. Proper environmental control is also necessary during the plating process itself.

CoWP AND ULK MATERIALS

When shrinking semiconductor devices, the RC-delay of the wiring layer increasingly limits the speed of such devices [3]. To overcome this speed limitation, the interconnects are embedded in ULK ILDs.

Figure 2 shows the impact of changing the dielectric from dense low-K (k=3.0) to ULK dielectric (k=2.4) while keeping the geometry of the copper wires and the etchstop layers unchanged. A 14% decrease in the total capacitance with such ultra-low-k dielectrics without changing the resistance is observed.

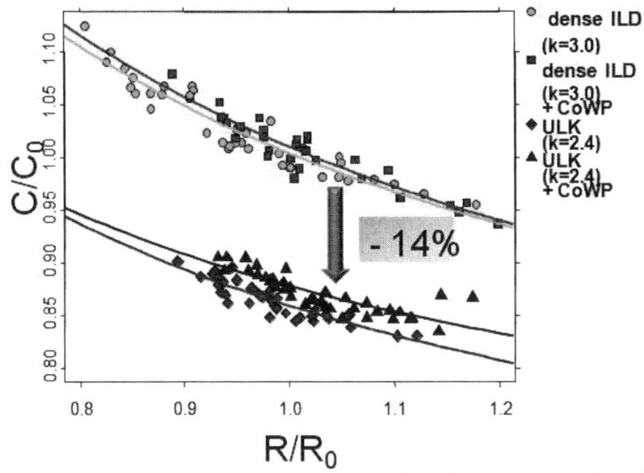

Figure 2. The addition of a COWP film does not significantly impact RC when deposit on either dense or ULK dielectric.

Despite the positive effects of ULK for RC, it is a major obstacle for reliability. The largest problems with ULK are dielectric breakdown (TDDB) and EM. The EM lifetime of copper wires in porous ULK dielectrics is reduced to about one half of the lifetime in dense low-k dielectrics (Figure. 3), which seriously impacts the EM robustness.

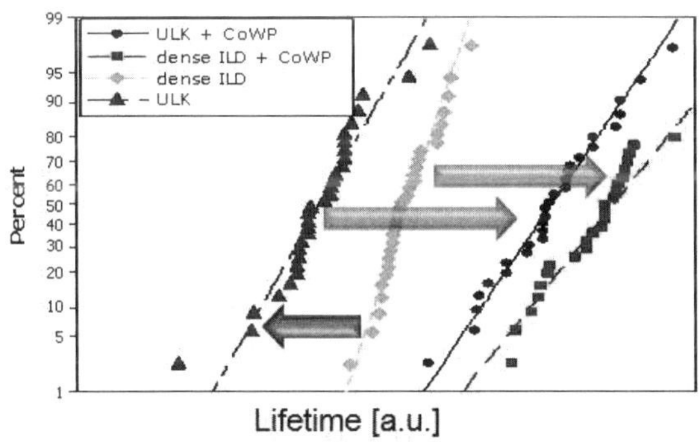

Figure 3. Comparison of EM lifetimes for dense low-k (diamonds) and ultra-low-k (triangles). In both cases, CoWP improves EM lifetime.

Interconnect capping layers such as CoWP are known to increase the resistance of copper lines against EM failure [4]. An EM increase by a factor of at least 500x is observed when a CoWP layer is present on top of the copper lines. This EM improvement enables the use of ULK in the tightest pitches at the 32nm node and beyond.

Using an optimized process for depositing the CoWP capping layer does not result in TDDB performance degradation. As shown in Figure 4, the introduction of ULK reduces TDDB significantly. The introduction of CoWP in conjunction with dense ILD does not impact TDDB performance. However, the introduction of CoWP in conjunction with ULK slightly improves TDBB performance. This behavior can be explained by the extra pre- and post-clean that are used as part of the CoWP deposition process.

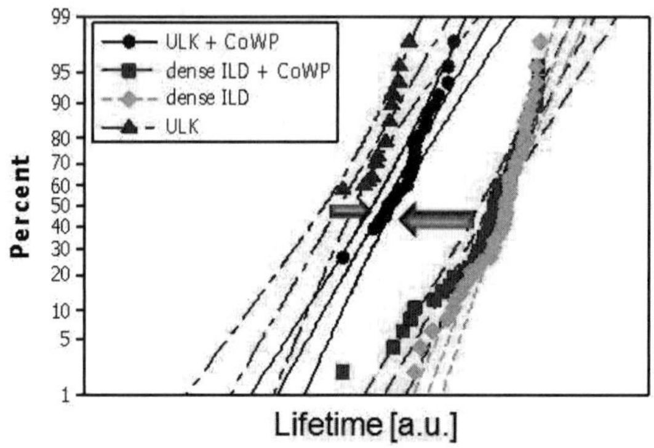

Figure 4. Comparison of TDDB lifetimes for dense low-k (diamonds) and ultra-low-k (triangles). CoWP layer samples are in squares (dense ILD) and circles (ULK).

HIGH-VOLUME MANUFACTURING (HVM)

For implementing CoWP into a high-volume manufacturing environment, the control of the applied chemistry is of great importance [5]. Therefore, all relevant chemical metrology parameters have to be closely monitored. Online metrology methods are preferred to allow automated operation. For CoWP deposition, the control of deposition solution pH, cobalt concentration, and reducing agent concentration are the key factors for the process. The typical control chart for a tool running a CoWP plating process is shown in Figure 5.

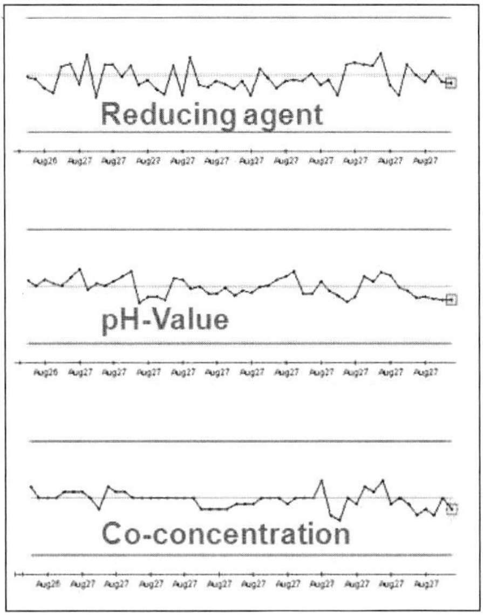

Figure 5. Statistical process control (SPC) chart of relevant chemistry parameters for CoWP deposition

For the success of the process, a scheduled qualification, which measures the particle performance, thickness, and nonuniformity or thickness range, is essential. A chart for thickness and range is shown in Figure 6.

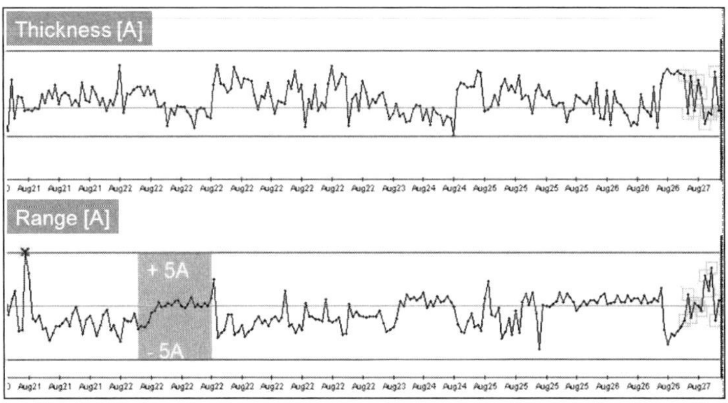

Figure 6. Thickness and range chart of CoWP deposition

INTEGRATION

The integration of CoWP into the manufacturing workflow requires the addition of several steps. In addition to the CoWP plating step, inline metrology steps such as CoWP thickness measurements and post CoWP defect measurements with defect review should be added.

The biggest challenge for integration is compatibility of the CoWP film with the post plasma etch wet clean [6]. The problem is schematically described in Figure 7. The physical illustration of this problem can be seen by TEM (Figure 8).

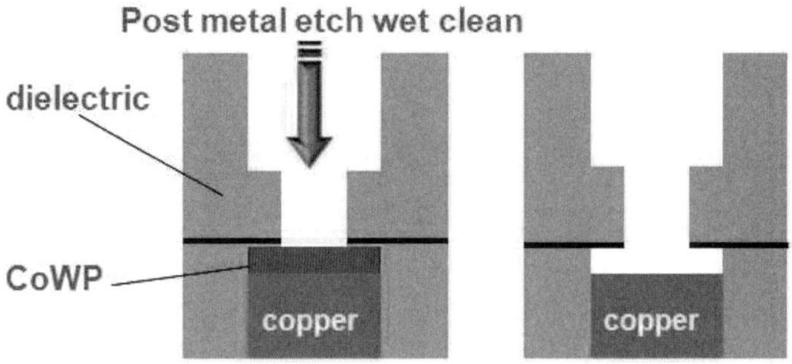

Figure 7. Schematic explanation of post plasma etch wet clean problem

Figure 8. TEM picture illustrating the post plasma etch wet clean problem. Strong undercut is visible (picture after metal-plating)

The post plasma metal solution removes the CoWP not only vertically in the trench, but also laterally at the interface between the copper and the dielectric. After testing different wet etch chemistries, a solution for avoiding lateral etching of CoWP was found. Figure 9 shows a TEM picture of a wafer post plasma etch wet clean with the newly developed chemistry.

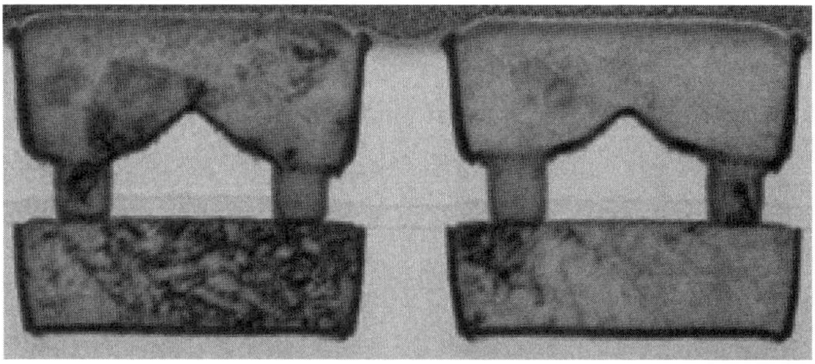

Figure 9. TEM picture taken after applying the newly developed post plasma etch wet clean chemistry. No undercut is visible.

After the post plasma etch wet clean problem, the most critical electrical parameter is line to line leakage. Figure 10 shows CoWP deposited using optimized process conditions does not increase line to line leakage.

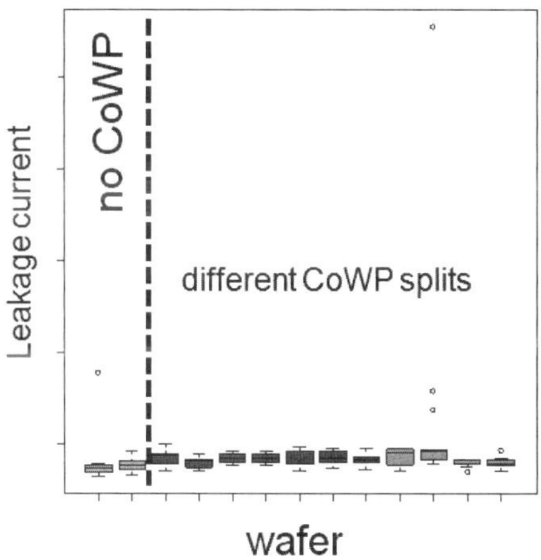

Figure 10. Leakage plot of a lot with and without CoWP

YIELD

CoWP does not cause yield loss in 45 and 32nm technologies. In fact, higher yields have been seen when using the CoWP capping process. The definition of yield here is via chain yield and full function yield on products. Yield data can be seen in Figure 11.

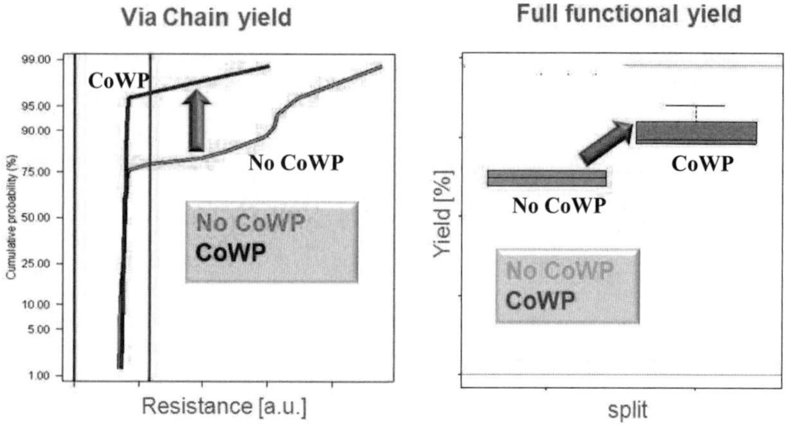

Figure 11. Via chain yield and full functional yield on a lot with and without CoWP

A reason for this improved yield could be the fact that the CoWP film acts as a sacrificial layer. This means that the removal of most of the CoWP film along with etch polymers during the post plasma etch wet clean process leaves a cleaner surface for barrier seed deposition. This theory is schematically explained in Figure 12.

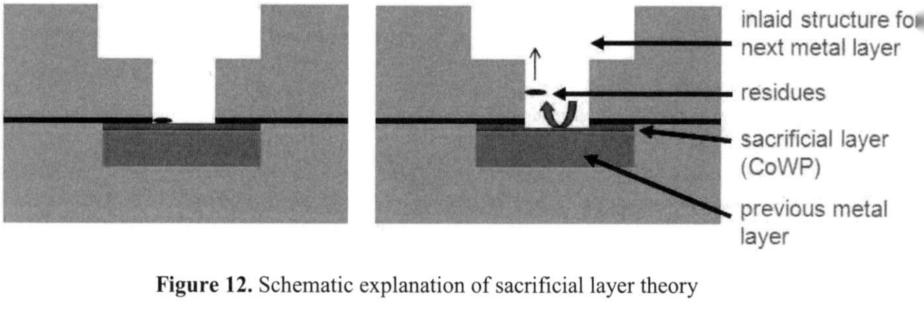

Figure 12. Schematic explanation of sacrificial layer theory

An additional benefit attributed to introducing CoWP into the process flow is that the CoWP film separates the copper from the plasma etch chemistry. When the copper is exposed to the plasma etch chemistry, detrimental copper-fluoro polymers can be created.

SUMMARY

The introduction ULK dielectric materials poses serious integration challenges, including significant TDDB and EM performance degradation. As described in this paper, the introduction of a CoWP capping layer can reduce the negative impact of these issues.

CoWP electroless deposition is ready for 45 and 32nm production. Process control is understood. Selectivity and film coverage is good. As expected, EM lifetime is improved with the addition of a CoWP capping layer on top of copper lines.

An unexpected additional benefit of adding CoWP to the process flow is that the additional cleans for CoWP and the changes to the integration flow result in increased TDDB performance and yield (via chain and full functional) relative to no CoWP at 45 and 32nm technology nodes.

ACKNOWLEDGEMENT

The authors would like to thank the GLOBALFOUNDRIES FAB1 CoWP team, Robert Seidel, Tobias Letz, Matthias Schaller, Carsten Peters, and Lam Research Corporation.

REFERENCES

[1] O. Aubel, *et al.* Proc of IRPS 2008, p. 675 (2008)
[2] S. Sankaran, S. Arai, R. Augur, et.al, „A 45 nm CMOS node Cu/Low-k/ Ultra Low-k PECVD SiCOH (k=2.4) BEOL Technology", IITC2006 (2006)
[3] G. Schindler, *et al.* Microelec. Eng. Vol.70, p. 7 (2003).
[4] C.-K. Hu *et.al.* Appl. Phys. Let.,vol. 83, p. 869 (2003).
[5] M. Nopper, *et al.,* Proc. of VMIC 2008, p. 243 (2008)
[6] R. Seidel *et al.*, Proc. of IITC 2008, p. 123 (2008)

Conference Proceedings AMC XXV © 2010 Materials Research Society

Advanced metallization developments for 32-nm node CMOS technology contact architecture

Doug H. Lee, Valli Arunachalam, Kuldeep Amarnath, Filippos Papadatos[1], Hao Zhang[2], Zhengwen Li[1], Keith Wong[1], Woo-Hyeong Lee[1], Shurong Liang, Michael Chudzik[1], David Brown, Scott Luning, Dan Mocuta[1], Doug Bonser, John Pellerin

GLOBALFOUNDRIES, [1]IBM Systems & Technology Group, [2]Chartered Semiconductors
at IBM Semiconductor Research & Development Center (SRDC), Hopewell Junction, NY 12533, USA
Corresponding Author: Doug Lee, Email: doug.lee@globalfoundries.com, Tel: +1-(845) 892-1116

Introduction

Advancement in contact resistance (Rc) reduction for CMOS process is crucial for logic/memory manufacturing in next-generation semiconductor technologies. In 32-nm node CMOS technology, the Rc increase due to scaling down of contact size imposes significant performance and yield degradation.

Ultra-thin contact liner metal combined with optimized W nucleation process allowed continuous use of conventional Tungsten metal scheme for 32 nm CMOS logic technology. The series of experiments varying contact liner metal and W nucleation processes resulted in Rc reduction while preventing WF6 attack of PVD Ti layer. A previous study showed the importance of contact profile fine tuning and use of optimized liner metal processes in reducing Rc mean value and variation.

This paper describes a newly developed atomic layer deposition (ALD)-TiN contact barrier metal process solution resulting in lower Rc for state-of-the-art sub-32-nm generation logic CMOS products. The optimized barrier metal stack process produced 50% lower median resistance while the variation was reduced 80% compared to a conventional CVD process.

Results and Discussion

Sub-40-nm scale contact patterns were generated by immersion-lithography and advanced plasma etch process on fully integrated silicon-on-insulator (SOI) wafers. The Ti/TiN/W metal stack was used to fill the contact holes either by CVD or ALD methods. The conventional dual-damascene metal process produced metal layer over the contacts. Kelvin tests were performed to measure contact resistances.

The contact metallization process consists of PVD Ti deposition, followed by TiN deposition and CVD W fill. Both the CVD and ALD TiN processes use TDMAT as a precursor. The CVD TiN process consists of sequential cycles of thermal decomposition of TDMAT followed by an *in situ*, directional N2/H2 plasma treatment. Due to the directionality of the plasma in the CVD process, the barrier film on the sidewall of the contact receives less plasma treatment than the film at the top and bottom. It creates positive step coverage, as indicated in Fig. 1(a). The ALD TiN process consists of significantly shorter deposition and plasma cycles, resulting in a net film thickness of only ~2Å per cycle. The deposition/cycle is significantly lower in the ALD process than in the CVD process. The thin film is then treated by non-directional remote plasma of nitrogen radicals, resulting in very uniform step coverage, as shown in Fig. 1(b).

The thickness and density of the TiN film and the thickness of the W nucleation influence the exposure of the underlying Ti film to fluorine attack during the CVD

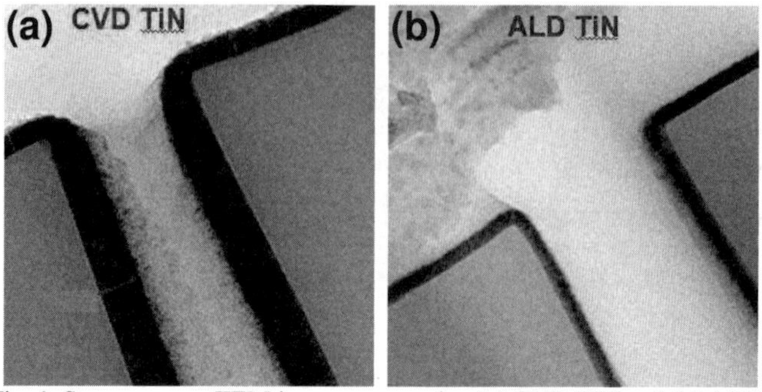

Fig. 1. Step coverage TEM images comparing (a) CVD-TiN and (b) ALD-TiN.

portion of the W fill. The TiF reaction is undesirable because it results in defects/unfilled contacts as shown in the top down optical image (Fig 2(a)). Reduction of the TiN process temperature resulted in a lower deposition rate, thereby improving plasma treatment/densification of the TiN film. Fig. 2(b) shows defect improvement by optimized ALD-TiN process.

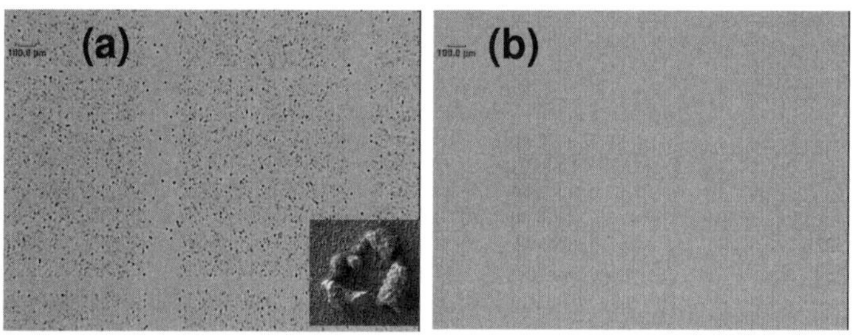

Fig. 2. Higher plasma treatment rate and reduced process temperature resulted in lower defects: (a) Without process optimization. (b) With ALD-TiN process optimization.

The conventional CVD-TiN process resulted in wider Rc variation (Fig. 3(a)), especially contact dimensions below 32 nm. ALD-TiN significantly improved median resistance and distribution with smaller contacts, as shown in Fig. 3(b). The contact resistance at various CD size was normalized to the resistance value at 40nm CD. We expect the effective aspect ratio of the contact opening is smaller with CVD than with ALD processing. The optimized Ti barrier metal process improved further with contact resistance. The Ti-2 process provided better normalized Rc distribution in contact dimensions below 32 nm (Fig. 3(c)).

The impact to CMOS transistor performance was estimated by looking at drive current at fixed leakage as a function of relative contact resistance. Fig. 4 shows a plot of relative NMOS FET Idsat (Vdd = 0.9 V) change at fixed leakage as a function of a single contact resistance. The x-axis represents the change in single contact resistance relative

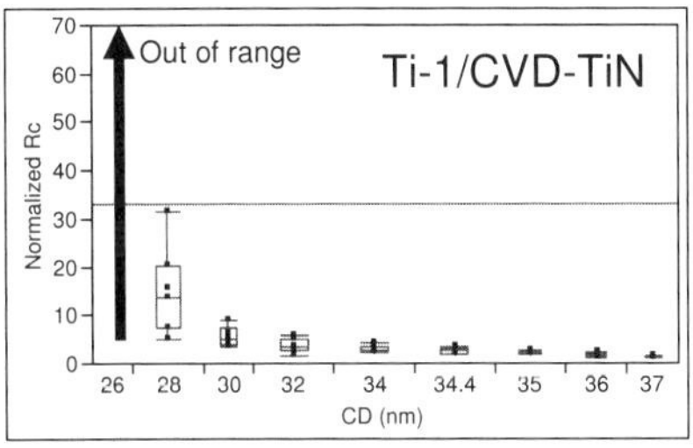

Fig. 3. (a) Normalized contact resistance with various contact sizes processed with CVD-TiN and Ti-1 barrier metals. The Rc was normalized to resistance at 40nm CA CD. 26-nm CA size showed very high resistance out of range in the graph.

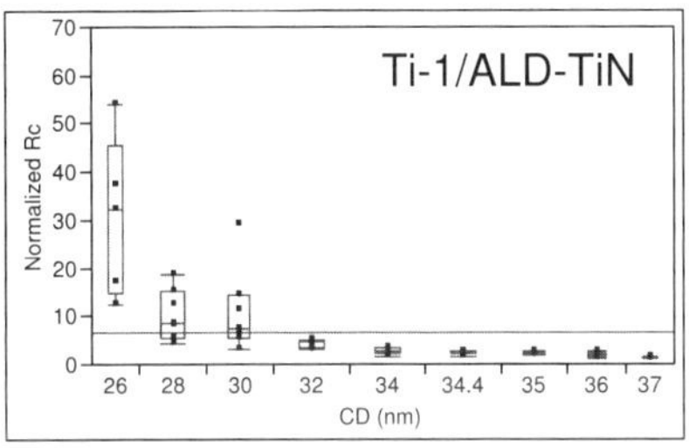

Fig. 3. (b) Normalized contact resistance with various contact sizes processed with ALD-TiN and Ti-1 barrier metals.

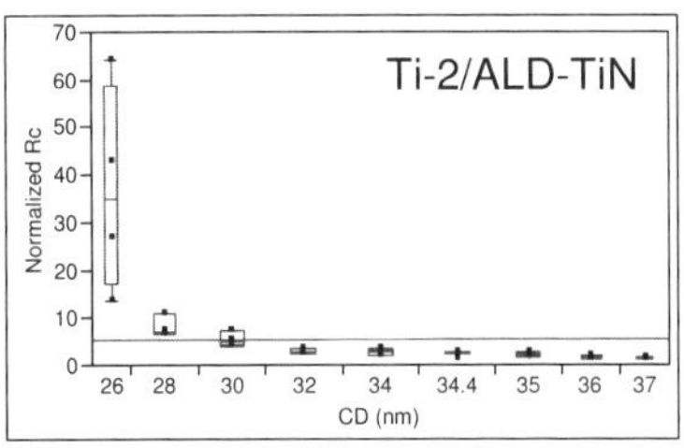

Fig. 3. (c) Normalized contact resistance with various contact sizes processed with ALD-TiN and Ti-2 barrier metals.

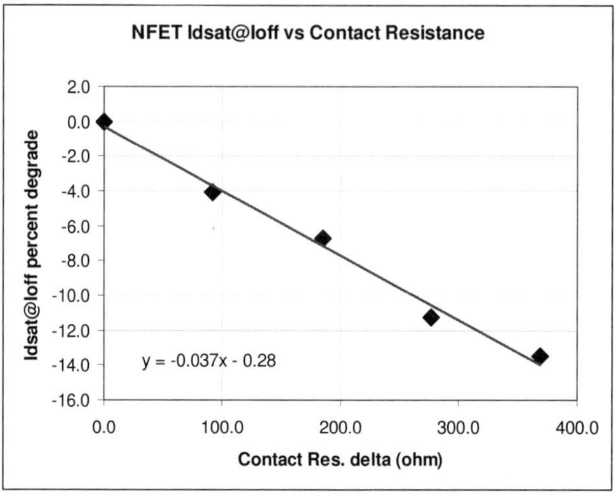

Fig. 4. Relative NMOS Idsat (at fixed leakage) vs. single contact resistance. The x-axis represents the change in single contact resistance relative to a reference case representing 0% drive change. 10 ohms change in contact resistance resulted in approx. 0.37% Idsat change for a 0.325-um wide NMOS FET with three contacts on either side.

to a reference case representing 0% drive change. The FET is 0.325 um wide and has three equally spaced contacts on either side. The estimated sensitivity is approximately 0.37% for a 10-ohm contact resistance change. While the impact is marginal for small median resistance changes of the contacts, there is significant impact to the CMOS circuit performance degradation due to the high end of the contact resistance distribution tails.

Conclusion

The contact resistance improvements by Ti and TiN barrier metal process optimization were electrically verified with Kelvin and CMOS device performance experiments. The high-end tail of the resistance distribution resulted in severe device performance degradation. More than 4% performance degradation was observed with 100-ohm Rc increase per contact.

Acknowledgement

This work has been supported by the independent Bulk CMOS and SOI technology development projects at the IBM Microelectronics Div. Semiconductor Research & Development Center, Hopewell Junction, N.Y.

Reference

[1] A. Topol, C. Sheraw, K. Wong, X. Shao, R. Knarr, S. Rossnagel, C.-C. Yang, B. Baker-O'Neal, A. Simon, B. Haran, Y. Li, C. Ouyang, S. Allen, C. Brodsky, S. Cohen, L. Deligianni, X. Chen, S. Deshpande, C. Y. Sung, and M. Ieong in Digest of Technical Papers VLSI Technology (Symposium on VLSI Technology, Honolulu, HI, 2006).

[2] K. Wong, F. Papadatos, Z. Li, H. Zhang, V. Arunachalam, J. Choi, J. Strane, J. Wallner, H. Zhuang, D. Lee, T. Graves-Abe, A. Knorr, D. Mocuta, S. Liang, S. Grunow, X. Chen in Advanced Metallization Conference 2008 (AMC 2008), edited by M. Naik, R. Shaviv, T. Yoda, and K. Ueno, (Mater. Res. Soc. Conf. Proc. AMCXXIV, Pittsburgh, PA, 2009) pp. 187-193.
[3] S. Park, S. Kang, G. Choi, U. Chung, J. Moon in Interconnect Technology Conference 2002, (IEEE Proceeding 2002) pp. 282-284.

Conference Proceedings AMC XXV © 2010 Materials Research Society

Integration of an organic ultra low-k (k=2.2) material

M. Pantouvaki[1], L. Zhao[2], C. Huffman[1], N. Heylen[1], A. Ferchichi[1], Y. Ono[3], M. Nakajima[3], K. Nakatani[3], H. Struyf[1], G. P. Beyer[1], and M. R. Baklanov[1]

[1]IMEC, Kapeldreef 75, 3001 Leuven, Belgium
[2]Intel Corporation, Kapeldreef 75, 3001 Leuven, Belgium
[3]Sumitomo Bakelite Co, Ltd, Yokohama, Kanagawa, 245-0052, Japan

ABSTRACT

The integration of an organic dielectric material of pristine k-value = 2.2 is demonstrated in 65 nm technology node interconnects. By varying process conditions an integrated k-value of 2.4 is achieved. For the same conditions, wafer-level time dependent dielectric breakdown measurements resulted in a mean time to failure of 10 years at ~1 MV/cm.

INTRODUCTION

Inter-metal dielectric materials with low dielectric constant (k-value) are interesting for use in future technology node interconnects of high electrical performance, because they have the potential to reduce RC delay and power consumption. For reduction of the k-value to < 3, porosity is introduced in such materials, which most commonly are Si-based or organic dielectrics. The integration of porous low-k materials is facing major challenges, especially with respect to patterning, plasma damage, clean and chemical-mechanical polishing (CMP). Such processing steps can result in increased integrated k-value and compromised interconnect reliability. Finding a low-k dielectric with suitable intrinsic properties and resistance to processing degradation is therefore of major significance for future interconnects.

The material properties of an organic low-k dielectric developed by Sumitomo Bakelite Co Ltd, pristine and after exposure to CMP or etch plasma were presented in [1] and are summarized in Table 1. Advantages of this spin-on material were the low k-value of 2.2, the small pore radius and its non-hybrid nature which increase its resistance to processing impairments, and the sufficiently low CTE in comparison with other organic low-k materials. In this paper we demonstrate the damascene integration of this material resulting in a k-value of 2.4 and we discuss its electrical performance.

Table 1: Material properties of Sumitomo polymer M

k-value	2.2
Porosity	27%
Pore radius	0.56 nm
Elastic modulus	5.01 GPa

RESULTS AND DISCUSSION

The integration of the organic low-k material was performed on 90nm half-pitch interconnects on wafers of 300 mm size. The processing flow is shown in Figure 1. A double hardmask scheme comprised of SiCN and SiO2 dielectrics is used for patterning. This scheme was selected to protect the organic low-k during resist stripping and to improve the CD control at low-k etch. Another advantage is the good selectivity of the etch chemistries used for each layer to the other materials exposed. After stack deposition and lithography, the pattern is printed on the SiO2 layer, followed by resist strip, SiCN hardmask etch and low-k etch using an N2/H2 chemistry. A relaxed TaN/Ta barrier thickness of ~10 nm was used in this feasibility study. After barrier / seed deposition and Cu plating, CMP was done using POR slurries for Cu and barrier polish. Good structural integrity was maintained after CMP, as shown in Figure 2. All wafers were then capped with passivation for isolation of the low-k from the ambient conditions during electrical characterization and they were sintered at 420 ^{0}C for 20 min.

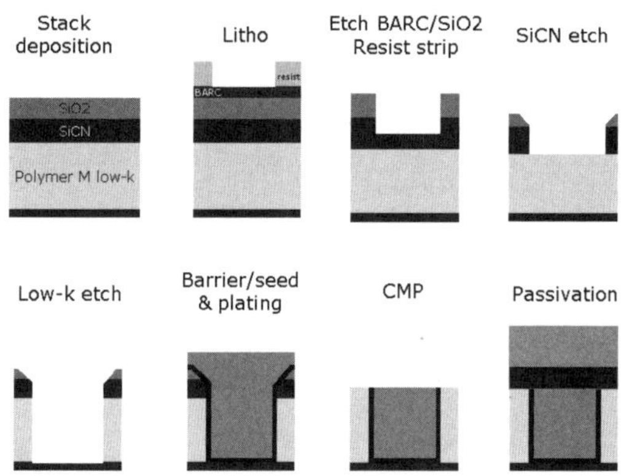

Figure 1: Integration flow

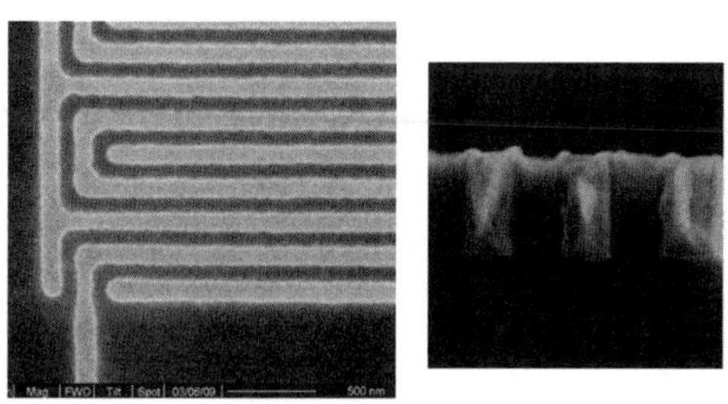

Figure 2: Top down and x-section SEM images after Cu CMP.

Two variations of the integration scheme of Figure 1 are examined in this paper. The differences are summarized in Table 2. In the first case (Process 1), we are using ~20% over-etch time to ensure that the bottom of the trenches has been reached across the wafer. In the second case (Process 2), the over-etch time is reduced to ~7%. In Process 1, we allow a 10 nm SiCN layer originating from the hardmask to remain on the low-k material after CMP. The purpose of this layer is to protect the organic low-k material from a diluted He + NH3 plasma treatment used before deposition of the passivation layer. Such a plasma treatment has been found to improve the time dependent dielectric breakdown (TDDB) on Cu interconnects due to reduction of the CuO to Cu on the post-CMP surface before deposition of the passivation stack [2]. However this plasma is also etching the organic low-k hence the use of residual SiCN for protection of the low-k. In Process 2, the SiCN hardmask material has been completely removed by CMP and no plasma treatment was applied before passivation.

Table 2: Process 1 versus Process 2

	Over-etch	SiCN hardmask after CMP	Plasma before passivation deposition
Process 1	20%	10 nm	He + NH3
Process 2	~7%	0	-

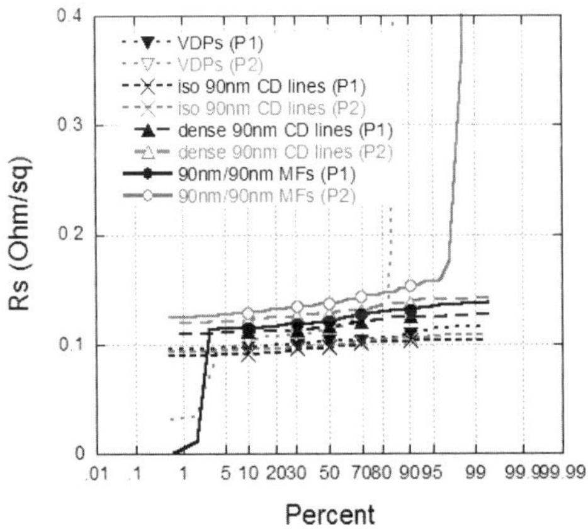

Figure 3: Cumulative plot of sheet resistance of various BEOL test structures.

Figure 3 shows the sheet resistance of various BEOL structures for the two processing flows. Good uniformity and yield are achieved in both cases. The RC comparison is presented in

Figure 4. The two process variations demonstrate a tight distribution of the resistance values within the wafers; however Process 2 has lower values and larger distribution in capacitance. This implies that the 7% over-etch time used in the second case could be marginal. TEM pictures from the integrated Cu / organic low-k interconnects are shown in Figure 5. The profile of the trenches with Process 1 conditions, i.e. ~20% over-etch time, has a trapezoidal shape, with the bottom of the trenches wider than the top. The negative slope makes the barrier deposition at approximately half-way of the trenches challenging, resulting in a possibly weak point for TDDB, especially if thinner barriers are necessary. The intensity profile of carbon at the CMP interface is also shown in Figure 5. Carbon depletion is observed within ~10nm from the low-k top surface in the case of Process 1. This is most likely due to the He+NH3 treatment done before deposition of the passivation: The plasma penetrates the ~10nm of SiCN remaining after CMP and attacks the low-k. When the He+NH3 treatment is omitted in Process 2, no carbon depletion is observed.

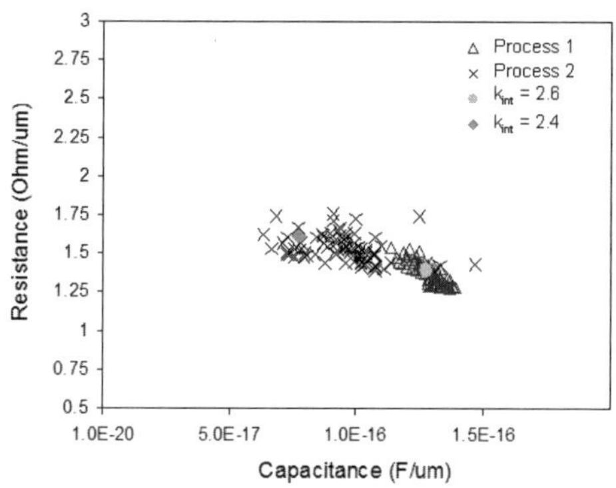

Figure 4: RC comparison of Process flows 1 and 2. The position in the RC plot of the dies used for TEM x-sections and integrated k-value calculation is also indicated.

The TEM x-sections were used to simulate the real geometry and the spacing of the Cu lines and to extract the integrated k-value of the material in each case. The geometry of the Cu / low-k x-section is reproduced by simulations and the k-value is optimized to fit the measured capacitance values [3]. The integrated k-values calculated using this technique were k = 2.6 for Process 1 and k = 2.4 for Process 2. This difference is reasonable when taking into consideration the carbon depletion in the case of Process 1 (Figure 5). The location of the RC values for the exact dies used for TEM with respect to the distribution across the wafer is indicated in the RC plot of Figure 4 as a cyan and a green dot for comparison. Even though the RC distribution may appear significant at a first glance, most of the values for both process conditions correspond to integrated k-values between 2.4 - 2.6.

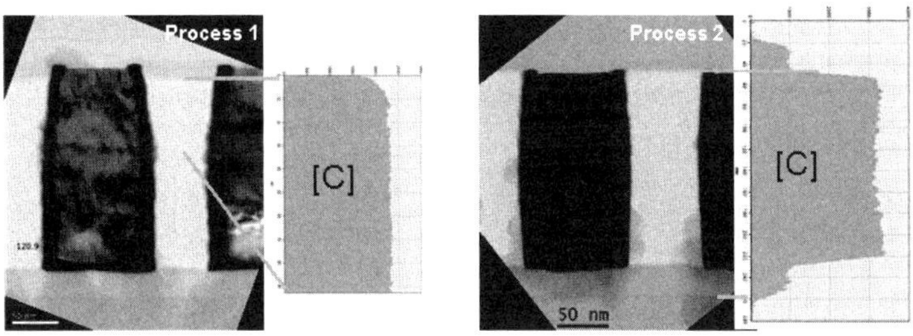

Figure 5: TEM pictures of Cu / organic low-k trenches using integration Processes 1 and 2. The carbon intensity profile at the CMP interface is also shown.

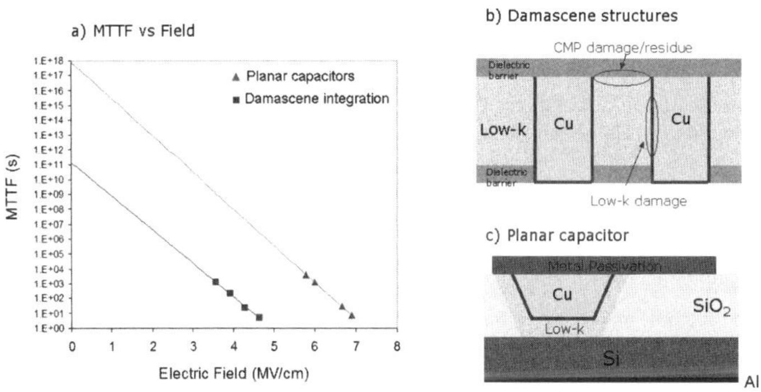

Figure 6: a) Mean time to failure ($t_{50\%}$) versus field of damascene structures (Process 2) as compared to planar capacitors. Schematic x-section of b) damascene structures versus c) planar capacitors is also shown.

Wafer-level time-dependent dielectric breakdown measurements were done on 1 cm-long meander-forks at high field conditions and at a temperature of 100^0C for Process 2. Figure 6(a) shows the mean time to failure measured on these damascene structures. 10 years lifetime is reached at an operating field of 1 MV/cm or lower assuming the E-model for extrapolation from the measured data. The TDDB lifetime of the damascene structures was compared to that obtained from planar capacitor structures as that shown in the schematic x-section of Figure 6(c), which was presented in [4]. In this test vehicle the low-k film is deposited after patterning of the capacitor, and it is followed by barrier-seed deposition, Cu plating, Cu CMP and passivation. The low-k therefore remains pristine if no intentional damage is introduced after deposition. Such planar capacitors allow for the characterization of the pristine low-k material, as opposed to damascene structures where damage may be introduced at different processing steps such as patterning or CMP (Figure 6(b)). TDDB measurements done on planar-capacitors using pristine organic low-k material and the corresponding lifetime extrapolation to user conditions are shown in Figure 6(a) for comparison with the results after damascene integration. The lifetime of the organic low-k material measured using the planar capacitor test vehicle shows that further optimization of the integration scheme could improve the TDDB performance of the Cu / organic low-k interconnects.

CONCLUSIONS

The integration of an organic low-k dielectric material in the 65nm technology node has been demonstrated, resulting in an integrated k-value of ~2.4 and exceeding 10 years wafer-level TDDB lifetime at user conditions. Further optimization of the damascene integration of this material is possible by optimization of all processing steps.

REFERENCES

1. A. Ferchichi et al, "Evaluation of an advanced organic ultra-low-k material", Advanced Metallisation Conference, pp. 587-592, 2008
2. J. Noguchi et al, "Effect of NH3 plasma treatment and CMP modification on TDDB improvement in Cu metallization", IEEE Transactions on Electron Devices **48** (7), pp. 1340-1345, 2001.
3. K. Maex et al, "Low dielectric constant materials for microelectronics", J. of Appl. Physics 93 (11), pp. 8793-8841, 2003.
4. L. Zhao et al, "A novel test structure to study intrinsic reliability of barrier/low-k", IEEE International Reliability Physics Symposium, 2009.

Conference Proceedings AMC XXV © 2010 Materials Research Society

CoWP Metal Caps for Reliable 32 nm 1X Cu Interconnects in Porous ULK (k=2.4)

E. Huang[1], M. Oh[3], S. B. Law[4], S. Petitdidier[5], T.-M. Ko[2], H. Sawada[6], T. J. Tang[4], C.-K. Hu[1], S. Cohen[1], E. Liniger[1], G. Bonilla[1], D. Rath[1], T. Shaw[1], L. Gignac[1], D. Edelstein[1], D. Permana[3], C. Child[3], R. Buengener[3], J. Fitzsimmons[2], P. Findeis[2], C. Taft[2], M. Angyal[2], O. Ogunsola[2], J. King[2], P. Flaitz[2], F. Baumann[2], S. Molis[2], C. Dziobkowski[2], J. Coffin[2], R. Davis[2], M. Zaitz[2], A. Kapur[2], L. Kermel[2], C. Truong[2], S. Grunow[2], X. Chen[2], M. Sankarapandian[2], B. Li[7], F. Chen[7], T. Lee[7], C. Christiansen[7], A. Kolics[8], N. Gilbert[8], O. Rigoutat[8], and N. Li[8]

[1]IBM T.J. Watson Research Center, Yorktown Heights, NY 10598, USA
[2]IBM Microelectronics Division
[3]GLOBALFOUNDRIES, Inc.
[4]Chartered Semiconductor Manufacturing
[5]STMicroelectronics
[6]Toshiba America Inc.
Hopewell Junction, NY 12533, USA
[7]IBM Burlington, Essex Junction, VT 05452, USA
[8]Lam Research Corporation, Fremont, CA 94538, USA

ABSTRACT

CoWP metal cap layers with good reliability (EM, SM, TDDB) for 32 nm thinwire interconnect structures in low-k (k=2.7) and porous ultralow-k (ULK) dielectrics (k=2.4) are demonstrated. Relative to structures not having CoWP, substantial EM benefits, comparable yield, and TDDB behavior are observed with the inclusion of CoWP metal caps layers.

INTRODUCTION

Electromigration (EM) involves metallic atom diffusion under an electric potential and has been extensively investigated[1]. For Cu interconnects, the top interface of the copper lines is critical as it may serve as a fast pathway for Cu diffusion leading to void formation which disrupts current flow[2]. EM data have indicated that reduced scaling leads to substantial challenges as current densities have increased and smaller voids are required to result in failure. These factors suggest that improvements in EM reliability for 32 nm technology node and smaller interconnects are critical. CoWP metal caps provide excellent EM improvements through an enhancement of the top interface of the Cu lines compared to the typical Cu/dielectric cap interface[3]. However, the implementation of CoWP metal caps can been challenging due to issues associated with leakage, shorts yield, and time-dependent dielectric breakdown (TDDB)[4,5].

EXPERIMENTAL

Interconnect structures comprised of ~50 nm wide Cu lines are fabricated with CoWP processed by electroless deposition. The M1 Cu lines are generated by single damascene in low-k SiCOH dielectric (k=2.7) while V1/M2/V2/M3 structures are built by dual damascene in ultralow-k porous SiCOH dielectric (k=2.4). The CoWP process involves a series of wet chemistries involving a preclean to clean the Cu surface after CMP, an electroless deposition of CoWP with thickness of ~4-5 nm, and subsequent post-plating clean steps.

.

DISCUSSION

Representative TEM images are shown in Figure 1 with the metal line parallel and perpendicular with the electron beam direction. Excellent CoWP coverage is observed atop the Cu line and near the landed via. Figure 2 shows lower magnificaton TEM and energy filtered TEM whereby the contrast is enhanced by highlighting the Co L3 edge electron energy loss region (779 eV).

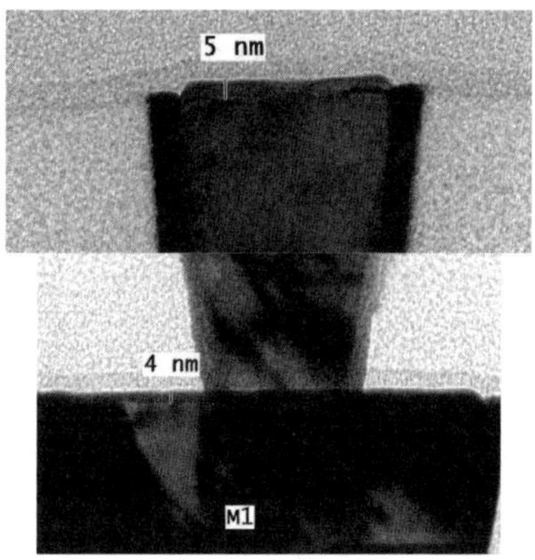

Figure 1. TEM image of a 50 nm wide Cu line having ~4-5 nm CoWP metal cap layer.

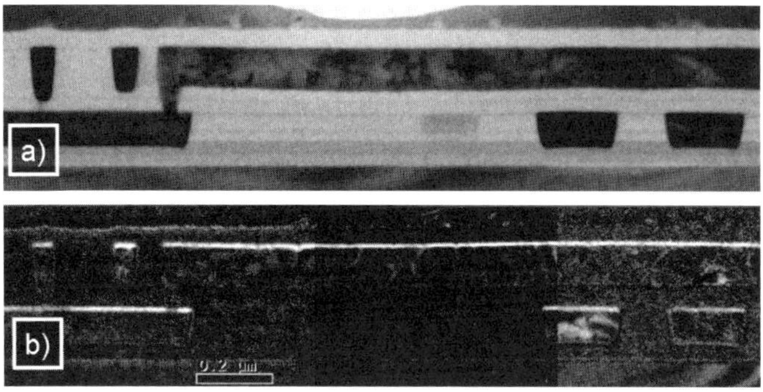

Figure 2. a) TEM and b) L3 Co EF-TEM of CoWP atop wide M1 and 50 nm M2 Cu lines.

As expected, significant improvements in electromigration lifetimes are realized in all directions (CA→M1, V1→M1, V1→M2, V2→M2) for thinwire interconnects with the inclusion of CoWP metal cap layers. Figure 3 shows the EM improvement for V2→M2 and V1→M2 whereby a ~100X improvement in EM lifetime is achieved. Greatly enhanced stress migration (SM) behavior has also been observed on thinwire interconnects having CoWP.

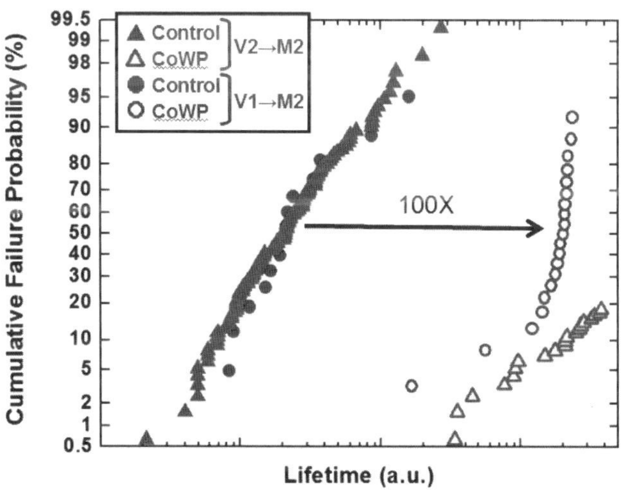

Figure 3. EM lifetimes for V1→M2 (circles) and V2→M2 (triangles) with and without CoWP metal cap.

Historically, metallic defects and/or residues leading to poor shorts yield or degraded TDDB behavior, have made implementation of CoWP metal cap layers extremely challenging. Consequently, focus has been placed on minimizing these issues especially for ULK dielectrics where the porosity may exacerbate these problems. It is demonstrated that CoWP can be processed with excellent shorts/leakage behavior on both "dense" and porous dielectrics and SRAM yields comparable to builds not having CoWP have been achieved.

Inline electrical test results on comb/serpentine/comb structure in porous SiCOH with serpentine length of 10 meters are shown below in Figure 4a for wafers having two different CoWP processes (1 vs. 2) and wafers without CoWP. The data demonstrates that wafers can be processed with CoWP having leakage currents that are comparable to uncapped structures. To better utilize inline electrical measurements and to understand the influences of CoWP processing on these structures, "bump" tests involving the measurement of leakage before, during, and after an elevated applied voltage of 15V is collected. Figure 4b shows the same structures after the elevated voltage bump. As shown, deteriorated leakages are observed after the bump test for one of process 1 while process 2 is comparable to wafers without CoWP.

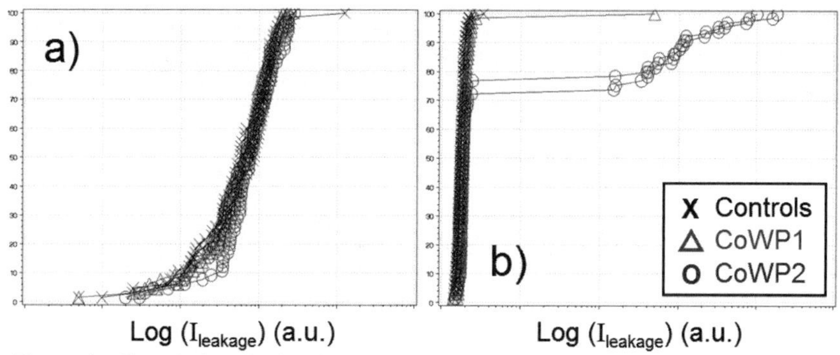

Figure 4. Cumulative distribution plots for shorts/leakage currents (a) before and (b) after an applied elevated voltage (15V) for two different CoWP processes and wafers not having CoWP.

To monitor contamination atop the dielectric layer, ToF-SIMS is conducted on 80x80 μm dielectric regions devoid of metal features on patterned wafers. This is shown below in Figure 5 for four different CoWP processes invoked on a M2 level having porous ULK. Varying Co intensities are observed while a consistent intensity is observed from the SiCOH peak. As the data indicates, interconnect structures having very low levels of Co contamination (*i.e.*, comparable to noise levels) can be achieved.

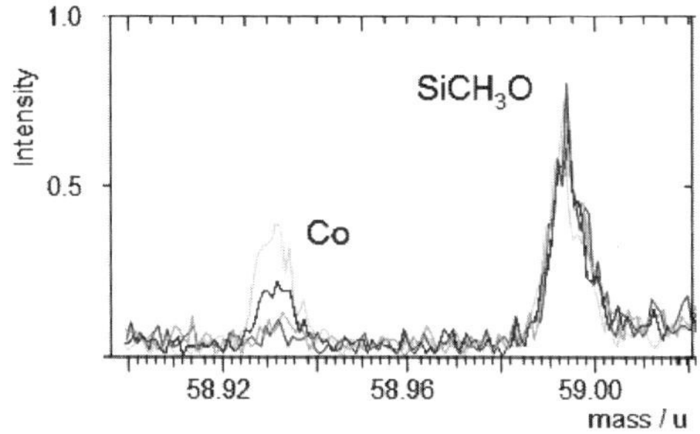

Figure 5. ToF-SIMS performed on 80x80 μm dielectric region (k=2.4) for four different CoWP processes.

Utilizing CoWP processes exhibiting excellent leakage behavior and low Co contamination levels from ToF-SIMS, good TDDB behavior can be realized. Figure 6 shows TDDB data from wafers with and without CoWP for (a) M2 and (b) M1 with varying applied voltages. As demonstrated, comparable or better TDDB lifetimes can be achieved with CoWP relative to structures not having CoWP. Using Weibull distributions and $E^{-1/2}$ modeling, voltage acceleration factors in excess of 10 are commonly observed with no influence on the presence of CoWP.

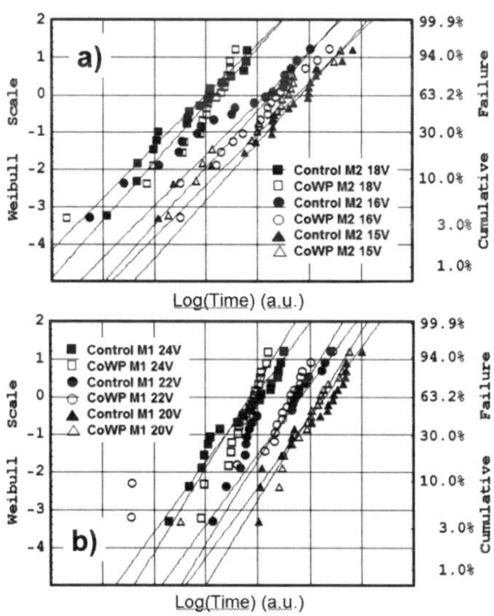

Figure 6. TDDB lifetime vs. Log(time) for control vs. CoWP for (a) M2 in ULK (k=2.4) (b) M1 in dense dielectric (k=2.7)

CONCLUSIONS

Reliable 32 nm interconnect structures having CoWP metal cap layers have been demonstrated with both low-k (k=2.7) and porous ultralow-k dielectrics (k=2.4). Substantial improvements in EM and SM behavior, comparable yield and TDDB behavior are observed with the inclusion of the CoWP metal caps relative to structures without CoWP. .

ACKNOWLEDGMENTS

This work has been supported by the independent Bulk CMOS and SOI technology development projects at the IBM Microelectronics, Div. Semiconductor Research & Development Center, Hopewell Junction, NY 12533. The authors acknowledge contributions from the following from LAM Research: Bill Lee, Fritz Redeker, Yaxin Wang, James Melley, Audrey Charles, David Byrnes, William Thie, Art Howard, Jared Lewis, Shawn Roberson, Alex Yoon, and Mike Maksymowicz.

REFERENCES

[1] C.-K. Hu, L. Gignac, R. Rosenberg, Micro. *Reliability* **46,** 213 (2006).
[2] C.-K. Hu, R. Rosenberg, H. S. Rathore, D. Nguyen, B. Agarawala,, *Proc. IEEE* **78**, 267, (1990).
[3] C.-K. Hu, L. Gignac, R. Rosenberg, E. Liniger, J. Rubino, C. Sambucetti, A. Domenicucci, X. Chen, A. Stamper, *Applied Physics Letters* **81**, 1782, (2002).
[4] A.. Preusse, R. Seidel, O. Aubel, M. Nopper, B. Freudenberg, M. Schaller, M. Fecher, T. Letz, C. Bartsch, A. Ott, M. Friedemann, F. Feustel, M. A. Meyer, P. Limbecker, IITC (2008) 123.
[5] O. Aubel, S. Thierbach, R. Seidel, B. Freudenberg, M. A. Meyer, F. Feustel, J. Poppe, M. Nopper, A. Preusse, C. Zistl, K. Weide-Zaage, Reliability Physics Symposium, IRPS (2008) 675.
.

Conference Proceedings AMC XXV © 2010 Materials Research Society

UV Curing-Induced Photoresist Poisoning in Advanced ULK BEOL Integration Schemes

Y. C. Ee[1*], D. Kioussis[2], M.S. Angyal[3], A. Kapur[3], Y. Feng[3], B.H. Engel[3], A. Passano[3], N. Klymko[3], A. Madan[3], J.L. Wang[3], J.A. Fitzsimmons[3], H. Yusuff[2], S.Y. Chong[3], D. Permana[2], O.O. Ogunsola[3], C. Child[2], T-J. Cheng[3], Z. G. Sun[1], Y. J. Choi[2], S.A. Sieg[3], S. Molis[3], C. Parks[3], J. Linville[2], S. Grunow[3], G. Biery[3]

IBM Semiconductor Research and Development Center (SRDC), Hopewell Junction, NY 12533, U.S.A.
[1]Chartered Semiconductor Manufacturing, [2]GLOBALFOUNDRIES, Inc., [3]IBM Microelectronics

*email: eeyc@charteredsemi.com ; phone: +1-845-892-2039

ABSTRACT

In today's advanced semiconductor technology, there are hundred of process steps that can significantly impact final yield and reliability performance. In our fabrication of advanced BEOL Cu integration schemes based on porous ultra-low dielectric constant (ULK) interlevel dielectrics (ILD), we found missing pattern defects after via lithography, resulting in 70% drop in SRAM and via chain yields. In the current work, we investigated the root cause of missing via pattern (MVP) in advanced BEOL Cu interconnects using porous ULK ILD materials. While it is widely known that photoresist poisoning can be caused by diffusion of amine (NHx) based compounds coming from the underlying low-k SiCHN Cu passivation and etch-stop layer (ESL), we also found that the diffusion of amines is exacerbated by the use of the UV curing process, which is used to create porosity (for lower dielectric constant, k) and enhances matrix cross-linking of the ULK films. Experimental results suggest that UV-induced resist poisoning in ULK/Cu BEOL integration schemes using SiCHN Cu passivation layer can be prevented by an N-free SiC layer, optimized hardmask properties, or exposing the oxide hardmask surface to a special plasma treatment.

INTRODUCTION

The introduction of low k ILD films with dielectric constant (k-value) less than 3.0 in thin wire (1X) metal levels at the 90 nm node and beyond implied the need to extend these type of films to lower k values in order to achieve higher device performance (RC) for advanced technology nodes. It is difficult to further reduce k-value by using fully dense materials. The solution to the limit is porous ULK materials with k=2.4 which are prepared by adding a porogen to the SiCOH skeleton and removing it after the deposition process. The removal of the porogen from the deposited films is done by UV irradiation, a method that has been generally adopted by the industry for the fabrication of SiCOH based ULK films. After the initial removal of most of the porogen from the deposited films, the UV curing generally results in an increase of the modulus and hardness of the films with curing time due to enhancement of the matrix cross-linking.

Significant challenges are encountered as device dimensions continue to shrink. More defects tend to surface when new integration schemes are applied in conjunction with porous ULK materials. Through careful partitioning of the integrated process flow we found that the UV cure process used in the fabrication of advanced ULK/Cu interconnect structures exacerbates the generation and diffusion of amine based species from the underlying SiCHN Cu passivation and etch-stop layer (ESL). Subsequently the amine species interfere with the photoresist, a phenomenon known as resist poisoning [1] leading to the imperfect transfer of the photoresist pattern to the underlying substrate, which in our case induced missing via pattern (MVP) defects.

Resist poisoning typically occurs due to neutralization of photogenerated H+ by nitrogen species from substrate materials (or by atmospheric trace amine compounds). To avoid resist poisoning, oxide films are commonly used to prevent diffusion of amine compounds. However, the present of pinholes in the oxide films become diffusion paths for amine compounds and subsequently neutralize photoresist. The self-catalyzed deprotection of resist polymer backbone cannot occur when the mobile deprotecting H+ species are neutralized by amine compounds [1]. Hence, resist is not rendered base soluble and remains intact in such locations, leading to "poisoning". The poisoned resist can result in a polysilicon micromask.

In the current work, we investigated the root cause of missing via patterns and propose three methods to suppress UV curing induced resist poisoning observed in advanced BEOL ULK/Cu integration scheme.

EXPERIMENT

A Cu/ULK based dual damascene interconnect test structure was fabricated on top of a standard single damascene metal 1 structure. After Cu chemical-mechanical polishing (CMP) process, a layer of Cu passivation and etch-stop film was deposited. In our process partition check to identify the source of resist poisoning, low-k SiCHN film, standard SiCHN film, and composite SiCHN/SiC film were used as Cu passivation and etch-stop layers. This was followed by deposition of a porous ULK film with k = 2.4 as an interlayer dielectric (ILD) material. A UV cure process is used to enhance matrix cross-linking of the ULK film and create porosity to obtain lower dielectric constant. Oxide hardmask (HM) film is then deposited on the ULK film. Two types of HM film, type I and type II, have been investigated as means to suppress missing via patterns. For evaluating missing via patterns, vias were patterned through standard lithography process with planarization film/anti-reflection coating/resist/topcoat stack.

Wafers were optically scanned to detect the presence of defects. Cross-section of the defects was characterized by scanning electron microscopy (SEM). Thermal desorption analysis (TDA) was used to characterize thermal outgassing profiles of Cu passivation and etch-stop films. Oxide hardmask film density, stress, hydroxyl (OH) content, and water outgassing were characterized using x-ray reflectivity (XRR), inline wafer bow scan, infrared spectroscopy, and TDA techniques respectively. Time-of-flight secondary mass ion spectrometry (ToF-SIMS) was used to analyze the composition of hardmask film with special plasma treatment.

RESULTS and DISCUSSION

Missing via pattern defects were found in the advanced porous ULK-based/Cu BEOL integrated scheme described in the experimental section after via lithography, resulting in approximately 70% drop in SRAM and via chain yields. A series of process partition experiments were carried out to investigate the root cause of missing via pattern defects as shown in Figure 1.

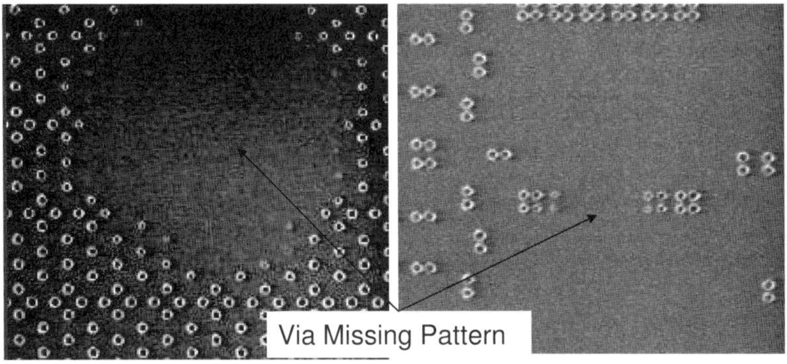

Figure 1. Missing via pattern defects are observed randomly after via lithography process for advanced porous ULK/Cu integration scheme.

Missing via pattern defects were observed randomly after via lithography and the affected area was found to be as large as one micrometer. Figure 2a shows scanning electron microscope (SEM) cross-section in X-direction across the affected area. From SEM cross-section micrograph (Figure 2b), missing vias were clearly observed and with evidence that there is no underlying defect which could cause localized region to be out of focus during the via lithography process.

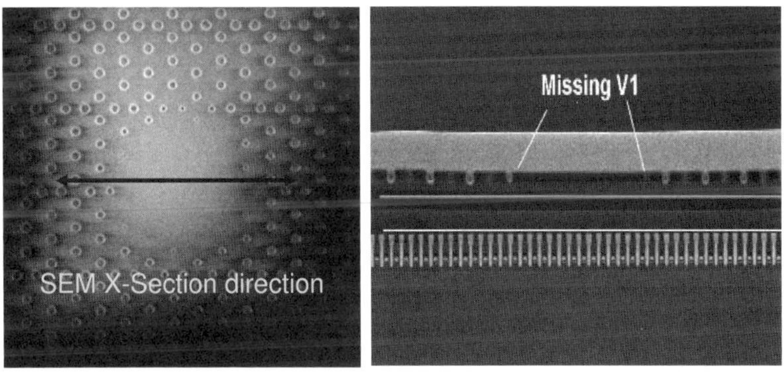

(a) (b)

Figure 2. (a) SEM cross-section direction of missing via pattern defect. (b) SEM cross-section micrograph indicates missing via pattern defect with no evidence of defect as the cause of focus issue.

These observations suggest that the photoresist properties are being modified, preventing subsequent patterning process. This implies that resist is poisoned by amine compounds. A partition check on the source of amine compounds was performed on Cu passivation and etch stop film: low-k SiCHN layer, which potentially generates amine compounds as poisoning agents. Three films were selected: 1) low-k SiCHN film, 2) standard SiCHN film, 3) composite SiCHN/SiC film. Scan data at post-via lithography showed massive missing via patterns for films 1 and 2. Conversely, no missing via pattern was detected on wafers using composite SiCHN/SiC film, as shown in Figure. 3, therefore implying that via resist poisoning is due to nitrogen-containing SiCHN films.

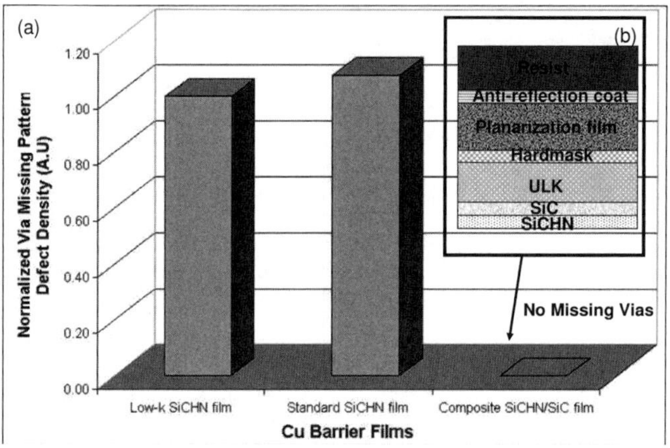

Figure 3. Scan data showed that no missing via pattern defects are observed on the wafer with composite SiCHN/SiC film.

Figure 4 shows thermal desorption analysis (TDA) thermal outgassing profiles from a reference oxide film and from the low-k SiCHN film. The thermal desorption profile from the nitrogen-free oxide reference film (Figure 4(a)) shows masses 17 and 18, which are in the ratio of 0.3 and represent the signal from water outgassing (mass 17 being OH and mass 18 being HOH). The profile of the low-k SiCHN barrier (Figure 4(b)) also shows evolution of masses 17 and 18, but in the ratio of 0.6. In the low-k SiCHN, mass 17 therefore derives not just from hydroxyl but also from another species with the same mass, which is ammonia (NH_3).

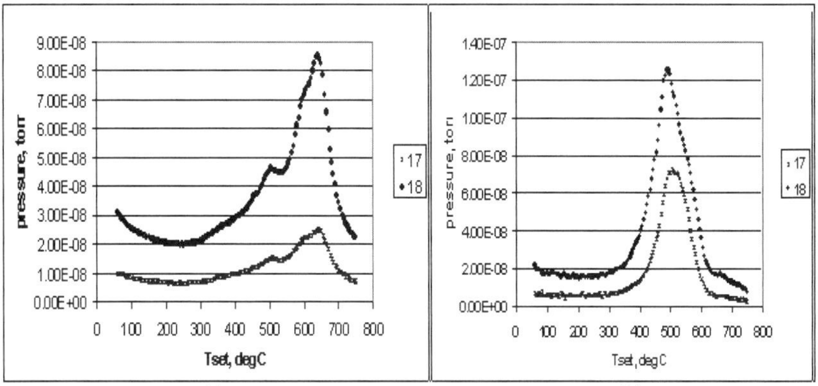

(a) (b)

Figure 4. TDA thermal outgassing profile of (a) SiO_2 reference film shows ratio of mass 17/18 = 0.3, from H_2O outgassing, (b) sample with low-k barrier shows ratio of mass 17/18 = 0.6, indicates presence of NH_3 in addition to H_2O.

The source of resist poisoning agents has been identified to be the SiCHN Cu passivation and etch stop films. To suppress missing via pattern defects in the ULK/Cu BEOL integration scheme, two sacrificial ULK oxide hardmask (HM) films, type I and type II, have been investigated. Type I versus Type II HM film density, stress, hydroxyl (OH) content, and water outgassing were characterized using techniques discussed in the experimental section. Type II film was found to be more compressive, denser, and have lower OH compared to type I HM, as shown in Figure 5. These are good indicators of the hermetic nature of the HM films.

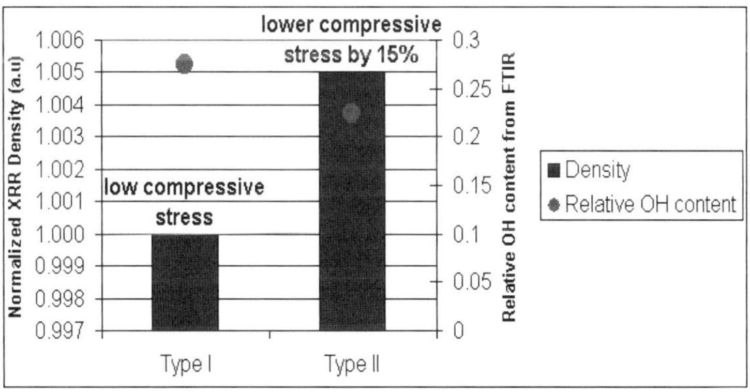

Figure 5. X-ray reflectivity (XRR), hydroxyl (OH) content, and as-deposited film stress comparison between type I and type II HM films.

In addition, different thickness splits were studied for both the HM type I and HM type II in order to access the ability to suppress missing via pattern defects. The scan data showed that type II HM with thickness X can effectively prevent resist poisoning. In contrast, 25% thicker type I film is needed to suppress the poisoning, as illustrated in Figure 6. This implies that compressive, denser and higher hermeticity film is favorable and able to prevent resist poisoning even with thinner thickness.

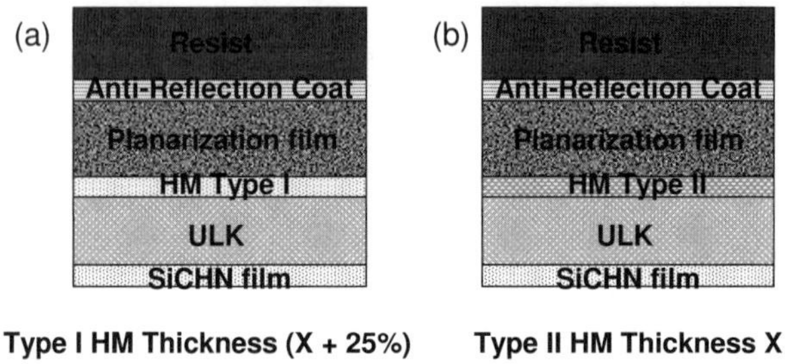

Figure 6. Type I HM needs an extra 25% thickness compared to type II HM to prevent amine compounds diffusion.

Further investigation was carried out to prevent the resist poisoning issues. Surface treatment A, which is designed to prevent amine compounds diffusion, was applied on the sacrificial HM surface prior to the planarization film. Scan data shows that there is no missing via pattern defects observed after via lithography. The surface treatment A was done in a Dual Frequency Capacitively coupled plasma chamber using a reducing chemistry of choice. ToF-SIMS analysis was conducted on experimental samples. When surface treated sample was compared to control sample that has not received surface treatment A, we observed fragments or species associated with tri-methylamine (C_3H_9N) or amine compounds. This suggests that a chemical reaction pathway in which the escaping amines species are either trapped or used as links in forming an organic film that enhance the absorption qualities of the sacrificial HM film, thereby protecting against amine diffusion. Figure 7a shows ToF-SIMS results from the control sample while Figure 7b shows that of the surface treated sample.

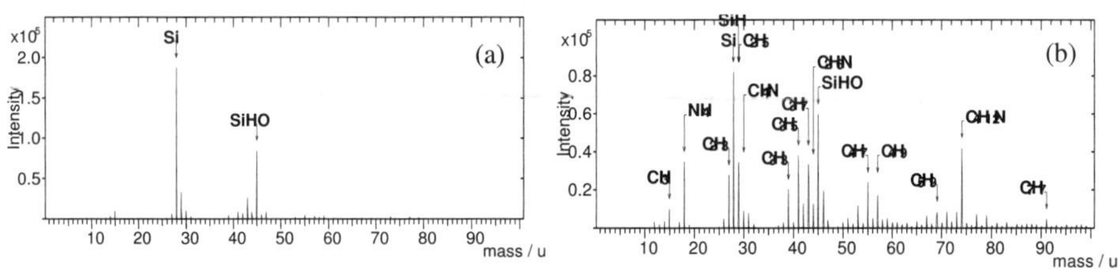

Figure 7. ToF-SIMS analysis of (a) untreated sample (control), (b) surface treated sample.

The hypothesis of a physical film property modification is further enhanced by confirmation using a D_2O SIM Analysis. D_2O vapor exposure was done on all samples either long (more than 24 hours) from all directions or confined by a suction cup to go only through the top for only 4 hours. As shown in Figure 8, data revealed that there is less Deuterium in the HM layer for treated sample when compared to the control sample. The reduction is to the magnitude of one order. This indicates less absorption due to a film modification from the plasma treatment.

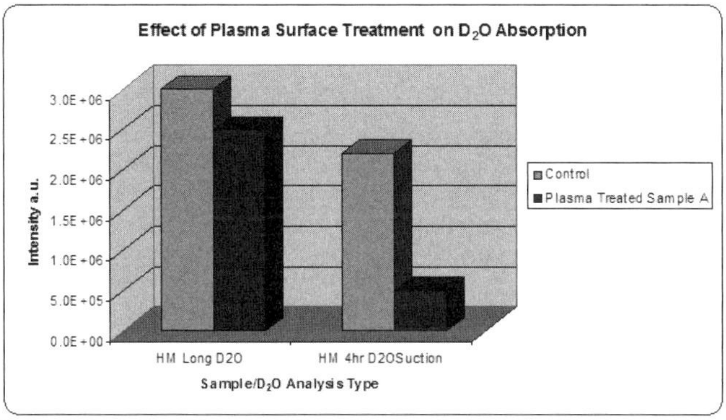

Figure 8. D_2O SIM analysis comparing control versus plasma treated sample.

Through the analysis of these observations, a potential mechanism can be proposed. In our ULK-based BEOL integration schemes, ULK ILD films need a subsequent high-temperature UV cure, which is necessary to increase porosity to reduce the dielectric constant for improved BEOL performance. This UV cure can chemically modify the underlying low-k SiCHN layer, promoting formation of amine species. The formation of these amine species is supported by an understanding of the thermodynamic enthalpy of formation data [2] related to these compounds. These amines can react with photochemical acid generators in the resist and, as a result, affect the overall resist sensitivity [3], resulting in the observed missing via pattern defect as illustrated in Figure 9.

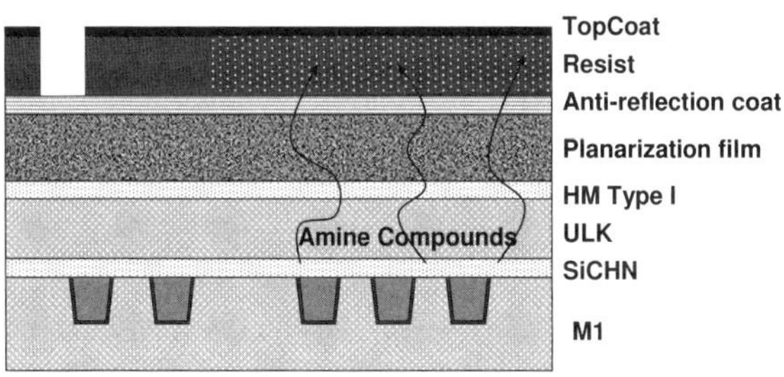

Figure 9. Illustration of the mechanism of missing via pattern by UV curing-induced photoresist poisoning.

Two possible mechanisms are proposed for amine diffusion through the SiCHN/ULK/HM film stack: 1) through pinholes present in the oxide HM when thickness is scaled below a certain threshold, as illustrated in Figure 6, and 2) classical membrane diffusion through the HM with the present of high temperature UV-cure.

Our results suggest that UV-induced resist poisoning in ULK/Cu BEOL integration schemes using SiCHN Cu passivation layer can be prevented by an N-free SiC layer, optimized HM properties, or exposing the HM surface to a special treatment to prevent amine compounds diffusion before subsequence lithography process.

CONCLUSIONS

UV curing-induced photoresist poisoning resulted missing via pattern defects in advanced ULK/Cu BEOL integration schemes has been shown and investigated. Missing via pattern defects were found to be related to the diffusion of amine (NHx) compounds coming from the underlying low-k SiCHN Cu passivation and etch-stop layer. The release and diffusion of amines is exacerbated by the use of the UV curing process. The possible mechanisms for amine diffusion are through pinholes in the oxide hardmask when film thickness is scaled below a certain threshold and classical membrane diffusion through the HM driven by the high temperature UV-cure. It is suggested that missing via pattern defects can effectively be prevented by using N-free SiC layer or composite SiCHN/SiC film, optimized hardmask properties to more compressive, denser and higher hermeticity, or special hardmask surface treatment to prevent amine compounds diffusion.

ACKNOWLEDGMENTS

This work has been supported by the independent Bulk CMOS and SOI technology development projects at the IBM Microelectronics, Div. Semiconductor Research & Development Center, Hopewell Junction NY 12533, USA.

REFERENCES

1. S. Dakshina-Murthy, S. A Bell, R. J Haungm R. C Nguyen, C. E. Tabery, "*Method of using amorphous carbon to prevent resist poisoning*", US Patent 6855627.

2. I. Barin, F. Sauert, E. Schultze-Rhonhof, and W.S. Sheng, *Thermochemical Data of Pure Substances*, 2nd ed., Weinheim, Federal Republic of Germany; New York: VCA (1993).

3. Joseph C. Salamone, *Polymeric Materials Encyclopedia*, 1st ed., Vol. 2 C., CRC Press Inc. New York (1996).

Conference Proceedings AMC XXV © 2010 Materials Research Society

Electroless CoWP Integration Scheme to Enhance Cu Interconnect Reliability at Ultra Narrow Line

Kwang-Jin Moon*, Jong-Ho Yun, Zung-Sun Choi, Hye-Kyung Jung, Jong-Myeong Lee, Gil-Heyun Choi, Siyoung Choi, Joo-Tae Moon
Process Development Team, Semiconductor R&D Center, Samsung Electronics
San#16 Banwol-Dong, Hwasung-City, Gyeonggi-Do, Korea, 445-701
*TEL: 82-31-208-2967 FAX: 82-31-208-0699 E-mail: honeymoon@samsung.com

ABSTRACT

Electroless cobalt tungsten phosphorus (CoWP) integration scheme on sub-50nm Cu interconnect of memory device was successfully demonstrated. Line-to-line leakage current and breakdown voltage (BV) was improved by removing contaminants with a specific cleaning treatment after CoWP deposition process and it could minimize a time-dependent dielectric breakdown (TDDB) degradation, which typically attributes to a selectivity loss of electroless deposition. In addition, electromigration (EM) lifetime of Cu line was increased significantly with a very thin electroless CoWP capping layer less than 5nm.

INTRODUCTION

Degradation of electromigration (EM) is a great concern in Cu BEOL interconnects. Many recent efforts have been addressed to develop an electroless CoWP self-aligned capping technology with the object of enhancing electromigration reliability[1-7]. All demonstrated that the metallic CoWP cap on Cu line remarkably improves EM reliability, which was explained by the increased activation energy of Cu electromigration for CoWP capped Cu and thereby suppressing void migration that favors its path on top of Cu line[4, 6-7]. However, as a design rule keeps shrinking the CoWP cap faces other challenges such as degradation of line-to-line leakage, breakdown voltage (BV), and TDDB. The reason for the degradation is that the self-aligned CoWP gradually losses its deposition selectivity for Cu lines as the spacing between Cu lines decreased. Some reported a comparable TDDB performance of electroless CoWP to that of comparison with no CoWP cap[4, 8] but no one yet showed an improved TDDB reliability by electroless CoWP.

In this work, we firstly demonstrated the application of electroless CoWP integration scheme on memory device below 50nm. After optimizing the electroless CoWP plating process and introducing specific chemical treatment after plating, we could obtain good electrical characteristics such as line-to-line leakage and breakdown voltage that are comparable to the one that has no CoWP capping, as well as prominently improved EM lifetime.

EXPERIMENTAL

Sub-50nm single damascene lines were prepared with PVD-TaN/Ta barrier, PVD-Cu seed and electroplated Cu. The electroless CoWP layer was selectively formed on top of the Cu line with $CoSO_4$ for the cobalt source, DMAB for the reducing agent, and additives such as complexing agent, surfactant and stabilizers which are typical chemistries found in other literatures[1,2]. To evaluate Cu interconnect reliability, electromigration tests on samples with and without CoWP were completed. The tests were done on dual via chain structure for via EM evaluation and on 1 μm Cu line structure for line EM evaluation. Test conditions for via EM and line EM were 330°C, 1.0mA and 300°C, 0.8mA, respectively. A preliminary test for time-dependent dielectric breakdown (TDDB) was carried out with sub-50nm spacing structure and under 39V and 85°C test condition. Contaminants such as Co, W and Al from electroless deposition were analyzed by using ICP-MS with different post treatments. CoWP capping profile was characterized by using TEM and Co contamination was analyzed by Energy Disperse X-ray Spectroscopy (EDS).

RESULTS and DISCUSSION

Selective Deposition of CoWP by electroless plating

One of advantages that the electroless deposition has is that metallic capping layers such as CoWP, CoWB, NiWP, or NiMoP can be selectively deposited on a conductive layer. This contributes to reducing integration costs by eliminating CMP process after electroless deposition process. We observed that our CoWP layer was selectively capped on Cu lines. Figure 1 illustrates that Cu grains and their boundaries were visible after Cu CMP process, but the following electroless CoWP selectively deposited on Cu lines without any noticeable defects and contaminants.

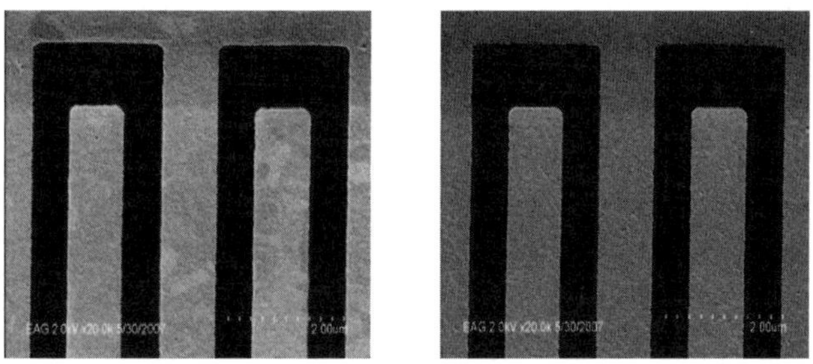

(a) (b)

Figure 1. Plain SEM analysis on (a) Cu line after CMP and (b) Cu after CoWP capping

Electrical Characteristics

Figure 2. showed the line-to-line leakage current and breakdown voltage (BV) at narrow Cu trench less than 50nm. Leakage current of CoWP capped sample was higher than reference with no CoWP capping. Breakdown voltage of CoWP capped sample was still less than that of reference. However, we were able to obtain noticeable improvements in both leakage current and breakdown voltage (BV) when the specific treatment with acidic solution and additives was performed at the cleaning phase.

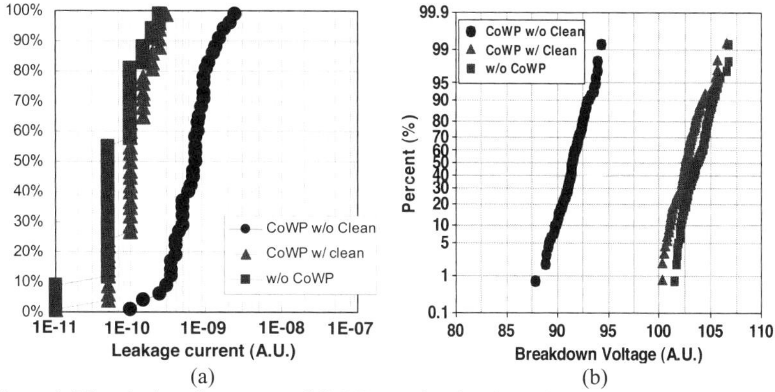

Figure 2. Electrical performances of CoWP capping by electroless deposition with and without cleaning treatment (a) Line-to-line leakage current (b) Breakdown voltage

We suspect that there are three main factors causing BV and TDDB degradations. First, metallic contaminants such as tungsten particles from plating chemicals and reduced cobalt or copper on interlayer dielectric (ILD) surface are likely to form a conducting path between Cu lines for TDDB degradation. Secondly, other type of contaminants such as Co oxide or unidentified particles may reduce TDDB lifetime. Lastly, CoWP can grow laterally toward ILD surface resulting in a decreased distance between Cu lines. All these factors can accelerate Cu ion diffusion across ILD and consequently will deteriorate TDDB performance[9]. We found that a thin and continuous CoWP layer was formed on Cu lilne but no lateral growth of CoWP was observed by TEM analysis as shown in Figure 3.

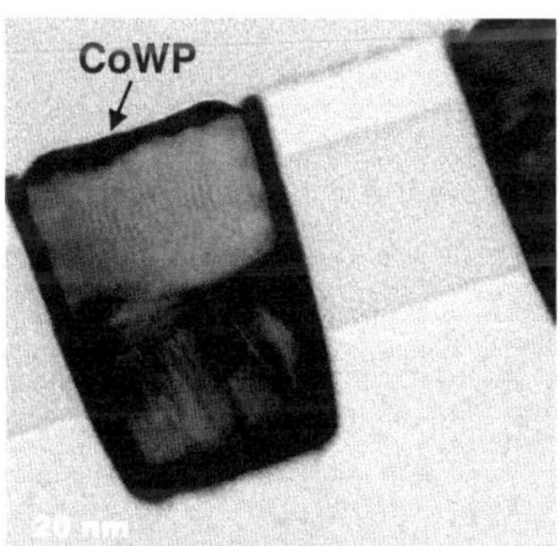

Figure 3. TEM image of CoWP capped Cu line. A thin and dense CoWP layer less than 5nm was formed on Cu line, however no lateral growth of CoWP was observed.

In order to minimize the ILD contamination, we focused on the development of several chemical cleanings that consist of a pre-clean treatment to remove Cu oxide on Cu surface and a following post-clean step to passivate CoWP and prevent further $Co(OH)_2$ formation after plating. In addition, final cleaning phase was applied to further remove residual contaminants on ILD surface thereby decreasing residues.

According to the ICP-MS analysis, Co and W contaminants on ILD was 8×10^{11} atoms/cm^2, 4×10^{10} atoms/cm^2 respectively and decreased to 5×10^9 and 2×10^9 atoms/cm^2 by following cleaning process. It indicated that metallic contaminants such as Co and W on dielectric were removed and finally current path was limited by specific cleaning treatment. It is also verified by TEM and EDS analysis. As shown in Figure 4, Co and Cu contaminants were remained on ILD after electroless CoWP deposition while no Co contaminant was detected with following cleaning treatment.

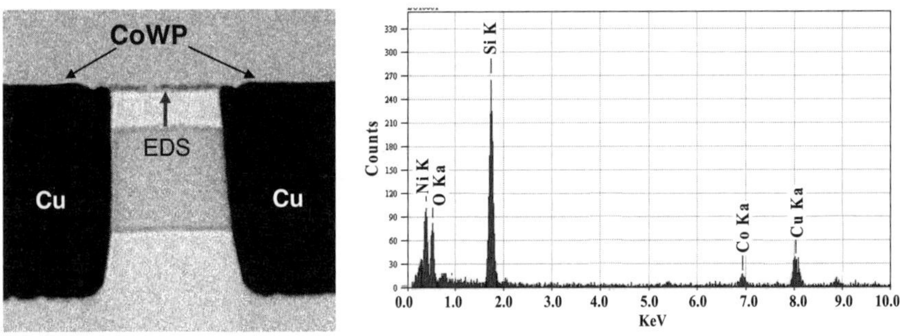

(a) (b)

Figure 4. STEM and EDS analysis on CoWP capped Cu single damascene without specific post treatment (a) STEM Image (b) EDS. Co and Cu contaminants were detected on ILD.

Reliability test

Via EM test revealed that CoWP cap sustained EM lifetime for about 50 times longer than that of reference with no CoWP cap as shown in Figure 5. Failure criterion was set as 2% resistance increase. Most failures of the uncapped Cu line happened at 10 hours while the CoWP capped one failed at 500 hours. TEM result on the sample with CoWP revealed that the failure location was in the second metallization Al segment which has no CoWP.

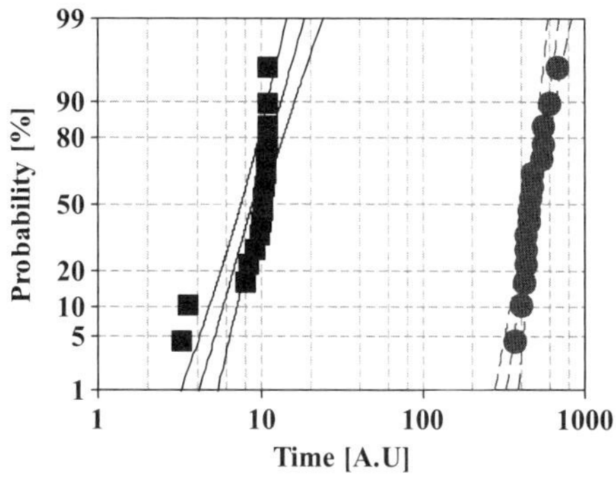

Figure 5. Plot of via electromigration time-to-failure (TTF) in Lognormal scale for Cu interconnect line, with and without CoWP capping layer. Test condition was 330°C and 1.0mA

In Figure 6, resistance of uncapped Cu lines (No CoWP) increased sharply above the set point of failure criteria which was 2% increase. The uncapped Cu lines failed after 8 hours but the CoWP capped Cu lines did not fail after 1,500 hours. Thus, time-to-failure of Cu lines were increased more than 150 times when they are capped with CoWP layer. In addition, it was shown that both CoWP capped Cu line and uncapped Cu line exhibited a comparable TDDB lifetime.

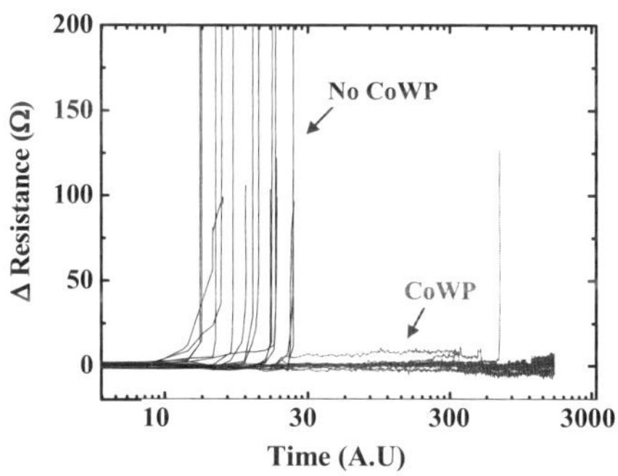

Figure 6. Graph showing the resistance increase of Cu Interconnect line, with and without CoWP capping layer as a function of time. Test condition was 300°C and 0.8mA

CONCLUSIONS

Selective CoWP capping integration by electroless deposition was demonstrated for reliability enhancement of Cu BEOL interconnect of memory device. Electromigration lifetime of Cu line was increased significantly with a very thin electroless CoWP capping layer less than 5nm. In addition, line to line leakage current and breakdown voltage (BV) were improved by removing contaminants with a special cleaning treatment after CoWP deposition process. It indicated that a time-dependent dielectric breakdown (TDDB) degradation which typically attributes to a selectivity loss of electroless deposition could be minimized by the post-clean treatment.

ACKNOWLEDGEMENT

This work has been collaborated with Lam Research Corporation. We would like to acknowledge the helpful support and equipment from Lam Research.

REFERENCES

1. J. Gambino, J. Wynne, J. Gill, S. Mongeon, D. Meatyard, B. Lee, H. Bamnolker, L. Hall, N. Li, M. Hernandez, P. Little, M. Hamed, I. Ivanov and C.L. Gan, Microelectronics Engineering 83, pp. 2059-2067 (2006)
2. J. Guillan, L.G. Gosset, R. Delsol, E. Ollier, Ph. Brun, E. Petitprez, R. Gras, V. Girault, M. Gall, M. Hauschildt and J. Torres, Microelectronics Engineering 84, pp. 2629-2633 (2007)
3. N. Kawahara, M. Tagami, B. Withers, Y. Kakuhara, H. Imura, K. Ohto, T. Taiji, K. Arita, T. Kurokawa, M. Nagase, T. Maruyama, N. Oda, Y. Hayashi, J. Jacobs, M. Sakurai, M. Sekine and K. Ueno, Proceeding IITC, pp. 152-154 (2006)
4. A. Preusse, R. Seidel, O. Aubel, M. Nopper, B. Freudenberg, M. Schaller, M. Fecher, T. Letz, C. Bartsch, A. Ott, M. Friedemann, F. Feustel, M.A. Meyer, Proceeding IITC, pp. 123-125 (2008)
5. T. Decorps, P.H. Haumesser , S. Olivier, A. Roule, M. Joulaud, O. Pollet, X. Avale and G. Passemard, Microelectronics Engineering 83, pp. 2082-2087 (2006)
6. C.-K. Hu, L. Gignac and R. Rosenberg, Microelectronics and Reliability Volume 46, pp. 213-231 (2006)
7. C.-K. Hu, D. Canaperi, S.T. Chen, L.M. Gignac, S. Kaldor, M. Krishnan, S.G. Malhotra, E. Liniger, J.R. Lloyd, D.L. Rath, D. Restaino, R. Rosenberg, J. Rubino, S.-C. Seo, A. Simon, S. Smith and W.-T. Tseng, Thin Solid Films Volume 504, pp. 274-278 (2006)
8. R. Daamen, A. Kolics, J. Michelon, J. Noiray, P.H.L. Bancken, V.H. Nguyen, G.J.A.M. Verheijden, R.G.R. Weemaes, Advanced Metallization Conference 2006, pp.45-51 (2006)
9. W. Wu, X. Duan, J.S. Yuan, Proceeding IRPS, pp. 282-286 (2003)

Reliability and Yield

Conference Proceedings AMC XXV © 2010 Materials Research Society

Electromigration Improvement for 40nm and below through Diffusion Barrier Interface Engineering using a new Precursor

Y. Le-Friec[1†], W. Ye[2†], V. Zubkov[2], G. Conti[2], M. Y. Shek[2], L.-Q. Xia[2]*, D. R. Witty[2], S. Chhun[1], L.Arnaud[1], E. Petitprez[1], L.Clement[1], C. Monget[1], D. Galpin[1],S. Hong[3], J. Fort[3]
[1]STMicroelectronics, Crolles France
[2]Blanket Dielectric Division, Applied Materials Inc.
3225 Oakmead Village Drive, Santa Clara, CA 95052 USA
[3]STM Account, Applied Materials France

ABSTRACT

The rate limiting factor for Electromigration (EM) enhancement for 40nm technologies and below becomes reducing the interface diffusion along the copper line. This study proposes to strengthen Cu-Si bonding by forming $CuSi_xN_y$ using the clustered molecule-trisilylamine $N(SiH_3)_3$, TSA, as an alternative to SiH_4. This approach consists in a single process step added to a conventional copper barrier deposition process. A factor of 3 EM improvement is obtained in comparison to the baseline while resistance degradation is kept minimal (<2.5%). Modeling of the process is detailed and surface characterization in accordance with EM results is exposed.

INTRODUCTION

In an industrial 40nm BEOL process built with Trench First Hard Mask architecture all problems of poisoning are removed by construction, ULK properties fully conserved, line/via profiles are favorable for void free Cu filling and the line to line space is preserved for TDDB[1]. After all these issues are solved, the rate limiting factor for Electromigration (EM) enhancement for 40nm technologies and below becomes reducing the surface diffusion along the copper line[2]. A previous technique to strengthen Cu-Si bonding was to form $CuSi_xN_y$ using SiH_4 soak followed by NH_3 plasma nitridation before copper barrier deposition[3]. However, this approach significantly increases the Cu line resistance (Rs) due to Si diffusion into the Cu bulk through the Cu grain boundaries. In this paper a new approach to form $CuSi_xN_y$ is proposed using the clustered molecule-trisilylamine $N(SiH_3)_3$, TSA, as the soak precursor on Cu. This molecule is known to be useful as a single source precursor for growth of silicon nitride films [4]. A new application of this molecule for the formation of a $CuSi_xN_y$ interface before copper barrier deposition is investigated. The advantages of this approach using TSA compared to SiH_4 are described and resulting significant EM improvement is exposed.

EXPERIMENT

In order to evaluate EM improvement, it is necessary to make the study on a stable baseline. Therefore samples were prepared using 45nm CMOS process with a Trench First Hard Mask Architecture in an ultra low k dielectric [1]. Prior to the metal 2 line cap, NH_3 pre-treatment was performed for copper deoxydation purpose, then various TSA process conditions were used and compared with SiH_4 approach as well as conventional baseline process (no TSA, no SiH_4). Copper dielectric barrier (BLOK) is a standard SiCN film (k=4.9) deposited in a PECVD system (Applied Materials Producer® GT Twin Chamber™). Process conditions are summarized in table1.

Table 1: process conditions for line 2 top treatment

Process conditions	Soak precursor	Nitridation step
BLOK reference	None	none
2 steps CuSiN BLOK	SiH4	Yes
TSA soak BLOK	TSA	none
TSA soak + nitridation BLOK	TSA	Yes

EM time to failure (TTF) was measured on 70nm width structures uniformly spread across the wafer to capture process variation within the wafer. A fail criterion of a 10% resistance change was applied.

After EM measurement, samples were submitted for Transmission Electron Microscopy (TEM) cross sections and Copper-SiCN interface was analyzed through TEM Electron Energy Loss Spectroscopy (TEM EELS).

Additional blanket samples were prepared for chemical analysis of the copper-barrier interface. Post-CMP blanket copper substrates were exposed to a conventional NH_3 pretreatment followed with various TSA conditions. X-Ray Photoelectron spectroscopy (XPS, V6 Scientific Theta 300) and angle-resolved XPS (AR-XPS) were carried out to determine interface bonding modification related to various surface treatment impact using TSA. Al k-α source with kinetic energy of 1486.68 eV was used. The spectrum energy resolution is 0.49 eV.

RESULTS AND DISCUSSION

<u>Electrical and EM Results:</u>

Sheet resistance (Rs) change as a function of line width with various TSA conditions is shown on Fig 1. It is clearly observed that TSA process conditions exhibit minimal Rs shift <2.5%, compared to the conventional SiH_4 CuSiN process.

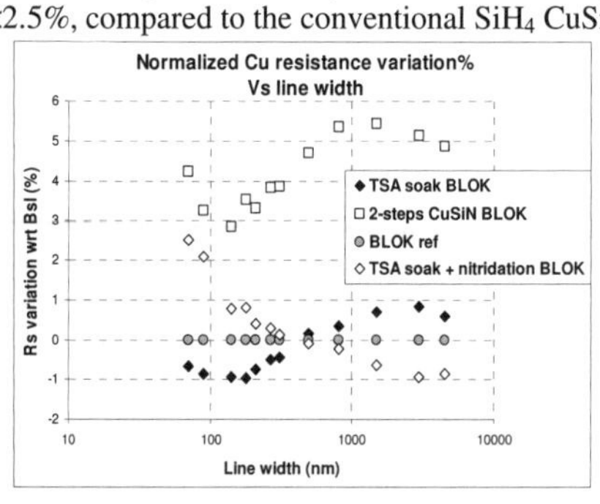

Fig. 1 Sheet Resistance shows no degradation with TSA

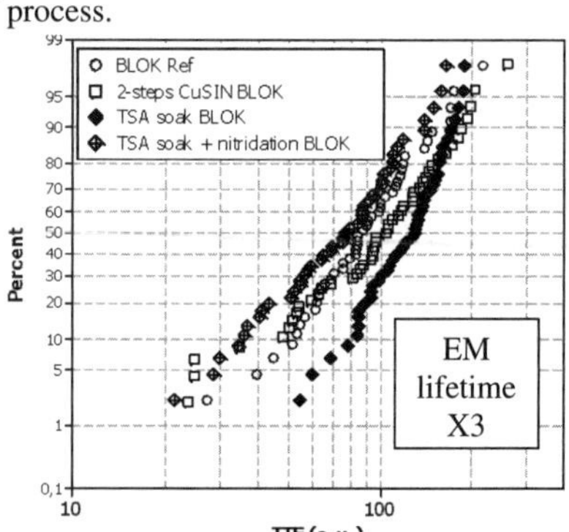

Fig. 2 EM results showing 3x improvement in product lifetime with TSA soak

Fig. 2 shows the percentage of failure accumulation as a function of lifetime testing. TSA process conditions exhibit two different behaviors regarding EM performance: when TSA soak is followed by a nitridation step, no improvement is achieved compared to the reference process. However the TSA soak only condition shows EM performance improved by a factor of 3 compared to baseline process. CuSiN with SiH4 process was tuned to have an acceptable Rs increase (5%) but the consequence is no EM improvement is observed, confirming the limitation of this previous technique at this node.

Discussion :

To understand the process fundamentals, Fig.3 presents *ab initio* modeling performed to simulate the adsorption and decomposition of TSA molecule on Cu, with a comparison to SiH_4.

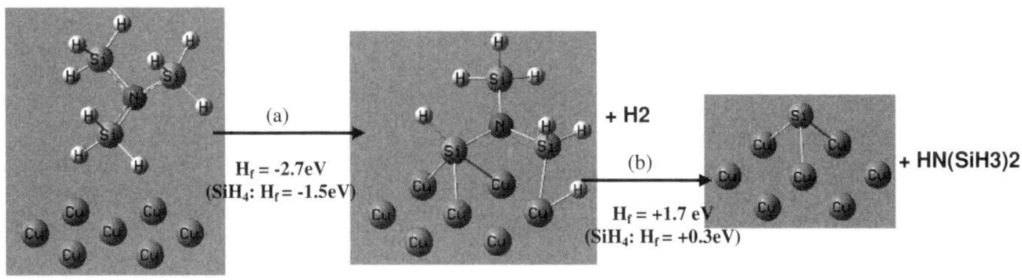

Fig. 3 Simulation results show TSA easily adsorbs on Cu (a) and with the Si-N bond difficult to break (b)

The proposed model indicates the H split from TSA and bonding of Si to Cu is very favorable (H_f=-2.7eV), which is similar to SiH_4 soak on Cu. However the subsequent split of Si-N bond in TSA is more difficult than the subsequent split of Si-H bond in SiH_4 (+1.7eV vs 0.3eV). Thus, the diffusion barrier of Si into Cu is significantly increased upon TSA soak, therefore preventing Cu line Rs increases.

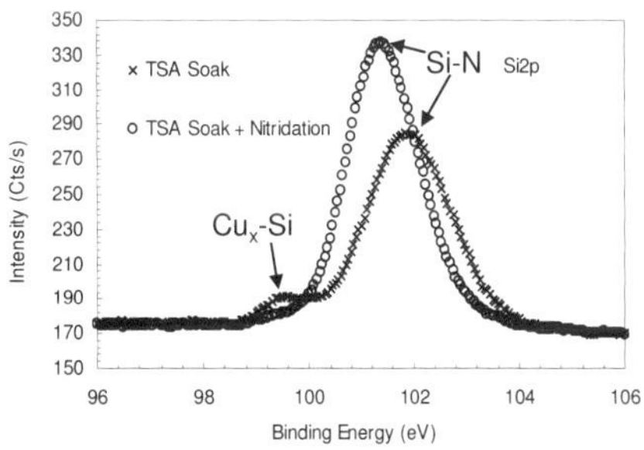

Fig. 4 XPS spectra showing Cu-Si bonding formation using a TSA soak, while no Cu-Si bonds form using a TSA soak followed by nitridation

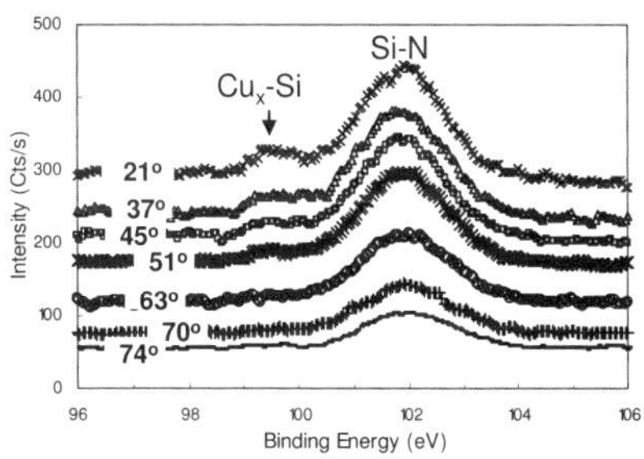

Fig. 5 Angle-resolved XPS spectra for a TSA soak-only process with various incidence angles shows Cu-Si bonding is only detected at small incidence angles

(a) (b)

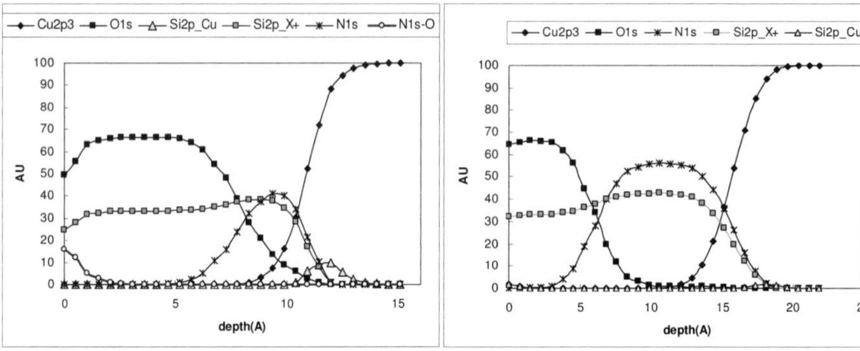

Fig. 6 Depth profiles from AR-XPS measurement on blanket Cu samples submitted to (a) TSA soak only (b) TSA soak + nitridation. For Si2p_$^{X+}$: X=3 for N; X=4 for O.

Post nitridation step was found to play a Si pinning role in previous 2-step CuSiN process with SiH_4 [5]. Fig. 4 shows the XPS spectra from processed copper blanket wafers using two different process conditions: with TSA soak only and with TSA soak followed by nitridation sted. The 99.5eV peak is related to the presence of Cu-Si bonding: this peak is observed with TSA soak only conditions, while no peak is observed at this binding energy for TSA soak followed by nitridation. AR-XPS spectra with various incidence angles are shown in Fig. 5 for a TSA soak-only condition. The Cu_xSi peaks are only observed at small angles, which confirms that the $CuSi_xN_y/CuSi_x$ layer formed at the Cu surface is very thin.

To get in-depth information, samples were then measured using Angle-resolved XPS. Spectra are shown in Fig. 6.

The copper surface of the sample treated with TSA soak only presents a very thin modified layer where Cu-Si and Si-N bonding can be identified. Due to exposure to air break, the sample presents an 8nm oxidized layer. The copper surface of the sample treated with TSA soak followed by a nitridation step doesn't exhibit any Cu-Si bonding but a thick Si-N layer. Because of this SiN layer, surface oxidation due to air break is limited to 5nm.

These results are in agreement with the EM data shown on Fig.2: copper surface modification that leads to EM improvement are observed with TSA soak but not with TSA soak followed by nitridation. The hypothesis is that such Si pinning step (NH_3 plasma) breaks the Cu-Si bonding that was formed during TSA soak, therefore converting the TSA adsorbed layer to a SiN like layer. In the TSA process approach, the post nitridation step is not required as for SiH_4 CuSiN process approach, making the process even simpler.

Fig. 7 shows for a product wafer a TEM-EELS analysis of a copper line treated with TSA soak only condition. Fig 7(b) confirms the presence of a thin modified layer on top of copper.

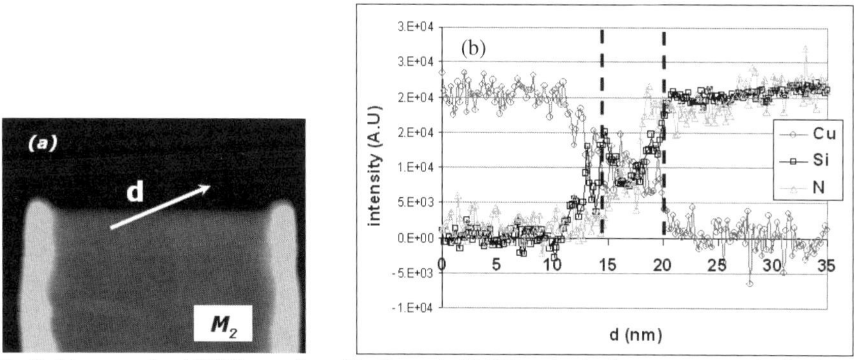

Fig. 7: (a) cross sectional TEM image of top of line2 treated with TSA soak only. d vector represents the direction of EELS collection: x axis in (b); (b) TEM EELS profile evidencing interface layer between copper and SiCN on top of a copper line treated with TSA soak only

All the analyses data (XPS, AR-XPS, TEM-EELS) are in agreement to show Si diffusion is limited to the very surface in accordance with the simulation data: when TSA is used to soak copper surface, the molecule structure itself prevents Si diffusion into Cu. Moreover, the study demonstrates that in the TSA process approach, the nitridation step is not required as for SiH_4 CuSiN process approach.

CONCLUSIONS

TSA soak process prior to SiCN barrier deposition is a new approach to enhance Cu/SiCN interface towards EM performance with minimal line resistance degradation on a stable baseline. The structure of the TSA molecule itself allows strengthening of Cu-Si bond while limiting Si diffusion in copper as observed when using SiH_4. It is thus believed that this simple approach is a very promising alternative to electrolessly deposited alloys in order to achieve Electro-migration target for 40nm node and below. Future work will address implementation of this process in manufacturing.

REFERENCES

1. D Galpin, C.Monget, E.richard, N.Posséme, Ph.Brun, M.Guillermet, C.Verovc, A.ostrovsky, Y Trouiller, M.Mellier, S.Petitdidier, V.Caubet, M.Rivoire, O.Hinsinger, G.Imbert, Y.Le Friec, L.Favennec, I.Thomas, J.Guillan, K.Haxaire, P.Caubet, P.Vannier, Ph.Normandon, S.Chhun, V.Arnal, P.Ferreira, T.Berger, C.Besset, L.Arnaud and D.Roy, in Proc. Advanced Metallization Conference 2008 (AMC2008), edited by M. Naik, R. Shaviv, T. Yoda, and K. Ueno (Mater. Res. Soc. Conf. Proc. AMCXXIV, Warrendale, PA, 2009) pp. 11-18.
2. M.H.Lin, Y.L Lin, K.P. Chang , K.C.Su, Tahui Wang Microelectronics reliability 45 (2005) 1061-1078
3. L.G. Gosset, V. Arnal, S. Chhun, N. Casanova, M. Mellier, J.-Ph. Reynard, M. Broekaart, Ph. Brun, X. Federspiel, J.-F. Guillaumond, L. Arnaud, J. Torres, in Proc. Advanced Metallization

Conference 2008 (AMC2003), edited by G.W. Ray, T. Smy, T. Ohta, and M. Tsujimura (Mater. Res. Soc. Conf. Proc. AMCXIX, Warrendale, PA, 2004) pp. 321-328.
4. JM.Girard, C. Dussarat, T.Kimura, Air liquid Patent IPC8 Class :AC23C1622FI
5. S. Chhun, L.G. Gosset, N. Casanova, D. Ney, D. Delille, C. Trouiller, M. Hopstaken, P. Chausse, M. Grégoire, B. Gautier, J.-C. Dupuy and J. Torres Microelectronic Engineering, Volume 82, Issues 3-4, December 2005, pp. 587-593

Conference Proceedings AMC XXV © 2010 Materials Research Society

Adhesion, Copper Voiding, and Debonding Kinetics of Copper / Dielectric Diffusion Barrier Films

Ryan Birringer[1], Roey Shaviv[2], Tom Mountsier[2], Jon Reid[2], Jian Zhou[2], Roy H. Geiss[3], David Read[3], and Reinhold Dauskardt*[1]

[1]Department of Materials Science and Engineering, Stanford University, Durand Building, Stanford, CA 94305-4034
[2]Novellus Systems Inc., 4000 North First Street, San Jose, CA 95134
[3]National Institute of Standards and Technology, 325 Broadway, Boulder, CO 80305

*Corresponding Author: 496 Lomita Mall, Durand Building, Office 157, Stanford, CA 94305-4034, dauskardt@stanford.edu, (650) 725 – 0679

ABSTRACT

Effects of the chemistry of electroplated copper and the composition of dielectric diffusion barriers on stress-induced voiding and adhesion are reported. The void density, as observed by scanning electron microscopy, decreased markedly with increasing Cu purity. The Cu / barrier interface fracture energy, measured using the four-point bend technique, showed that Cu / barrier adhesion improves with increased Cu purity and is highly dependent on dielectric diffusion barrier composition. For the Cu / SiN interface, increases in the activity of both oxidizing and reducing environments increase the kinetics of interface debonding. Increases in both temperature and current density through the Cu film resulted in increased fracture resistance.

INTRODUCTION

It is widely accepted that electromigration (EM) in modern Cu interconnects is closely related to adhesion and bonding between the Cu and the top capping layer and is a major reliability concern for decreasing Cu feature size in advanced technology nodes [1]. While higher purity Cu may allow for easier fill of narrow and high aspect ratio features, it has also been shown to be more susceptible to stress migration [2], and stress induced voids may serve as preexisting nuclei for EM void growth. In a previous communication we discussed a novel approach to study the effect of current density, environment, and temperature on debonding kinetics of the interface between Cu and the dielectric barrier [3]. In the present study, stress induced voiding and Cu / barrier adhesion are characterized for a variety of Cu electroplating chemistries and dielectric diffusion barrier (DDB) compositions. Additionally, the kinetics of Cu / barrier interface debonding are characterized in a variety of oxidizing and reducing environments, and the effects of temperature and current density through the film are explored.

EXPERIMENTAL DETAILS

Cu was electroplated (500 nm) onto a 50 nm physical vapor deposited Cu seed layer and then reduced down to 200 nm using chemical mechanical polishing. The Cu electroplating was performed using one of three different plating chemistries (low purity, high purity #1, or high purity #2) as discussed elsewhere [4]. A H_2 or NH_3 reducing pretreatment was performed to reduce the surface of the copper film immediately prior to deposition of one of four dielectric diffusion barrier films (SiN, N doped SiC, dual layer SiC, or O doped SiC). This produced a total of twelve different sample splits consisting of three different Cu film chemistries and four different barrier films. All samples were subjected to similar thermal histories, and X-Ray diffraction residual stress measurements confirmed the stresses in the Cu films were consistent with expectations for electroplated Cu films [5].

The well-known four-point bend thin-film adhesion technique was used to measure critical fracture energies of the Cu / barrier interfaces [6,7]. Additionally, a modified version of the standard double cantilever beam (DCB) technique was used to probe the debonding kinetics of the copper / SiN interface in the presence of varying current density through the film, environment, and temperature [6,8]. The DCB technique was used to measure debonding kinetics because it allows for the monitoring of debond growth rates as low as 10^{-11} m/s [9]. The four-point bend and DCB techniques result in a vastly different mode mixity at the crack tip (approximately 43 degrees and 0 degrees, respectively), and DCB measurements will generally result in lower measured interface fracture energies than four-point bend measurements [6,10]. Lastly, in order to clarify the fracture path and rationalize the measured adhesion values, the resulting fracture surfaces were characterized using X-Ray Photoelectron Spectroscopy.

Voids visible at the Cu / barrier interface were characterized using scanning electron microscopy (SEM). All SEM images were taken at 20kX to 60 kX magnification with an acceleration voltage of 20 to 30 kV. The areal density of voids was determined by averaging a manual count of 3 to 4 images. Voids greater than about 20 nm in diameter were counted in the images. Grain size distribution was also measured using the backscatter detector.

RESULTS AND DISCUSSION

The interface fracture energies varied widely between samples, ranging from 4 to 13 J/m^2. Figure 1 shows fracture energies for the three different Cu chemistry films capped with SiN. Cu plating chemistry has a noticeable effect on DDB adhesion. Barrier composition has a much more pronounced effect on DDB adhesion, with SiN barriers exhibiting the highest interface fracture energies in this study. X-Ray photoelectron spectroscopy was performed on the resulting fracture surfaces after testing and confirmed failure at the Cu / DDB interface in all cases.

The grain size distributions for all the Cu films were similar, widely scattered over a range of 0.1 μm to 3 μm. Figure 2 shows a plot of the areal density of Cu voids vs. the measured fracture energy of the Cu / SiN interface. There exists an inverse relationship between the areal density of voids and adhesion of the SiN film for varying

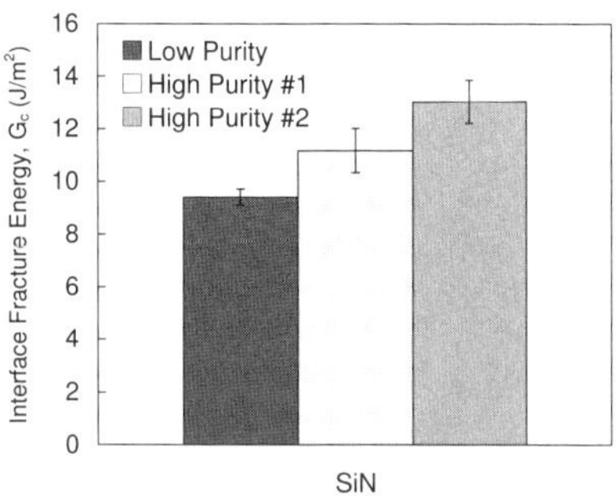

Figure 1. Interface fracture energies for three different Cu plating chemistries capped with a SiN barrier.

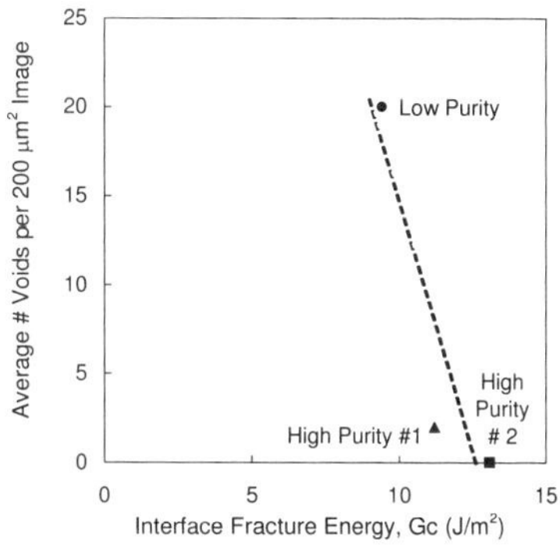

Figure 2. Average areal density of voids vs. interface fracture energy for the SiN samples.

Figure 3. Representative SEM image of Cu voiding in the SiN low purity samples.

Cu chemistries. Figures 3 and 4 are representative SEM images which show the different voiding behavior of the SiN low purity and high purity #1 samples, respectively.

The kinetics of Cu / SiN interface debonding are characterized by performing load relaxation tests and calculating the resulting v-G curves, which are plots of the velocity at which the debond is advancing versus the applied driving force for debond advancement [6]. One fact to note is that the driving forces quoted in these tests are less than the critical fracture energies of the SiN samples measured using the four-point bend technique. This is expected and can be attributed to the different mode mixity states present in the two different measurement techniques used (four-point bend versus DCB) [6,10].

Figure 4. Representative SEM image of Cu voiding in the SiN high purity #1 samples.

The kinetics of Cu / SiN interface debonding are largely influenced by environment. The rate of debond growth in a wet oxidizing environment (lab air, 30% relative humidity) is nearly three orders of magnitude larger than in a dry oxidizing environment (5 mol% O_2 in N_2), as shown in Figure 5. This striking behavior may be attributed to oxidation kinetics at the tip of the advancing debond. Surprisingly, the debond kinetics in reducing environments (H_2 in N_2) are accelerated as the activity of the reducing species is increased (not shown). The growth curves in Figure 6 show an acceleration of growth rates, likely due to increased reaction kinetics, and a large increase in debond driving force with increasing temperature. The increased G values may be due to increased plastic energy dissipation in the region surrounding the crack tip with increasing temperature. Increased current flow through the film was found to inhibit debond growth for current densities greater than 130 kA / cm^2 and was independent of polarity. We believe these subcritical debonding metrologies, together with the role of current flow, provide a rapid and promising new method to characterize the effectiveness of novel barrier materials.

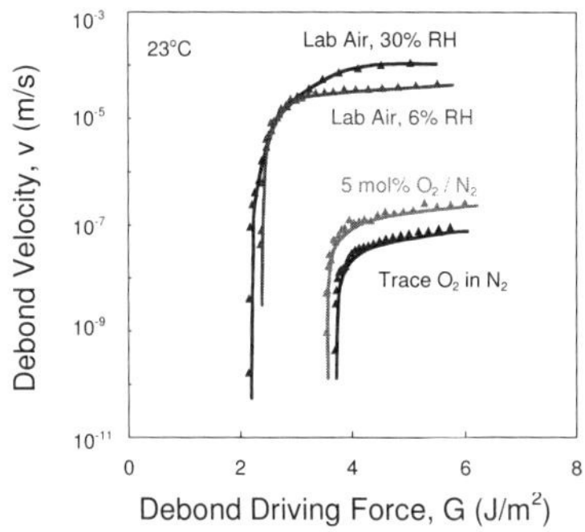

Figure 5. Subcritical debonding curves in oxidizing environments for the SiN high purity #1 samples.

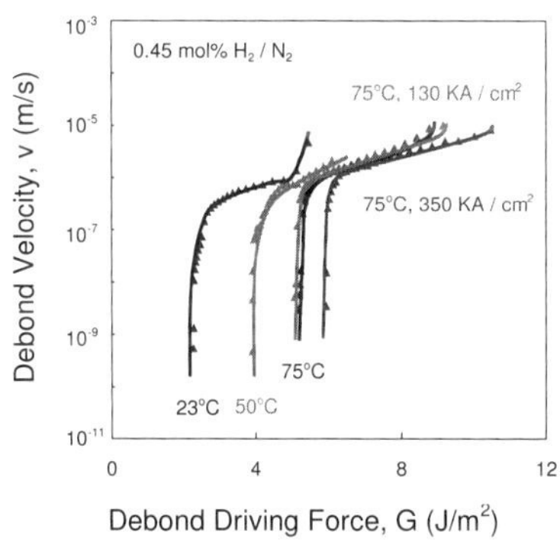

Figure 6. Subcritical debonding curves in a reducing environment showing the effects of temperature and current density through the film for the SiN high purity #1 samples.

ACKNOWLEDGMENTS

This research is supported by SRC Task 1292.052 (SRC Monitor: Dr. Scott List). This research is a partial contribution of the US National Institute of Standards and Technology. Not subject to copyright in the US.

REFERENCES

1) M. Lane, E. Linger, and J. Lloyd, J. Appl. Phys. **93**, 1417 (2003).
2) A. Uedono, T. Suzuki, T. Nakamura, T. Ohdaira, and R. Suzuki, Jpn, J. Appl. Phys. **46**, 1938 (2007).
3) R. Birringer in Advanced Metallization Conference 2008 (AMC 2008), edited by M. Naik, R. Shaviv, T. Yoda, and K. Ueno, (Mater. Res. Soc. Conf. Proc. AMCXXV, San Diego, CA, 2008) pp. 331-336.
4) M. Marshall in Advanced Metallization Conference 2009 (AMC 2009), in press, (Mater. Res. Soc. Conf. Proc. AMCXXVI, Baltimore, MD, 2009).
5) A. Volinsky, M. Hauschildt, J. Vella, N. Edwards, R. Gregory, and W. Gerberich, Mat. Res. Soc. Symp. Proc. **695**, L1.11.1 (2002).
6) R. Dauskardt, M. Lane, Q. Ma, and N. Krishna, Engng. Fract. Mech. **61**, 141 (1998).
7) R. Shaviv, S. Roham, and P. Woytowitz, Microelectron. Eng. **82**, 99 (2005).
8) M. Kanninen, Int. J. Fracture. **9**, 83 (1973).
9) E. Guyer, and R. Dauskardt, J. Mater. Res. **19**, 3139 (2004).
10) A. Evans, M. Ruhle, B. Dalgleish, P. Charalambides, Metall. Trans. A. **21**, 2419 (1990).

Conference Proceedings AMC XXV © 2010 Materials Research Society

Electrical and microstructural investigation of the Reliability of Power Devices after Electro-thermal Fatigue

D. Martineau[1], T. Mazeaud[2], M. Legros[1], Ph. Dupuy[2], J.-B. Sauveplane[3], P. Tounsi[3], C. Levade[1,4] and G. Vanderschaeve[1,4]
[1]CEMES-CNRS, 29 rue Jeanne Marvig, 31055 Toulouse, France
[2]Freescale Semiconductor, Inc., avenue du Général Eisenhower, 31023 Toulouse, France
[3]LAAS-CNRS, 7 avenue Roche, 31077 Toulouse, France
[4]Université de Toulouse, INSA, 31077 Toulouse, France

ABSTRACT

Extreme electro-thermal fatigue tests have been performed to failure on power MOSFET devices that were later observed using acoustic, electronic and ionic microscopy. At variance from the former generation of devices, fatigue-induced ageing of these components affects only the source metallization zone. Ageing is characterized both by a local increase in resistance and by an alteration of the metal structure of the wire bondings and the metallization. The increase in drain source resistance may originate from a loss of contact between the wire bondings and the Al layer but an extensive decohesion between the metal grains of the top metal is also investigated. Complete failure modes were also probed and revealed local melting of the Al and the subsequent creation of Al/Si eutectic alloys.

INTRODUCTION

Because the number of power devices integrated in automobiles keeps increasing, a good knowledge of their reliability is required. This implies to have a deeper insight in their modes of ageing, both from the electrical properties and structural evolution points of view [1]. The components investigated here were recently developed by Freescale Semiconductor to support high electrical loads [2]. Due to the many different types of power devices that enter a vehicle fabrication, and the various specifications they have to comply with, there is no current standard test procedure to assess for their ageing properties [3,4]. The procedure adopted in this study is an association of extreme cyclic electrical fatigue, finite element (FE) calculations, and multi-scale structural analysis. The tests have been performed beyond the operating limits to assess the limit of device's robustness in extreme conditions.

The studied devices gather on a single frame a MOSFET power device and a control chip which is able to monitor the electric fluctuations and to detect errors. Their purpose is to command the on/off states of light bulbs for cars. The first device under test (DUT) contains 4 areas with 15 mΩ drain-source resistance (R_{dson}) at 25°C [5] while the second one contains two 10 mΩ R_{dson} areas and two 35 mΩ R_{dson} areas at 25°C [6]. The devices operate under a tension of 12 V and a nominal current of 20 A (5 A per area).

In the former generation devices, the main failure mode was the delamination at the solder that attaches the silicon substrate at the back of the die to the copper heatsink [7]. The origin of the delamination lies in the large mismatch between the thermal expansion coefficients of copper and silicon [8]. In the new generation devices, it is shown that this failure process is no longer operative and that the fatigue-induced ageing mainly affects the active region of the MOS, that is the source metal and the wire bondings [9]. Detailed evidences of the ageing modes of this region are given and are discussed in terms of possible physical processes leading to structural changes in the metal layer and bondings.

EXPERIMENTAL PROCEDURE

The electrical bench has the capacity to test 10 coupons at a time. Each coupon contains one DUT. The whole procedure has then been established to carry out batches of 10 samples. The life time (or number of cycles to failure) of a given DUT for a given set of testing parameters is taken when 5 out of 10 switches cease to operate. Prior to electrical ageing, the 10 devices were inspected by acoustic microscopy to evaluate potential defects and initial delamination on top and beneath the power die. SAM (Scanning Acoustic Microscopy) and SAT (Scanning Acoustic Transmitted microscopy) were respectively used to inspect the mold/power die and the power die/heat sink potential delamination. X-Ray analysis was also performed to analyze voids inside the different layers. As in the preceding generation, small voids were shown to persist throughout testing without evolving, which allowed us to rule out any implication of these defects in the ageing of the DUT. After these initial non-destructive tests, batches of coupons carrying these power devices were electrically tested in two similar and thermally regulated electrical benches. Among all the different tests, the most harmful one consisted in trains of 90 A, 6 ms square electrical pulses followed by a 9 times longer pause at 90°C. The pulse intensity was high enough to activate an overload protection that turns off the MOS. Stressed devices were re-examined by X-Ray, SAM and SAT for comparison with initial component structures. Improved analyses were made on failed devices. The mold compound was removed by chemical etching, making possible the Scanning Electron Microscopy (SEM) observation of the metallization and Al wires. Energy Dispersive X-Ray (EDX) analyses were carried out to reveal a potential evolution of the composition of the different Al alloys. We mainly focused on the active MOS area: Al wires and metallization on top of the silicon substrate. Transmission Electron Microscopy (TEM) cross-sections were mechanically prepared using a tripod and diamond lapping films while final electron transparency (foil has to be thinner than 400 nm to be observed in a TEM) was attained with Ar ion bombardment. In regions hardly accessible with tripod preparation such as Al bonding and Al metallization, TEM cross-sections were made using a Focused Ion Beam (FIB) apparatus equipped with a micromanipulator. In parallel, Finite Element (FE) calculations using ANSYS® and COMSOL® Multiphysics softwares were run to simulate and map the temperature increases inside the device during electrical pulses.

RESULTS AND DISCUSSION

The most relevant electrical parameter to characterize the ageing of a power MOSFET is the drain-source resistance R_{dson}. This resistance has to remain low to prevent local heating of the switch, which in turn may accelerate its ageing. Fig. 1 shows the evolution of R_{dson} for a batch of 10 switches that underwent 842 kcycles before 5 ceased to operate. The first step consisted in monitoring the evolution of R_{dson}. As seen in Fig. 1, this parameter always increased during electrical ageing, but the rise was different between components of a same batch. In our tests, the larger augmentation of R_{dson} was 30% in functional components, but failed components showed much larger increases. More detailed electrical investigations were then carried out on the remaining 5 functional components, focusing on the active transistor region (see below, Figs. 4, 5).

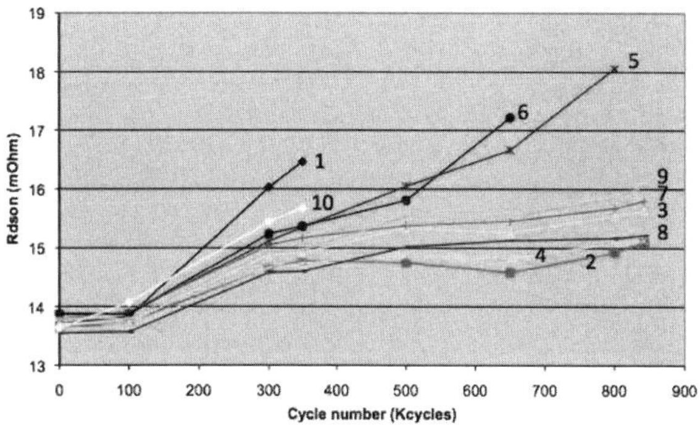

Fig. 1. Distribution of R_{dson} evolution for ten different stressed devices which have undergone cycles of 100 mΩ short circuit at 40°C. After 842.000 cycles, five components (3, 4, 7-9) were still operative but their R_{dson} increased between 11% and 16%. Components 1, 2, 5, 6, 10 failed during the test.

The second step was to compare SAM and SAT results before and after electrical tests. We investigated both the mold compound/die and leadframe delamination by SAM and the die attach delamination by SAT. As illustrated in Fig. 2, delamination was not observed in these components neither under the mold (fig. 2b) nor at the die attach (Fig. 2c). Delamination was identified as the main failure mode in devices from the former generation [7]. It arises from the very different coefficient of expansion between the Silicon of the die and the Copper heat sink it is attached to. The higher the temperature rises in the switch, the higher the stresses leading to delamination go.

This result may underline the better ability of devices from the new generation to maintain a constant evacuation of heat between the silicon and the copper heat sink. To gain further insight on that point, FE simulations were run using the same parameters as in the actual test. Fig. 3 represents partial results obtained with ANSYS® simulations.

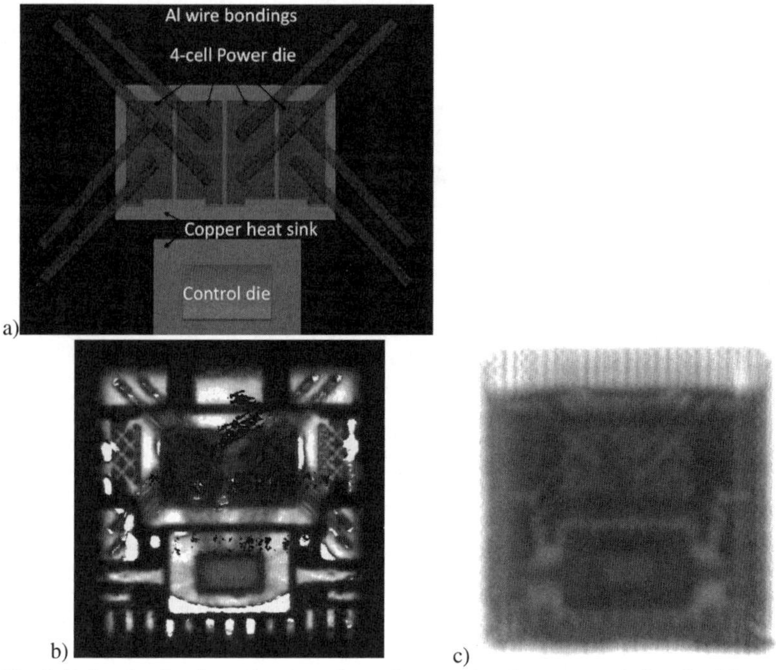

Fig. 2. Delamination investigations through acoustic microscopy. a) sketch of the DUT. b) SAM images showing little delamination under the mold (white). c) SAT images indicate no delamination at the power and control die attach.

In Fig. 3a, the initial mesh used for the simulation can be seen. Finer mesh has been used in interfacial zones while low current areas (control die and connectors, on the left) supported larger grids. Fig. 3b displays the temperature map of the device when a 90 A current pulse is injected in the front cell after 5.7 milliseconds. The Al metallization is taken to very high temperatures, even far away from the wires. The Al wires are affected very locally. The gradient between the bottom of the Si die and the copper heat sink is limited to less than 30 °C, which is less than in the previous generation devices [7-10]. The inactive cell in the back is very slightly affected. These simulations explain why the delamination between the heat sink and the die is lower in these new generation power switches. On the other side, it points out the strong heat gradients seen by the metallization and stresses the need of further analysis of this area.

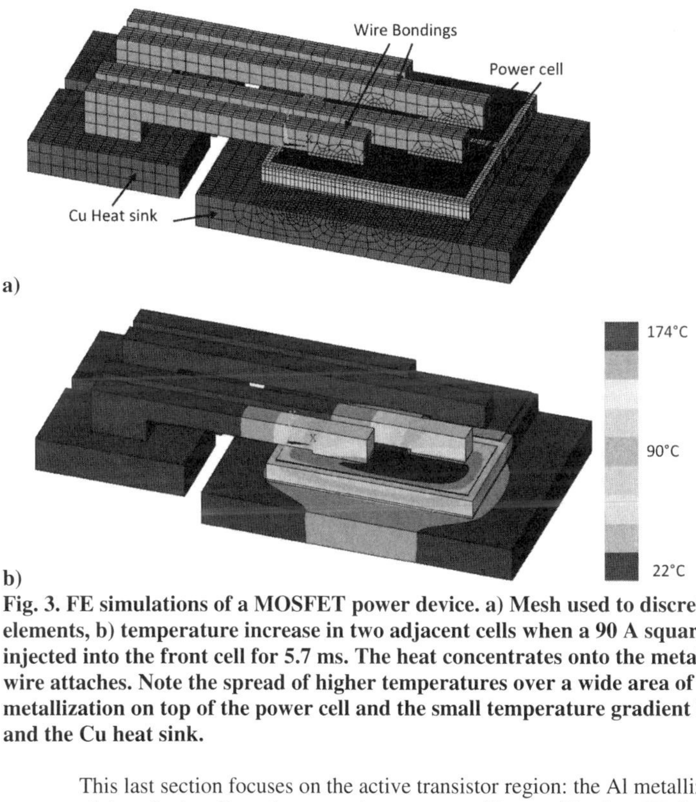

Fig. 3. FE simulations of a MOSFET power device. a) Mesh used to discretize the different elements, b) temperature increase in two adjacent cells when a 90 A square pulse current is injected into the front cell for 5.7 ms. The heat concentrates onto the metallization and at the wire attaches. Note the spread of higher temperatures over a wide area of the Al metallization on top of the power cell and the small temperature gradient between the Si die and the Cu heat sink.

This last section focuses on the active transistor region: the Al metallization of the source and the wire bondings that carry the current to this metallization. We indeed found that most of the degradation occurred in this specific site of the device. Local electrical measurements also showed a 30% increase of R_{dson} between wire bonds and Al metallization in one aged device. It further confirms that the overall R_{dson} augmentation probably roots in the upper metal degradation/wire attach. Fig. 4 regroups SEM images of the wire bondings and metallization. Close observations of the wire bondings reveal that cracks were present across the wire section (Fig. 4c). Most of these cracks span over a small section of the wire and are unlikely to cause such an increase of resistance in the wire

Other cracks were however observed below the wire, at the interface with the Al layer (Fig. 4b). It is difficult to estimate how far these cracks go in plane view SEM. If they crawl underneath the wire from one side to the other, they could lead to a complete detachment of the wire from the metallization. Even incomplete, such a delamination could result in a substantial increase of the resistance between the wire and the active cells (Fig. 4b). So far, we have not found a way to image these cracks. Comparisons between electrical measurements and undergoing simulations may help to evaluate this effect.

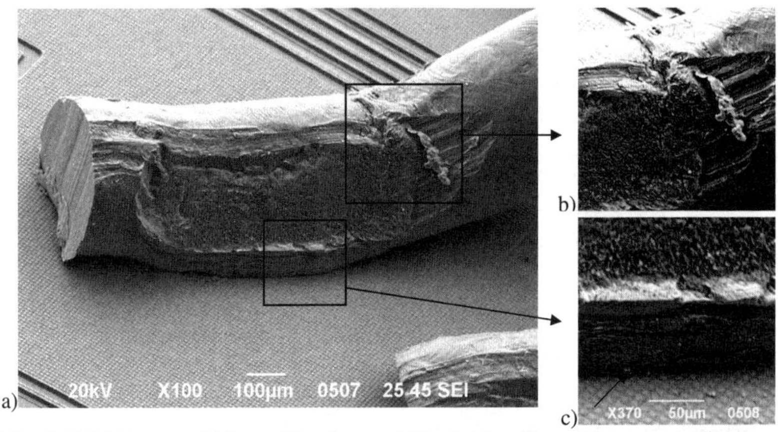

Fig. 4. SEM images of Al metallization and Al wire bonding. a) Low magnification of an Al wire bonding. In (b) black arrow indicates mold debris present in cracks parallel to the interface (those can be the result of a local heating or of the post mortem chemical etching of the mold). c) Al wire partially cracked along its section.

SEM observations made away from the wire bondings clearly point to the degradation undergone by the metallization. Fig. 5 represents the surface of the device before testing (a) and after 250 kcycles at 85°C with an 85 A square electrical pulse (b). Fig. 5c is a close-up of a degraded surface where inter-granular cracks are easily seen (arrows).

Square structures in Fig. 5a correspond to transistor cells. The metallization underwent a complete degradation (seen in Fig.5b), characterized by many inter-granular cracks in the stressed metallization (Fig. 5c). This signifies that, as suggested by the FE simulations, the whole metallization area undergoes a local heat gradient between the Si die and Al metallization. This gradient is at the origin of inter-granular decohesion. The cracks between the Al grains are probably formed by stress and temperature-induced accelerated diffusion at the grain boundaries. Temperature cycles generate, through the thermal expansion coefficients difference between Si and Al, some tension compression sequences that are sufficient to activate the plastic deformation of the metal layer [11]. However, at intermediate temperatures, dislocations cannot accumulate in the Aluminum layer because of their evacuation at the free surface and at the Al/ oxide interfaces [12-14]. On the contrary, grain boundaries perpendicular to the film/substrate interface can carry Al in and out of the film to accommodate the stress [15]. Such a diffusive mechanism may not be completely reversible in Aluminum as a native oxide forms very fast at exposed surfaces. The grooving initiated at the grain boundary/free surface intersection may therefore increase as the thermal cycles go on and lead to a decohesion of the grains. A severe loss of contact between Al grains, which is not complete here, could again induce an increase of the total source resistance, but this effect has not been validated by local electrical measurements.

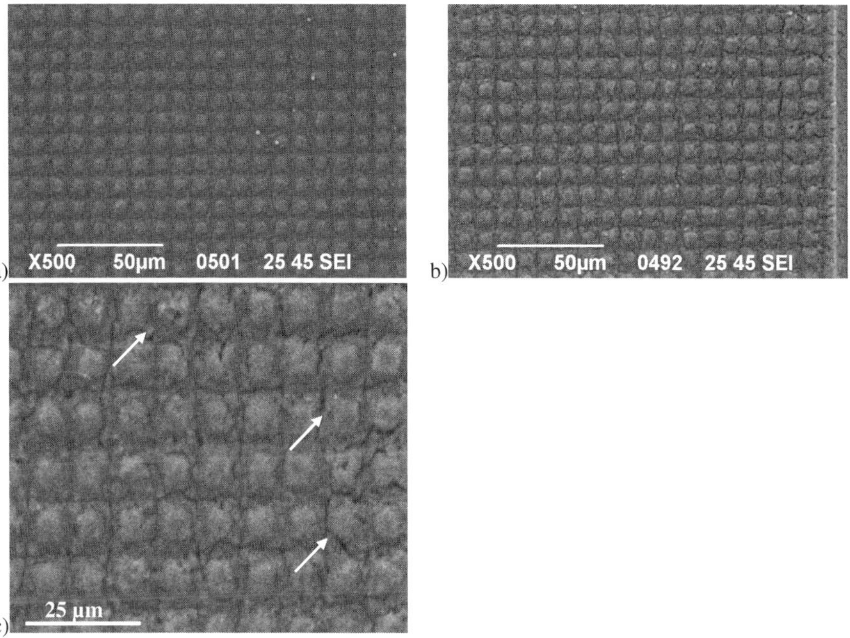

Fig. 5. SEM images of the Al metallization before (a) and after (b) tests. In (a), the Al surface is smooth and individual grains cannot be detected. The destructured metallization in (b) corresponds to a stressed device after 250 kcycles (85 A square electrical pulse) at 85°C. This component was still operative but its R_{dson} increased about 30%. c) is a close-up of image (b) ; white arrows show some inter-granular cracks in the metallization layer.

We also investigated components that were no longer operative after electrical tests. Some of them burned, probably because the resistance drastically increased in a very small spot, thus leading to an abrupt temperature upsurge. Regular TEM preparation was not possible in case of burned out Al wires so FIB lamellae had to be extracted from the damaged region and then analyzed by means of EDX and TEM (Fig. 6). Before test (Fig. 6a), Al wires consist of randomly oriented Al grains, separated by boundaries that eventually contain Si precipitates. In the burned wire (Fig.6b), the metal stands in a characteristic eutectic structure where all the Al grains have the same orientation (white patches) that are separated by Al/Si eutectic and probably Si precipitates (grey inter-granular zones and dark dots, respectively). The richer Si content of the burned wire was revealed and quantified at 10% of Si in Al wire by EDX. Formation of such structure reveals both a migration of Si from the active area of the MOS to the Al wire and a complete reorganization of Al grains in the wire. This structure also indicates that a local melting of the Al metallization occurred in this region (temperature larger than 660°C). Further according to Si-Al phase diagrams, it is necessary to reach a temperature of 577°C to form an eutectic compound (12,2% Si-Al) [16].

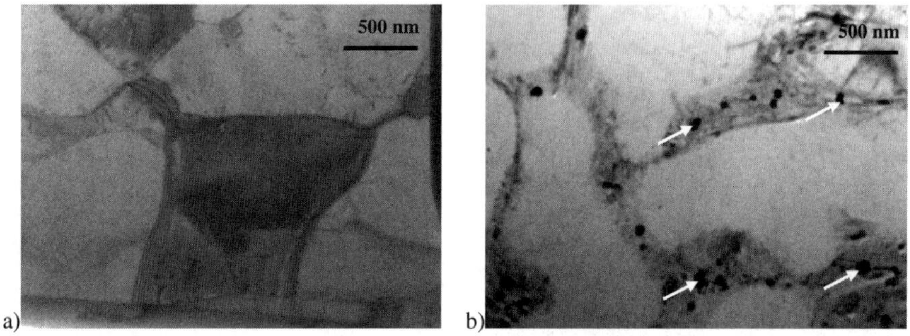

Fig. 6. TEM images of Al wires. Al grains with various crystalline orientations are observed in an untested Al wire (a). In a burned out Al wire (b), a single crystallographic orientation is found for all Al grains and sub-grains. White arrows in (b) indicate probably Si grains at the eutectic Al/Si regions (grey zones) between Al sub-grains (white zones). This microstructure is typical of a high temperature eutectic alloy formation between Si and Al.

Local micro-structural changes are helped both by the rise of temperature caused by an increased resistance and by the associated stresses. This resistance increase could be itself caused by the decohesion of the Al metallization or some cracks appearing between the wires and the underlying Al layer.

ACKNOWLEDGMENTS

The authors would like to thank the Regional Council of Région Midi-Pyrénées for its financial support.

REFERENCES

1. Muller-Fiedler R., Knoblauch V., Reliability aspects of microsensors and micromechatronic actuators for automotive applications. Microelectronics Reliability, 2003. 43(7), 1085.
2. Khong B., et al., Innovative Methodology for Predictive Reliability of Intelligent Power Devices Using Extreme Electro-thermal Fatigue. Microelectronics and Reliability, 2005. 45(9-11), 1717.
3. Lefebvre S., Khatir Z., Saint-Eve F., Experimental Behavior of Single-Chip IGBT and COOLMOS Devices Under Repetitive Short-Circuit Conditions. IEE Transactions on Electron Devices, 2005. 52, 276.

4. Belaid MA., et al., Reliability study of power RF LDMOS device under thermal stress. Microelectronics Journal, 2007. 38(2), 164.
5. Freescale Data Sheet MC15XS3400: http://www.freescale.com/webapp/sps/site/prod_summary.jsp?code=MC15XS3400&fsrch=1
6. Freescale Data Sheet MC10XS3435: http://www.freescale.com/webapp/sps/site/prod_summary.jsp?code=MC10XS3435&webpageId=M98459&nodeId=01435979968459&fromPage=tax
7. Khong B., et al., Alterations induced in the structure of intelligent power devices by extreme electro-thermal fatigue. Physica Status Solidi(c), 2007. 4, 2997.
8. Ciappa M., Selected failure mechanisms of modern power modules. Microelectronics Reliability, 2002. 42, 653.
9. Arab M., Lefebvre S., Khatir Z., Bontemps S., Experimental investigations of Trench Field Stop IGBT under repetitive short-circuits operations. Power Electronics Specialists Conference 2008. 4355.
10. Sauveplane J.B., et al., Smart 3-D Finite-Element Modeling for the Design of Ultra-Low On-Resistance MOSFET. IEEE Transactions on Advanced Packaging, 2007. 30, 789.
11. Nix W. D., Mechanical properties of thin films. Metallurgical and Materials Transactions A, 1989. 20(11), 2217.
12. Gao H., et al., Crack-like grain-boundary diffusion wedges in thin metal films. Acta Materialia, 1999. 47(10), 2865.
13. Legros M., et al., Microstructural evolution in passivated Al films on Si substrates during thermal cycling. Acta Materialia, 2002. 50, 3435
14. Legros M., et al., Plasticity-related phenomena in metallic films on substrates. Multiscale Phenomena in Materials Experiments and Modeling Related to Mechanical Behavior, 2003, San Francisco.
15. Legros M., Cabie M., Gianola D.S., In situ deformation of thin films on substrates. Microscopy Research and Technique, 2009. 72(3), 270.
16. Murray J.L. and McAlister A.J., Bull. Alloy Phase Diagrams, 5(1), 74-84 (1984)

Conference Proceedings AMC XXV © 2010 Materials Research Society

Impact of Impedance Mismatch of On-Die Interconnects and Logic Cells on Device Reliability and Functionality

Pavel Livshits[1], Moshe Gurfinkel[2], Alexander Rysin[2, 3], Sergey Sofer[3], Yoram Shapira[2] and Yefim Fefer[3]

[1] School of Engineering, Bar Ilan University, Ramat Gan 52900, Israel
[2] School of Electrical Engineering, Tel Aviv University, Tel Aviv 66978, Israel
[3] Freescale Semiconductor Israel Ltd., Herzlia 46120, Israel

ABSTRACT

The impact of impedance mismatch of on-die transmission lines and CMOS logic cells upon device reliability was studied. The experimentally obtained results from a test chip performed on a 45 nm CMOS technology process were compared with SPICE simulations. The results reveal that the impedance mismatch manifested by voltage overshooting and slowed down voltage transition may accelerate the device aging as well as lead to increased power consumption and logic faults.

INTRODUCTION

The system-on-a-chip (SoC) design has been making huge strides forward in recent years. New technology advents have required increasing on-die data transmission distance and decreasing signal rise-times [1]. The latter became comparable with the times for signals to propagate from one end to another of the long on-die lines called global interconnects. Therefore, these lines cannot be treated as lumped, but as distributed lines. Today, global signaling is mostly facilitated through the insertion of active repeaters aimed at splitting the distributed lines into a number of small lumped lines. However, active repeaters introduce significant propagation delays, cause time uncertainty (jitter), consume power, waste scarce on-die routing resources and introduce noises into the system. These drawbacks are a driving force to study alternatives. On-die transmission lines are emerging as an alternative to long interconnects split by large numbers of repeaters. However, for maximum accuracy signal transmission, the driver (a CMOS logic cell) and the driven line impedances have to be matched. The impact of the impedance mismatch, for a given case, upon device reliability and functionality has not been discussed. This work is aimed at elucidating these hitherto untouched issues. Our study was performed on a 45 nm test chip utilizing on-die transmission lines. Waveforms, obtained experimentally using micro-probing techniques were compared with SPICE-like simulation results. Our results reveal that the impedance mismatch may indirectly affect wearout mechanisms, such as HCI (hot carrier injection), TDDB (time-dependent dielectric breakdown) and NBTI (negative bias temperature instability) and consequently accelerate the device aging by tens of percents. In addition, it may increase power consumption and cause logic faults and soft errors.

RESULTS AND DISCUSSION

The line driver output impedance Z_{Dr} (conventionally denoted as driver strength), in first approximation, depends on the transistor parameters and naturally may be higher, lower, or equal to the transmission line characteristic impedance Z_C (i.e., underdriven, overdriven, or matched case, respectively). The input impedance of the next stage CMOS logic cell (far end of the line) is very high (high resistance and low capacitance MOSFET gate). Hence, each wave that reaches the far-end will be fully reflected (i.e., the reflection coefficient equals unity). Reflection from the near-end occurs when Z_{Dr} does not match Z_C. The reflection coefficient is positive (negative) in the underdriven (overdriven) case. The initial voltage wave is based on a voltage division between Z_{Dr} and Z_C and is below (above) a half a supply voltage in underdriven (overdriven) case. Figure 1 shows SPICE simulated signal waveforms at the far-end of an on-die transmission line having dimensions close to typical on-die interconnects.

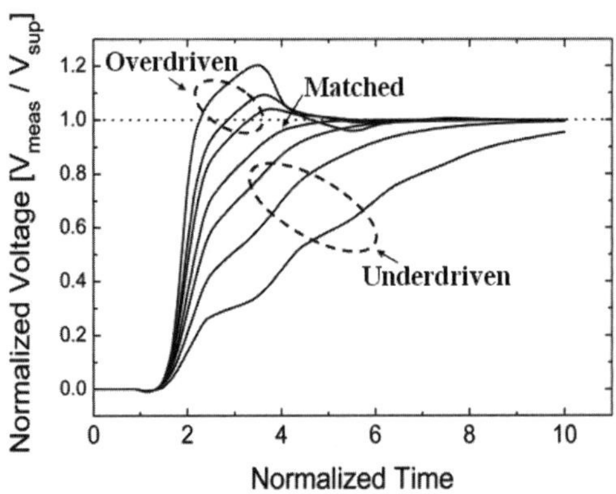

Figure 1. Simulated signal waveforms (45 nm CMOS technology) at the far-end of a typical on-die transmission line. The different curves correspond to different driver strengths.

The 3 cases may be clearly distinguished. Overdriving the line causes overshooting and ringing back. The overshoot amplitude strongly depends on the extent of the overdrive and may attain 20% of a supply voltage. The typical for overdrive overshoot peak in Figure 1 is caused by constructive interference of the incident (initial) voltage wave and the totally reflected wave at the far end. At the near end the reflection coefficient is negative, which results in phase inversion of the reflected waves. This explains the ringing back (the voltage, in the time domain, is the sum of all the incident and reflected waves). The voltage fluctuations quite rapidly decays due in large part to high active losses (Ohmic losses) in the line and also to partial absorption of the energy by the driver (the reflection coefficient is <1 at the near end). The voltage finally levels off at the nominal value.

The underdriven case is distinguished by a low voltage transient time, which increases with a further decrease in a driver strength. The signal becomes "undulating" due to slope changes resulting from bouncing of waves with different amplitudes back and forth along the line. Each slope change stems from the next reflection. The higher the impedance mismatch, the lower is the initial voltage, the higher is the energy absorption by a driver and the higher is the number of slope changes, consequently leading to the slower signal rise times).

The qualitative comparison with experimentally obtained signal waveforms (using micro-probing techniques) at the far end of a 4 mm on-die transmission line (presented in Figure 2) confirms the trends observed in SPICE simulated results i.e., three aforementioned cases are plainly discerned. The small differences in signal waveforms pertain to numerous small line impedance discontinuities, line length and extent of Ohmic losses, which cannot be accurately enough taken into account due to modeling limitations and process variations.

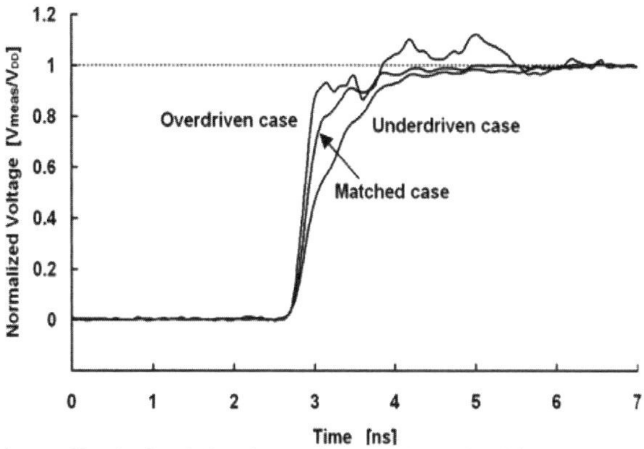

Figure 2. Experimentally obtained signal waveforms (45 nm CMOS technology) at the far-end of a 4 mm typical on-die transmission line. The different curves correspond to different driver strengths.

Reliability issues

The voltage overshoots (observed in an overdriven case) at the transistors gate aggravate the gate oxide wearout. Even though immediate damage is unlikely, the continuous voltage oscillations may lead to accumulated device degradation. A number of reliability models and simulation methodologies, such as the Berkeley Reliability Tools (BERT) [2], have been developed during the past decade for the modeling of device wearout. They are based on the physical failure mechanisms and include the major device wearout models for HCI, NBTI and TDDB. In our work we use an empirical function called Age from a BERT simulator in order to qualitatively asses the extent of impedance mismatch impact upon device degradation. The Age parameter is calculated for each failure mechanism separately using a physics-based model. The device degradation may be estimated as a function of device aging, but this left beyond the scope of the work and we discuss device aging only. In the general case, the Age parameter is calculated according to:

$$Age = C\int_0^t \frac{dt}{lifetime(V,E)} \qquad (1)$$

Where *V*, *E*, and *C* are voltage, electrical field and constant, respectively. NBTI, HCI and TDDB have an exponential or a power law relation to the oxide electric field, drain, and gate voltages [3-4]:

$$lifetime_{NBTI} \propto e^{-\beta \cdot E_{OX}} \tag{2}$$

$$lifetime_{HCI} \propto e^{\frac{\beta}{V_D}} \tag{3}$$

$$lifetime_{TDDB} \propto V_G^{-n} \tag{4}$$

Where n and β are experimentally obtained constants, V_G, V_D and E_{OX} are gate voltage, drain voltage and electric field, respectively. As a result, even small fluctuations in the gate voltage give rise to large changes in the device lifetime. In our work we analyze the TDDB wearout mechanism only. The aging simulation results shown in Figure 3 indicate a substantial additional degradation due to TDDB. Due to the exponential relationship between the gate voltage and the TDDB lifetime, the large voltage overshoots are responsible for the faster oxide degradation in the overdriven case. Similar results are expected for any other mechanism that is super-linearly dependent on the electric field or the drain/gate voltages.

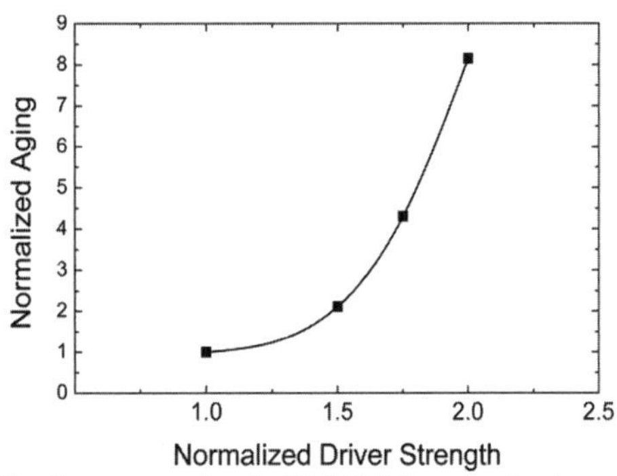

Figure 3. Simulated "Age" parameter for the TDDB failure mechanism as a function of normalized driver strength. Driver strength matched to the transmission line impedance is aligned to 1 to accentuate the acceleration of the device degradation due to overdrive.

The underdriven case is intuitively considered safe contrary to the overdriven case. This is definitely true for NBTI and TDDB failure mechanisms. However, the HCI mechanism strongly depends on the time when both NMOS and PMOS are open and in a pinch-off mode (also known as a velocity saturation mode). The impact ionization events occurring due to hot carriers during this time give rise to an impact ionization current, and therefore this time is called a device stress time. The device aging due to HCI is described by Eq. 5, where I_D, I_{ii} and W refer to drain current, impact ionization current and transistor width, respectively while H and m are determined experimentally for a given process technology [2].

$$Age = C\int_0^t \frac{I_D}{H \cdot W}\left[\frac{I_{ii}}{I_D}\right]^m dt \qquad (5)$$

Slower slew-rate at the CMOS inverter input (i.e., stronger underdrive of a previous stage) substantially increases the device stress time as shown in Figure 4. Consequently, our mathematical calculations of Age parameter (equation 5) show that the device life time becomes shorter by 4%, 9%, 12% and 18% for cases where driver is 1.5, 2, 3, and 5 times weaker with regard to a matched case, respectively.

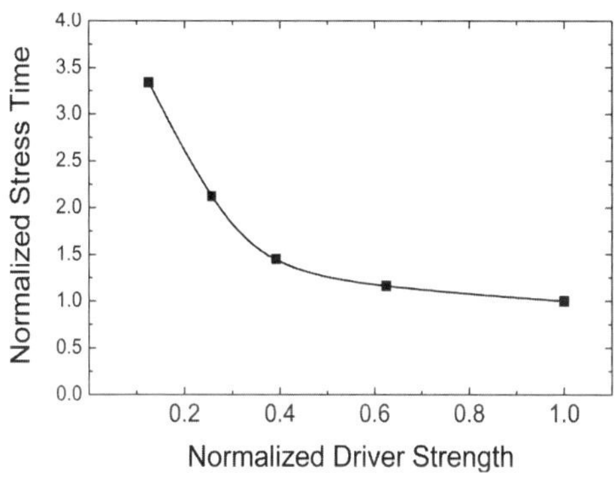

Figure 4. The SPICE calculated stress time (i.e., the time when both PMOS and NMOS are conductive and in a pinch-off mode) as a function of driver strength. The time was normalized so that it equals to 1 for an impedance matched case. The driver strength was normalized to matched driver.

Apart from device aging acceleration, the impedance mismatch, in both cases, may lead to excessive power consumption. Figure 5 compares the power consumption (calculated using a SPICE-like industrial simulator) for overdriven and underdriven cases with regard to a matched case.

In addition, voltage overshoots cause excessive cross-talk noise in neighboring lines and power supply noise, transmitted to other active elements, utilizing the same power distribution network. This may lead to functional failure of latch devices. In the underdriven case the longer transition of the signal voltage during logic cell switching process leads to signal jitter and even multiple switching.

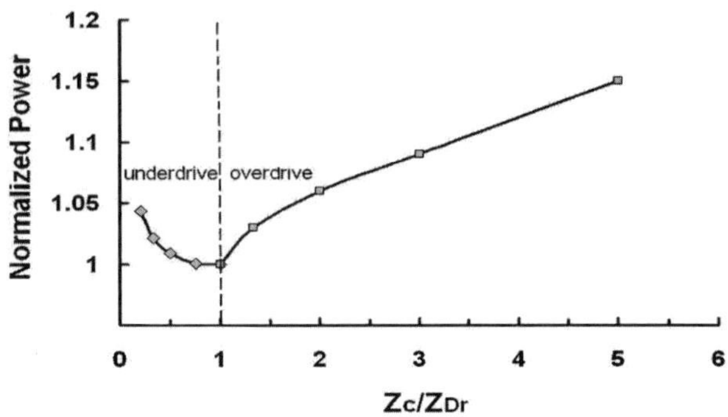

Figure 5. The power consumption in an overdriven case as a function of the ratio between line characteristic impedance and driver output impedance.

CONCLUSIONS

The on-die impedance matching may play an essential role in on-die circuit performance and reliability. It has been shown that the device aging of receiver transistors due to wearout mechanisms, such as TDDB and HCI may be considerably accelerated as a result of even a small impedance mismatch between driver output impedance and characteristic impedance of driven line. In addition, it has been demonstrated that impedance mismatch leads to power waste and the mismatch-induced functionality problems may lead to system failure. We conclude that the impedance matching has to be resolved in order to utilize transmission lines for on-die applications.

ACKNOWLEDGEMENT

The authors would like to thank Gadi Pinkovich from Freescale Semiconductor Israel for his assistance and support in devices' preparation for test.

REFERENCES

1. H. Johnson and M. Graham, "High-Speed Digital Design: A Handbook of Black Magic," Prentice Hall PTR, 2004.
2. R.H. Tu, E. Rosenbaum, W.Y. Chan, C.C. Li, E. Minami, K. Quader, P.K. Ko, C. Hu, "Berkeley reliability tools-BERT," IEEE Transactions on Computer-aided Design of Integrated Circuits and Systems, vol. 12, pp. 1524-1534, 1993.
3. S. Mahapatra, P. Bharath Kumar, M. A. Alam, "Investigation and Modeling of Interface and Bulk Trap Generation During Negative Bias Temperature Instability of p-MOSFETs," Transactions on Electron Devices, vol. 51, pp. 1371-1379, 2004.
4. E. Takeda, N. Suzuki, "An Emprical Model for Device Degradation Due to Hot Carrier Injection," IEEE Electron Device Letters, vol. 4, pp. 111-113, 1983

3D Integration

Conference Proceedings AMC XXV © 2010 Materials Research Society

CMOS Compatible Insulated Through Silicon Vias for 3D Silicon

A. K. Stamper[4], M. Shapiro[1], R. Herrin[4], V. Carlos[4], E. White[4], M. Brigham[4], B. Porth[4], J. Levy[4], D. Dang[4], A. Gupta[4], M. Interrante[2], P. Andry[3], B. Dang[3], C. Tsang[3], R. Liptak[2], J. Griffith[2], E. Sprogis[4], L. Guerin[5], V. Truong[5], D. Berger[2] and J. Knickerbocker[3]

[1]IBM Microelectronics, 1711 Burnet Rd., Austin, TX 78758 USA
[2]IBM Microelectronics, 2020 Rte. 52, Hopewell Junction, NY 12533 USA
[3]IBM Research, 1101 Kitchawan Rd., Yorktown Heights, NY 10598 USA
[4]IBM Microelectronics, 1000 River St., Essex Junction, VT 05452 USA
[5]IBM Canada, 23 Boulevard De L'aeroport, Bromont, QC J2L 1A3 CANADA
phone: 802.769.8466 email: astamper@us.ibm.com

ABSTRACT

This paper discusses the early technology development of a 90nm generation 3D chip stacking silicon carrier technology. The silicon carrier is an integrated CMOS wafer with insulated through silicon vias (TSV's), Pb-free solder bumps on the back of the mechanically thinned wafer under the TSV's, up to 7 levels of on-chip redistribution wiring, and one or more chips flip chip mounted to its surface. This paper focuses primarily on the silicon carrier wafer and not the flip chip packaging processes. The insulated TSV's are used to connect the front side of the wafer including the mounted flip chips to Pb-free solder bumps on the thinned wafer backside. To minimize TSV resistance and inductance and to maximize yield and reliability, the TSV's are placed in redundant arrays. The TSV process is compatible with standard 90nm high performance CMOS processing. CVD W with a CVD TiN liner is employed for the damascene TSV metallization and the TSV's are insulated from the substrate using a high-temperature (i.e. > 800C) thermal oxide. TSV arrays produced with this technology had excellent yield and voltage breakdown characteristics, did not change the CMOS FET or deep trench capacitor electrical parameters, and did not significantly degrade the overall defect density.

INTRODUCTION

As semiconductor performance, power, and cost improvement through device scaling has become more difficult, 3D chip stacking technology has become an attractive option [1-5]. The advantages of 3D technology include increased bandwidth, lower power, improved form factor, and lower cost [6]. The silicon carrier used in this 3D technology not only employs insulated TSV's to connect flip chips to the underlying substrate but is also contains up to 7 levels of signal redistribution wiring, high capacitance density decoupling capacitors, and digital CMOS logic circuits.

The development of IBM's 200mm wafer 0.35μm generation grounded TSV technologies laid the groundwork for developing insulated TSV's [7]. The key integration challenges of grounded TSV processing included deep silicon etch, wafer edge etch profile, TSV metallization, wafer bending/warpage, and reliability. The insulated TSV technology employs similar or identical processes as the grounded TSV technology including the deep silicon etch process with an enlarged TSV top opening and CVD TiN/W damascene metallization. The additional challenges of insulated TSV's involve electrically isolating them from the silicon substrate and integrating them into the base-CMOS without impacting active and passive device parameters.

IBM qualified high performance digital CMOS and low power RF CMOS 90nm generation technologies in 2003-4 and quickly ramped them into high volume 300mm manufacturing [8-10]. These 90nm generation silicon wafer technologies were transferred to 200mm and qualified for manufacturing in 2006-7 [11]. The 90nm generation 200mm wafer insulated TSV silicon carrier technology discussed in this paper uses the digital features of IBM's high performance 90nm digital CMOS combined with the analog wiring features of IBM's 90nm RF CMOS technology.

EXPERIMENT

IBM's 90nm high performance digital CMOS technology employs three gate oxide thickness MOSFET's with nominal and low threshold voltage, high capacitance density deep trench capacitors, and a low-k SiCOH intermetal dielectric BEOL wiring hierarchy using 1X, 2X, and 4X wiring pitches. IBM's low power 90nm RF CMOS technology, in addition to using low off current MOSFET's and RF passive devices, employs analog Cu and AlCu (i.e. super thick) wiring in the BEOL to improve RF performance [11]. The 90nm silicon carrier technology being discussed in this paper uses the high performance digital CMOS technology as a base and adds the analog wiring used in the low power RF CMOS technology. This fusion of the high performance digital and low power RF CMOS technologies gives optimal performance for silicon carriers, which require the high capacitance density (~40fF/μm^2) of deep trench capacitors, the high performance of low threshold voltage thin gate oxide transistors, the low parasitic capacitance of the tight pitch BEOL wires, and the low sheet resistance of the relaxed pitch super thick analog wiring. Table 1 summarizes the 90nm silicon carrier technology BEOL.

The 100μm deep TSV's are insulated with a high temperature (>800C) thermal oxide and are metalized with a CVD TiN/Tungsten process similar to the one employed for IBM's grounded TSV technologies [7]. A furnace grown thermal oxide is used for the TSV's instead of a lower temperature CVD oxide to improve the dielectric integrity between the TSV and the silicon substrate. After the wafer front side processing is complete, the wafers are attached to a temporary handle wafer, backside ground to expose the TSV's, solder bumps are formed on the exposed TSV's on the wafer backside, chips are diced and mounted to a laminate package, and one or more chips are flip chip mounted to the active silicon carrier front side. Figure 1 shows a simplified cross-sectional drawing and SEM of the laminate package, the silicon carrier chip with solder bumps on the backside, and flip chips mounted to the surface of the silicon carrier.

The silicon carrier technology discussed in this paper employed ~9cm^2 chips (~ 3cm by 3cm) . Large chips are typically used for 3D chip stacking, to maximize performance and minimize cost. The TSV array and thinned wafer backside solder bump pitch is 185μm, which translates into about 25,000 backside solder bumps per chip. The TSV bars are placed in parallel ~ 60μm by 60μm arrays, with seven TSV's per array.

DISCUSSION

The insulated TSVs are bar shaped and are placed in farms of seven TSV bars to minimize inductance and resistance; and to enhance reliability by provide redundant conduction paths (Figures 1-3). The bars are ~5μm wide at the top, ~3μm wide at the bottom, ~60μm long and 100μm deep; and the surrounding oxide insulator is ~1μm thick. The insulated TSV farm resistance has been measured for TSV's with direct contact to M1 at ~4 mΩ, or about 25 mΩ per single insulated TSV bar; and the other electrical properties were simulated using Ansoft [6], as shown in Table 2. The insulated TSV farm inductance and resistance values are comparable to similarly sized grounded TSV farms [7]. As mentioned above, the placement of TSV's in parallel arrays not only reduces the resistance but also reduces the inductance. TSV arrays composed of ~60μm long, ~3μm wide, ~150μm deep TiN/W filled TSV's grounded to the silicon substrate had measured inductances at 1GHz of 27, 15, and 5 pH for 1, 3, and 6 parallel TSV's, respectively [7].

Wiring level	Pitch	Thickness	Dielectric
1X M1 Cu	240nm	190nm	SiCOH (Keff=3)
1X Mx Cu	280nm	250nm	SiCOH (Keff=3)
2X Cu	560nm	500nm	SiCOH (Keff=3) or FSG (Keff = 3.6)
Analog Cu	2.4μm	3μm	FSG (Keff=3.6)
Analog AlCu	4.8μm	4μm	SiO_2 (Keff=4)

Table 1: 90nm generation Silicon Carrier BEOL wiring overview.

Parameter	Value
Resistance	25 mΩ
Inductance	21 pH
Capacitance	27 fF
Characteristic impedance	28 Ω

Table 2: Electrical properties of a single insulated TSV [6].

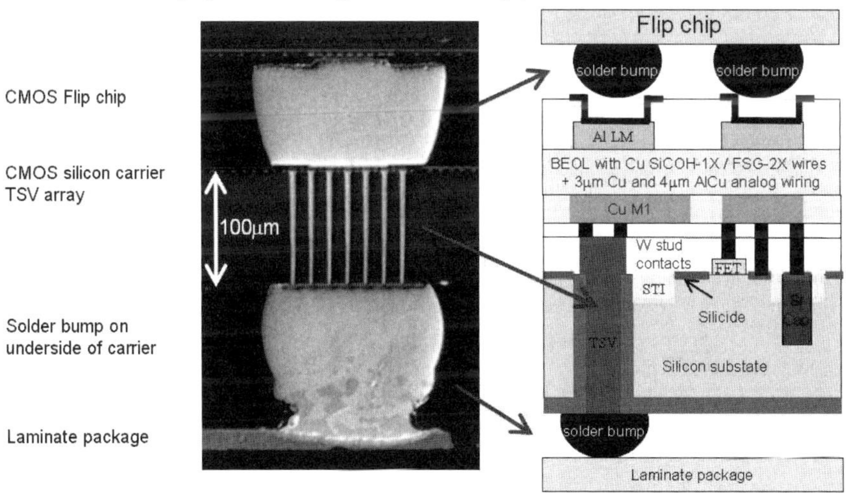

Figure 1: Simplified drawing and SEM [2] cross-section of silicon carrier technology.

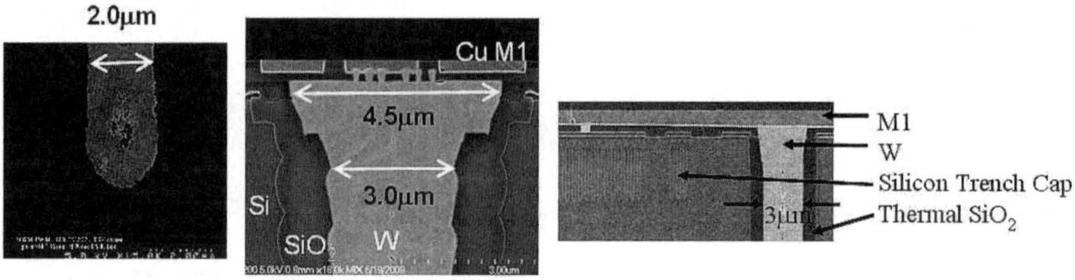

(a) (b) (c)

Figure 2: (a) TSV bottom, (b) TSV-C1-M1 top, and (c) details of TSV and 40fF/μm^2 silicon deep trench capacitor [2].

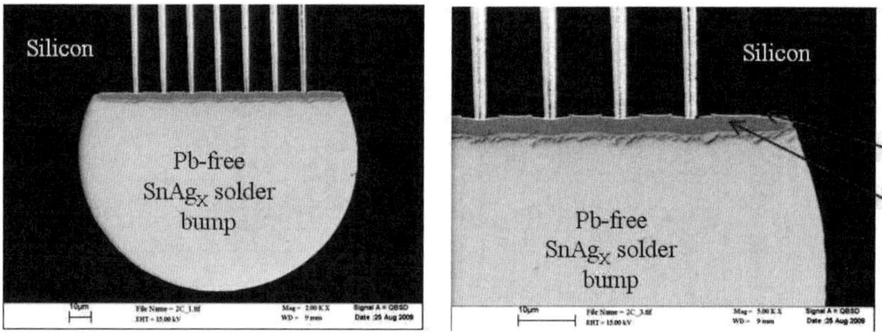

Figure 3: SEM cross-sectional images of solder bumps formed on thinned silicon carrier wafer backside connected to an insulated TSV array.

IBM's grounded TSV technologies employ a two mask approach for patterning the TSV's [7]. This two mask approach creates a ~5μm dielectric opening above the ~4μm wide TSV top opening, resulting in better CVD TiN sidewall coverage and CVD W TSV fill. We employ a similar integration scheme for insulated TSV's, resulting in a ~5μm opening in the dielectric above the ~4μm wide TSV top opening (Figure 2b). An added benefit of this approach is that the TSV insulator is not exposed to the damascene CMP processes and chemicals and instead is recessed ~0.6μm below the surface. This eliminates TiN/W smearing or metallic contamination of the TSV insulator with resultant improved TSV breakdown voltage and TDDB reliability.

Figure 3 shows SEM's of the silicon wafer backside after solder bump formation. After the silicon wafer front side processing is complete, including the formation of landing pads for flip chip mounting above the silicon carrier (Figure 1), the wafer is attached using an adhesive to a glass handle wafer, the wafer backside silicon is removed to expose the bottom of the insulated TSV (Figure 4a), the wafer backside is covered with CVD SiO_2, and Pb-free solder bumps are formed using standard processing under the exposed TSV's on the wafer backside (Figure 3). At this point, the silicon carrier chips are diced, the glass handle is detached, one or more flip chips are attached to the silicon carrier front side, and the solder bumps on the silicon carrier backside are attached to a laminate package. The final 3D stacked chips are shown in Figure 1.

The integration challenges of combining tungsten metalized insulated TSV's, silicon deep trench capacitors, high performance CMOS transistors, and damascene copper low-k SiCOH BEOL wiring include wafer stress control, wafer breakage, tungsten fill of the TSV's, integrating the TSV tungsten CMP metallization with the CMOS 90nm generation tungsten stud contacts and copper wires, and lithographic alignment of both new and existing levels.

As mentioned above, the TSV's are metalized with a damascene tungsten process similar to the one employed for grounded TSV technologies [7]. The TSV tungsten CMP process can severely recess or dish the ~5μm wide tungsten surface of the TSV bars during tungsten CMP and generate several 100nm of topography (Figure 5). This recessed topography above the TSV's is not a problem for subtractive-etch AlCu M1, but leads to residual liner/copper over and adjacent to the TSV during the Cu M1 CMP process (Figure 6a), which can cause M1 shorts.

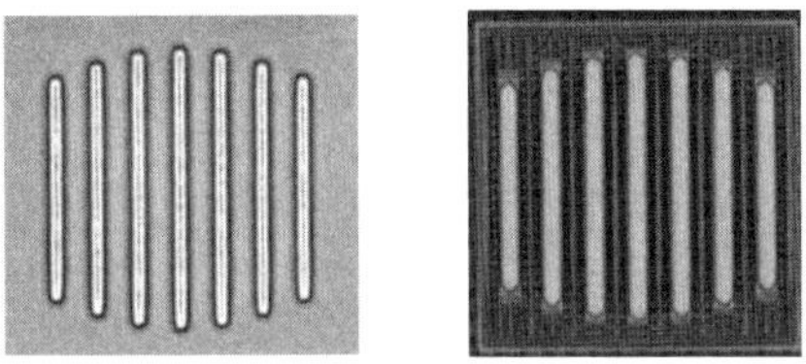

(a) (b)

Figure 4: 7 TSV farm (a) post backside grind showing the wafer backside and (b) post TSV tungsten CMP showing the wafer front side.

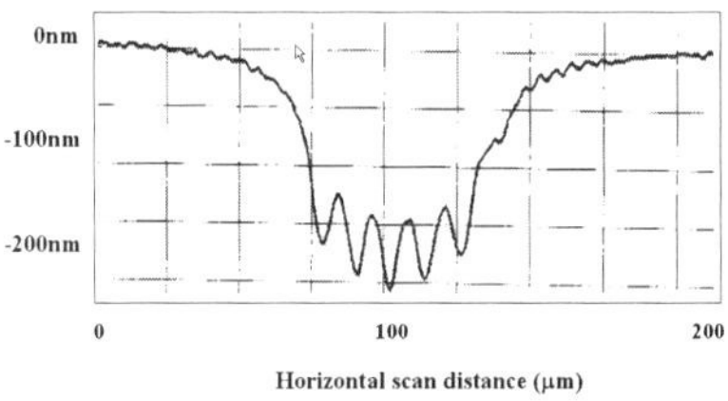

Figure 5: Surface scan across 5 TSV array after W CMP showing ~250nm surface recess [5].

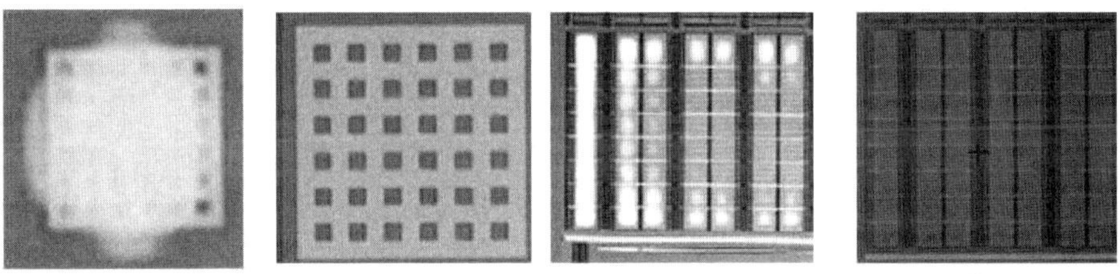

(a) (b) (c) (d)

Figure 6: (a) Unoptimized and (b) optimized TSV W CMP processing and resultant residual TSV pad M1 Ta/Cu; SRAM array M1 Ta/Cu residuals with (c) unoptimized and (e) optimized tungsten CMP processing.

Although we reduced the tungsten CMP TSV dishing or recess from ~500nm to ~150nm by changing from a 1-step high rate tungsten CMP process to a 2-step high/low rate process (Figure 7), the 150nm TSV tungsten recess still caused large areas of residual Ta liner and Cu post damascene Cu M1 CMP (Figure 6a). To further reduce this topography over the TSV incoming to M1 Cu CMP, we added a dielectric planarization step prior to forming the C1 studs. Adding the C1 dielectric planarization step further reduced the recess above the TSV to under 50nm (Figure 7), which eliminated the M1 Cu/Ta residuals, as shown in Figure 6b. Although our initial 2-step optimized CMP process dramatically reduced the TSV surface recess, it created >50nm of recess or dishing in SRAM arrays with closely spaced C1 contacts, which led to residual Ta/Cu post Cu M1 CMP (Figure 6c). Figure 8 shows a yield trend of a dense contact

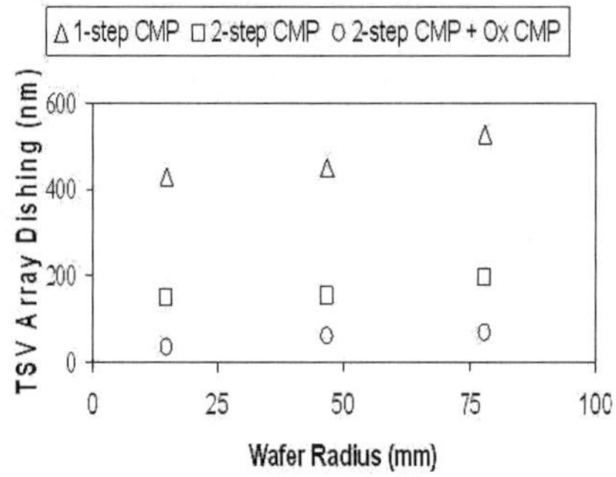

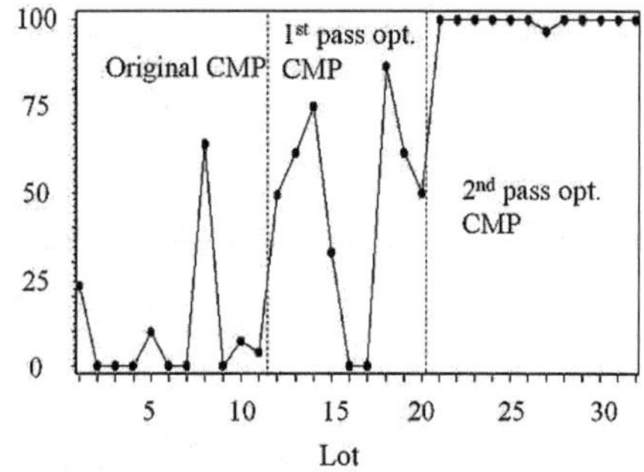

Figure 7: TSV array recess post TSV CMP processing incoming to damascene M1 process.

Figure 8: Dense SRAM-like C1-M1 W stud contact array to Cu M1 short yield trend.

stud C1 to damascene Cu M1 test structure which emulates the layout of a dense SRAM. Initial C1-M1 short yields with the 1-step TSV CMP process were ~0% and moving to the 2-step TSV CMP process boosted C1-M1 shorts yields to 10-90%. We further optimized the CMP processes to reduce the C1 stud recess to under 50nm in both TSV and dense SRAM areas of the chip and eliminate residual M1 Cu/Ta over both TSV's and dense SRAM arrays (Figure 6d). This final TSV CMP process change pegged C1-M1 shorts yields at 100%, as shown in Figure 8.

The addition of the TSV to the base 90nm technology affected both the random defect density of the C1 contact and M1 wire levels and the C1 stud contact resistance variability. Table 3 summarizes simple dicing channel comb-serp maze M1 opens and shorts; and M1-C1-silicided silicon-C1-M1 contact chain defect density. The addition of the metalized TSV under M1 and C1 degraded the defect densities, as expected, due to the added deposition, etching, and CMP processing associated with adding TSV's. Since the base technology yields were > 99.9%, the defect density degradation induced by adding insulated TSV's was minimal and we expect that the difference will diminish as the silicon carrier technology ramps into manufacturing. The C1 resistance, measured in M1-C1-silicided silicon substrate chains, showed increased variability for the silicon carrier technology (Figure 9). This increase in variability is due to an increase in C1 stud height variability, caused by the added dielectric erosion variability introduced by the TSV W CMP process, and the addition of dielectric CMP planarization above the TSV.

To minimize the C1 contact resistance in TSV arrays, we place dense arrays of C1 studs between the TSV and M1. Figure 10 shows a lateral M1-C1-TSV-C1-M1 test-site resistance chain, with 30 C1 studs in each link, used to measure the resistance of these added ~150nm tall ~120nm diameter C1 studs. Figure 11 shows that the measured resistance for these chains is ~1.5Ω per M1-C1-TSV-C1-M1 link for chains, independent of whether the chain contains 28, 56, or 72 TSV's. Assuming 15 sq. of TSV, 5 sq. of M1, TSV and M1 sheet resistances of 0.001 and 0.2 Ω/sq., respectively, and two parallel arrays of 30 C1's to connect the TSV to M1, we measured about 7Ω per C1 in the TSV-C1-M1 chains. 7Ω per C1 stud over the TSV is roughly the expected value since ~600nm tall C1's in M1-C1-silicided silicon chains have resistances of about 18 Ω per C1. Since we place ~1500 C1's between each TSV and M1, we expect that the added C1's will increase the single TSV bar resistance about 20%.

Parameter	90nm Base Technology	Silicon Carrier
C1 opens	1	3.2
C1 shorts	1	1
M1 opens	1	1.3
M1 shorts	1	1.9

Table 3: Normalized defect density data from ~30 silicon carrier lots and ~100 digital CMOS lots (3 waf/lot, ~1mm^2 M1 maze area).

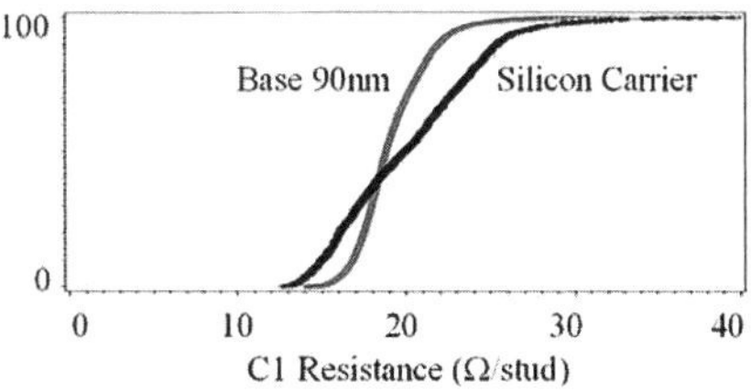

Figure 9: CDF of C1 resistance chip (~30 silicon carrier and ~100 digital CMOS lots, ~10 chips per wafer).

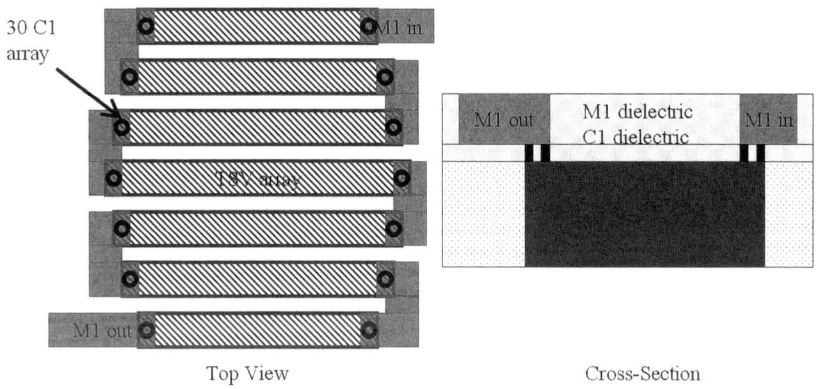

Figure 10: Simplified drawings of M1-C1-TSV-C1-M1 chains to measure C1 resistance.

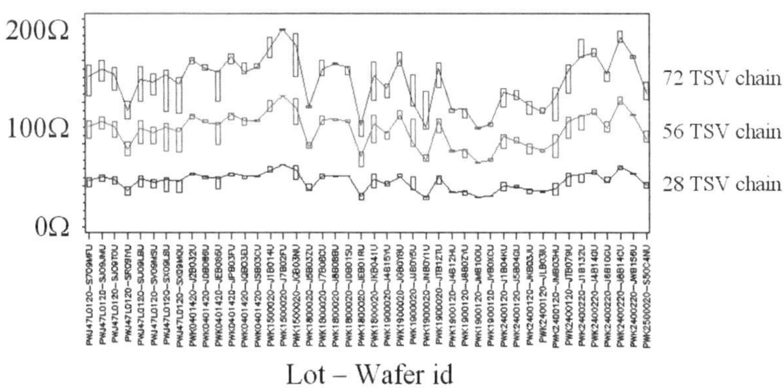

Figure 11: M1-C1-TSV-C1-M1 chain resistance for chains with 28, 56, and 72 TSV's

Table 4 summarizes measured substrate leakage currents for TSV farms, silicon trench capacitors, and CMOS transistors, respectively. The addition of insulated TSV's did not impact the electrical properties of the active 90nm generation devices; and the isolation between the TSV's and silicon substrate is excellent. We measure the breakdown voltage (VBD) of 7 bar TSV arrays, using an 1μA fail criteria, by ramping the voltage between the TSV conductor and silicided silicon adjacent to the TSV. TSV array breakdown voltage to substrate (VBD) is typically greater than 400V, which is the maximum tester voltage. Although we have not yet analyzed any VBD fails, we expect them to primarily occur at the top of the TSV/silicon interface, since the process used to flare out the top opening of the TSV reduces the oxide thickness on the sidewall of the TSV (Figure 2b).

Device	Base 90nm	90nm with TSV
TSV array leakage to substrate at 1.2V	-	< 10 pA per farm
TSV median breakdown voltage to substrate	-	> 400V
Silicon trench capacitor leakage at 1.5V (fA/μm2)	20	20
1.4nm Tox NFET gate leakage at 1.2V (na/μm^2	2.7	3.0

Table 4: TSV, 1.2nm gate oxide FET, and silicon trench capacitor median data.

CONCLUSIONS

IBM's 90nm generation 200mm wafer silicon carrier technology is being developed using a digital CMOS base combined with RF CMOS analog wiring and insulated TSV's. The silicon carrier is an integrated CMOS wafer with insulated through silicon vias (TSV's), Pb-free solder bumps on the back of the mechanically thinned wafer under the TSV's, with one or more Pb-free solder bump chips flip chip mounted to it's surface. To minimize TSV resistance and inductance and to maximize yield and reliability, the TSV's are placed in redundant arrays. The TSV process is compatible with standard 90nm high performance CMOS processing. CVD W with a CVD TiN liner is employed for TSV metallization and the TSV's are insulated from the substrate using a high-temperature (i.e. > 800C) thermal oxide. TSV arrays produced with this technology had excellent yield and voltage breakdown characteristics; and did not significantly change the CMOS FET, deep trench capacitor, contact resistance, or defect density parameters.

ACKNOWLEDGEMENTS

The authors wish to acknowledge the numerous engineers at IBM who contributed to the ground silicon via, silicon carrier, 90nm generation digital CMOS, RF-CMOS and damascene copper programs. Regarding this paper, we gratefully acknowledge the assistance of Felix Anderson, John Benoit, William Bosworth, William Cote, Timothy Dalton, Dinh Dang, Jim Dunn, Daniel Edelstein, John He, Patrick Hughes, Alvin Joseph, Jenifer Lary, Bob Leidy, Jessica Levy, Xuefeng Liu, Stephen Luce, David Meatyard, Paul Pfeiffer, Brett Philips, Rob Rassel, Daniel Vanslette, and Frank White..

REFERENCES

1. J. Knickerbocker et al., IBM Journal of Research and Development, v52, n6, pp. 553-569, 2008.
2. B. Dang et al., IBM Journal of Research and Development, v52, n6, pp. 599-609, 2008.
3. M. Kawano et al., in Technical Digest International Electron Device Meeting (IEDM 2006), San Francisco, CA, pp. 1–4.
4. G. Humpston et al., 2008 10th International Conference on Electronic Materials and Packaging (EMAP 2008), pp. 51-54.
5. E. J. Vardaman, Electronic Device Failure Analysis, v10, n4, pp. 30-32.
6. M. Shapiro et al., in Proceedings of the 2009 International Interconnection Technology Conference (IITC 2009), Sapporo, Japan, ISBN # 978-1-4244-4492-2, p.63.
7. A. K. Stamper et al. in Advanced Metallization Conference 2008 (AMC 2008), edited by M. Naik, R. Shaviv, T. Yoda, and K. Ueno, (Mater. Res. Soc. Conf. Proc. AMCXXIV, Pittsburgh, PA, 2009) pp. 495-500.
8. S. Parihar et al., in 2001 Technical Digest International Electron Devices Meeting (IEDM 2001), p249-255.
9. D. Edelstein et al., in 2004 International Interconnection Technology Conference (IITC 2004), p214-218.
10. L. A. Clevenger et al., in Advanced Metallization Conference 2004 (AMC 2004), edited by D. Erb, P. Ramm, K. Masu, and A. Osaki, (Mater. Res. Soc. Conf. Proc. AMCXX, San Diego, CA, 2005) p27-34.
11. A. K. Stamper et al. in Advanced Metallization Conference 2007 (AMC 2007), edited by M. Naik, R. Shaviv, T. Yoda, and K. Ueno, (Mater. Res. Soc. Conf. Proc. AMCXXIII, Albany, NY, 2008) pp. 363-368.
12. A. K. Stamper et al. in Advanced Metallication Conference 2004 (AMC 2004), edited by D. Erb, P. Ramm, K. Masu, and A. Osaki, (Mater. Res. Soc. Conf. Proc. AMCXX, San Diego, CA, 2005) pp. 37-43.

Conference Proceedings AMC XXV © 2010 Materials Research Society

Electrochemical deposition of reactive nanoscale metallization systems for low temperature bonding in 3D integration

L. Hofmann[1], J. Braeuer[2], M. Baum[2], S. E. Schulz[1, 2], T. Gessner[1, 2]
[1]Chemnitz University of Technology, Center for Microtechnologies, 09126 Chemnitz. Germany
[2]Fraunhofer ENAS Chemnitz, Technologie-Campus 3, 09126 Chemnitz, Germany

ABSTRACT

In the face of the demand for low temperature bonding processes in the 3D Integration of microelectronic or micromechanical components we want to introduce a method that uses a specific form of heat production. It is based on reactive material systems that consist of several layers of two different materials with nanoscale thicknesses. These layers show a self propagating and exothermic reaction during their joint intermixing. The resulting heat can be used as source for bonding processes such as solid liquid interdiffusion bonding. By arranging the reactive system in between the joining partners together with a solder layer the heat generation can be limited only to the joint interface while preventing heating of the whole device. Whereas others showed the fabrication of foils of such systems, that have to be placed between two joining parts by pick-and-place handling, we focused on the direct deposition of such reactive multilayer systems. The principle of this method is demonstrated by sputter deposition. The main part of this paper deals with the development of the electrochemical deposition of such reactive multilayers.

INTRODUCTION

The system integration and packaging of microelectronics and micromechanical systems is increasingly affected by three dimensional chip stacking [1]. Major roles therein play the fabrication of Through Silicon Vias (TSV) and the wafer bonding on wafer or chip scale [2]. When different functionalities such as electronics, mechanics or optics are to be combined the bonding process has to fulfill several conditions. The most important requirements on the joint interface are the mechanical strength, electrical conductivity and in most cases also hermetical tightness. Typical bonding processes today are silicon direct bonding, anodic bonding, glass frit bonding, eutectic bonding or adhesive bonding [3]. For most applications however, the bonding process has to be limited regarding to temperature (<400°C). Since the process temperatures of the mentioned standard bonding processes are above that temperature limit, new solutions must be found. One possibility is to use materials that need low temperatures to form a joint connection. Hereby several processes are under investigation in the recent years such as solid liquid interdiffusion bonding (Cu-Sn, Cu-In) or thermo compression bonding (metal-metal) [4]. Another promising method is to confine the applied heat only to the joint interface. This local heating can already be done for instance by the use of lasers.

In this work we would like to introduce a new method of local heating for wafer bonding processes that is based on reactive nanoscale metallization systems. Such systems consist of several (up to hundreds) alternating layers of two different thin metallic films. The principle of operation of such a reactive system is shown in Fig. 1.

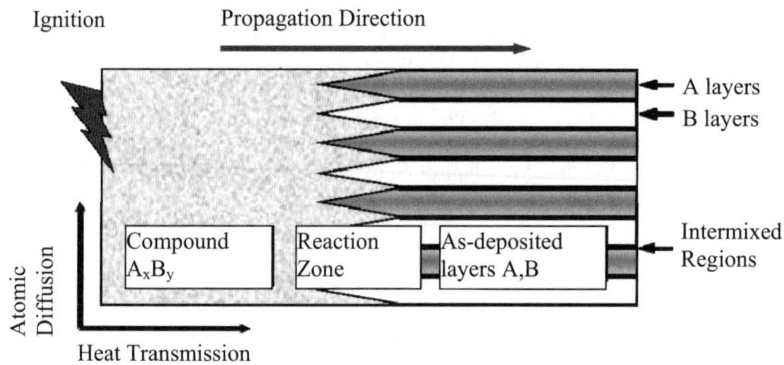

Figure 1. Principle of operation of a reactive multilayer system

Each single layer thickness is in the range of 10-1000 nm. The overall thickness of the system is given by the number of alternating layers, which is in the range of 20-10000 [5]. By means of applying an initial energy pulse (ignition) at one point of the multilayers the system starts to form inter metallic phases by interdiffusion of adjacent material layers. The reaction kinetic is promoted by the low thickness of the single layers due to a reduced diffusion path length. By choosing material combinations with negative enthalpy of formation this reaction runs exothermic. Thereby, the generated energy from the interdiffusion serves as initial energy for the reaction of the adjacent layers. If the rate of thermal energy generation is higher than the rate of thermal energy loss, the reaction is isotropic self propagating in lateral direction through the whole layer system [6].

Conventionally, commercial available foils of such reactive multilayer systems (NanoFoil©) are used for the assembly of macroscopic components. Hereby, the NanoFoil© is placed between the two joining parts in combination with solder pre-forms (Fig 2.a). Subsequently this assembly gets compressed and afterwards get ignited [7].

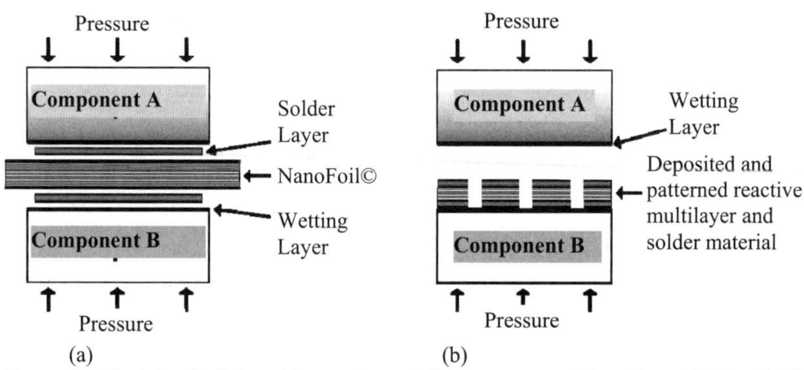

Figure 2. Principle of joining with reactive multilayer systems: a) bonding with NanoFoil© and b) new approach of bonding with direct pattern deposited multilayer

We already transferred this bonding method to micro scale applications and demonstrated the principle for instance by joining IR-emitting sensors onto conventional TO-packages. However, for bonding techniques in microelectronics and MEMS with much smaller dimensions (few ten microns) of bond frames, this method is not applicable in the same way. The reason therefore is because of handling issues and the limited ability of creating patterns with small dimensions. Our new approach is the direct deposition of the reactive multilayer system as well as the solder material onto the substrates (Fig. 2b). In this paper first results of sputter deposition are shown as demonstration of the principle feasibility of the direct deposition method. One drawback when using sputter deposition is the long process time and the complex process conditions due to the permanent alternation of the used targets. Moreover, the patterning of the blanket deposited multilayer systems is not straight forward and quite complicated. One way to overcome this problem is to use pattern plating, i.e. electrochemical deposition (ECD) over a photo resist mask. Focus in this paper is thus laid on the ECD of reactive multilayers.

EXPERIMENT

Feasibility of the direct deposition of reactive multilayers

In the first instance we used sputter deposition to demonstrate the feasibility of the multilayered deposition as well as the principle of the reactive effect. Although this method is very time consuming and thus expensive, it provides a good opportunity for layer-by-layer deposition of the desired thin and homogeneous metal films. For the choice of the materials a database was generated that includes all relevant material properties (e.g. energy of formation) for several material combinations. The selection was made on dependent on both the energy of formation as well as the availability at our institute. Those criteria were best satisfied by the material combinations Al-Ti as well as Ti-Si. The fabrication was done by magnetron sputtering onto standard 100 mm silicon substrates with a number of 114 layers for the Al-Ti system and a number of 50 layers for the Ti-Si system. The process parameters (namely the deposition rate) were adjusted in such a manner, that a stoichiometric composition of 3:1 for the Al/Ti system respectively 5:3 for the Ti/a-Si system should be reached. The deposited multilayer structures were first analyzed by SEM to investigate the achieved alternating structure with the corresponding single layer thicknesses. TEM and NanoSpot-EDX were used to investigate the material composition respectively the purity of each single layer. The reactive behavior of the structures was tested with the help of an electric spark from a simple battery. For the thermal and kinetic behavior of the intermixing reaction the Differential Scanning Calorimeter (DSC) technique and the observation with a high speed camera were used.

Reactive multilayer deposition by ECD

Electrochemical deposition (ECD) was also investigated as method for the fabrication of reactive multilayers. The motivation therefore is on the one hand the possibility of pattern plating what prevents the complex etching of structures after a blanket multilayer deposition. On the other hand the ECD process helps to reduce the process time and process complexity compared to sputter deposition. In principle there are two ways for ECD multilayer deposition. One way is to use a two bath method, which means an alternating deposition in two different plating baths. One drawback of this method is the process time, because of the permanent necessary intermediate process steps (rinsing, cleaning, pre treatment). Moreover the minimal possible thickness of each single layer is limited for this method. During the deposition of each single layer there is an initial deposition phase (wetting, adsorption of species, nucleation). For the dual bath method this phase is expanded since the surface condition (e.g. naturally grown oxide films) of the substrate after each intermediate step is slowing down steps such as nucleation. Moreover, the most likely present island growth leads to an extended initial phase until the surface of the substrate is completely covered. Thin film deposition (i.e. films in the range of a few tens of nanometers) is thus hard to achieve.

A more promising method is a one bath method, whereupon our primary interest is laid on. Hereby, the plating chemistry of both film components is present in one bath. The resulting composition of both components in the deposited film is dependent on several process parameters such as the concentration of the components, the current density, the mass flow or the pH value. By adjusting those parameters it is possible to achieve the preferred deposition of either component compared to the other. In most cases this adjustment is done by controlling the concentration of the components and by the control of the current density. Hereby, a noble and a less noble metal is used, whose electrochemical standard potentials lay sufficiently apart from each other. The more noble metal is present only at low concentrations, whereas the less noble metal is present at higher concentrations. The deposition of each single metal is then controlled by a pulsed current respectively potential signal. At low current densities the deposition of the noble metal prevails, caused by the electrochemical potential characteristic. At higher current densities normally both metals would be deposited. Since the concentration of the noble metal is very low compared to the concentration of the less noble metal, the mass transport of the noble metal ions is limited to diffusion. Hence, the deposition of the noble metal is suppressed during the higher current density phase. The number of materials that are suitable for ECD of reactive multilayer systems is limited. In the first instance the materials have to fulfill the requirements for reactive systems, i.e. the energy of formation must be below zero. Another criterion is that the used materials must be able to be deposited by ECD from electrolytes under maintainable conditions (e.g. the deposition of Al or Ti is only possible under very complex conditions such as high pressures, high temperatures in the ranges of 700°C in closed process chambers or by the use of ionic liquids instead of easy to use water based electrolytes). Furthermore, the selected electrolytes for both materials should be mixable and finally both materials must have electrochemical standard potentials that are sufficiently apart from each other.

The development of complete two component electrolytes is relatively complex. Therefore, we decided to begin the first experiments with an existing alloy bath and adapt the parameters such as the concentration of each chemical component and the current densities.

From the limited number of possible material combinations we identified materials from the platinum group (Pt, Pd) in combination with materials such as Sn, Zn or In as promising candidates, that fulfill the abovementioned requirements. For the initial experiments we choose the material combination Pd-Zn, because of the knowledge about an existing alloy bath with those components.This combination is expected to fulfill the abovementioned requirements regarding the reactive behavior, because the energy of formation of the desired reaction product $PdZn_2$ ($-50\ kJmol^{-1}$) is in the range of those material combinations that show a reactive behavior when used for reactive multilayer foils ($AlNi_3$: $-35\ kJmol^{-1}$, AlTi: $-60\ kJmol^{-1}$). All experiments were carried out in a beaker and with an inert platinized titanium grid as anode. For the generation of a convective mass transfer a magnetic stirrer with included hot plate was used. A DC power source *pe1010* from plating electronics was used for current control. The use of a pulse generator was abandoned for the initial experiments, because the necessary relative long cycle times would have needed further modification of our existing tool. As substrates standard 100 mm Si-wafers with a 500 nm thermally grown oxide film as insulating layer were used. A sputter deposited 10/20 nm Ti/TiN layer served as adhesion promoter respectively diffusion barrier and a 150 nm sputter deposited Cu film served as seed layer. The substrates were cut into 4 pieces and were immersed into the beaker via a holder with electrical clips. The anode cathode distance was held constant at 4 cm. For a better wetting behavior and a better nucleation those pieces were immersed into diluted sulfuric acid to remove the naturally grown copper oxide film. After the deposition the surface was blow-dried with nitrogen.

Prior to the actual investigations of the alloy bath we started plating experiments with the same electrolyte but containing only the less noble component Zn in order to better understand the electrolyte behavior and to find adequate plating parameters as starting conditions. Hereby the concentration of Zn was varied as well as the applied current density and the operating temperature. The influence of the buffer electrolyte, which is contained in the chosen alloy electrolyte by default, was also tested. Subsequently, we investigated the deposition characteristic of the alloy electrolytes with different concentrations of Pd in the electrolyte. The concentration of Zn was taken from the best results of the preliminary experiments. For each electrolyte composition different sets of current densities were used to find out the deposition ratio of Pd to Zn. Thus, for all parameter sets the mass concentration of Pd and Zn, respectively, in the deposited films were derived from EDX measurements. By means of this analysis we selected the parameters, i.e. the concentration of the noble metal and both current densities (higher and lower value) for the subsequent multilayer depositions. The multilayer structures were then observed by taking SEM images of the substrate cross section. TEM investigations including EELS analysis was used to investigate the single layer compositions.

DISCUSSION

Feasibility of direct deposition of reactive material systems

Via SEM imaging the cross section of the sputter deposited multilayers were analyzed to obtain the structural information. The SEM image of a Ti/a-Si system is shown in Fig. 3a. Herein the darker layers represent the Ti and the brighter layers the Si, respectively. TEM and NanoSpot-EDX analysis showed that no noticeable intermixing (about 3 nm) occurred during the deposition for the Al-Ti system. The intermixing region is depicted in Fig. 3b and 3c, respectively.

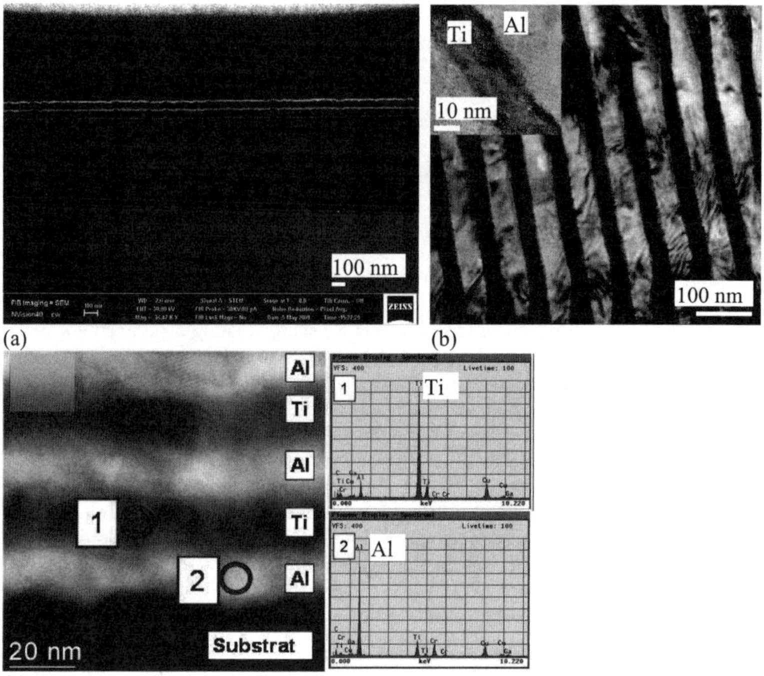

(c)

Figure 3. SEM/TEM/EDX-analysis of sputter deposited reactive multilayer: a) Ti/a-Si system with 12 bi-layers (30 nm Ti / 45 nm Si), b) TEM image of the Al-Ti system and c) NanoSpot EDX of the single layers in the Al-Ti system showing the intermixing regions [8]

The total thicknesses of the Al-Ti structures were identified as up to 4.5 µm with 114 Al and Ti layers. Single layer thicknesses of 14-120 nm could be measured. For the Ti/a-Si system with 50 layers a multilayer overall thickness of 2 µm and a corresponding single layer thickness of 15-45 nm were determined. For the latter system the intermixing zone was considerably larger. This effect is most likely due to a non optimized sputter process. The intermixing during the deposition may be minimized with adapted process parameters and a better Wafer backside cooling.

Analyzing the structures by the DSC technique showed a typical exothermic behavior. However for the Al-Ti system the reaction had only a self propagating characteristic after peeling the multilayer system off the substrate. For Al-Ti systems ignited direct on the substrate the reaction stops shortly after the applying of the initial energy. This effect is due to a absent thermal isolation layer between the substrate and the multilayer. Thus, all produced heat is directly transferred into the substrate. The Ti-Si system in contrast shows a much higher reactive behavior. Hereby the reaction in peeled off layers is highly explosive. Moreover, self-propagating exothermic reactions directly on the substrate could also be detected for the Ti/a-Si system. Hereby two aspects could be found out. On the one hand a minimal number of layers of

50 are required for this reaction. For systems with less single layers the reaction stopped immediately after the ignition. On the other hand the substrate temperature during the reaction plays an important role. To sustain the self propagating reaction over a wide range of the substrate (>5cm^2) the substrate had to be heated up to 300°C. A reacted Ti/a-Si multilayer is shown in Fig. 4 for two used substrate temperatures.

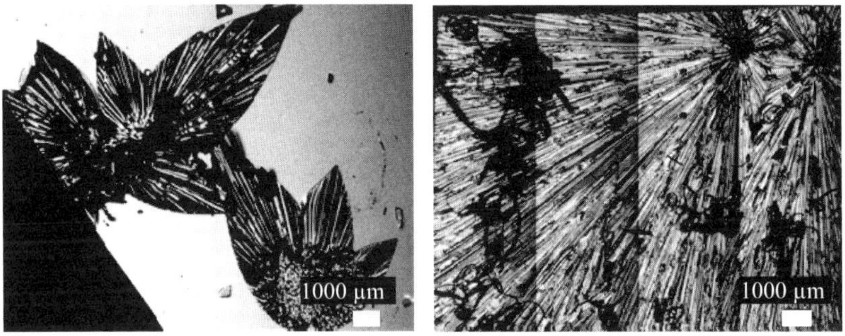

(a) (b)

Figure 4. Top view of a Ti/a-Si multilayer system after the self propagated reaction for two applied substrate temperatures: (a) 150°C and (b) 300°C

One reason for the short self propagating reaction at lower substrate temperatures is believed to be the large intermixing zone of the Ti/Si layer. This larger intermixing zone reduces the fraction of pure material in each single layer and moreover increases the diffusion length for the corresponding species (i.e. Ti and Si, respectively) between two layers. This induces a decelerated interdiffusion and a shortened reaction and in return this leads to a decrease in the generated heat. Hence the available initial energy for further reactions is reduced. Therefore, a heating of the substrate up to 300°C promotes the sustaining of the reaction.

With an optimized sputter process and hence with multilayers that have smaller intermixing zones, it should be possible to achieve reactions that generate more heat on the one hand and on the other hand to get reactions that are self propagating without any additional substrate heating.

Reactive multilayer deposition by ECD

The preliminary plating experiments with the electrolyte containing only the component Zn showed a strong dependence on the concentration of that component. This aspect is illustrated by means of the SEM surface views in Fig. 5 for the example of deposited Zn-films for three different Zn concentrations. For the lowest concentration the deposition character is only flake like and very porous. The films become more continuous and more even with increasing concentration; only for a concentration of 10 g/l the film is smooth and uniformly covers the complete substrate surface.

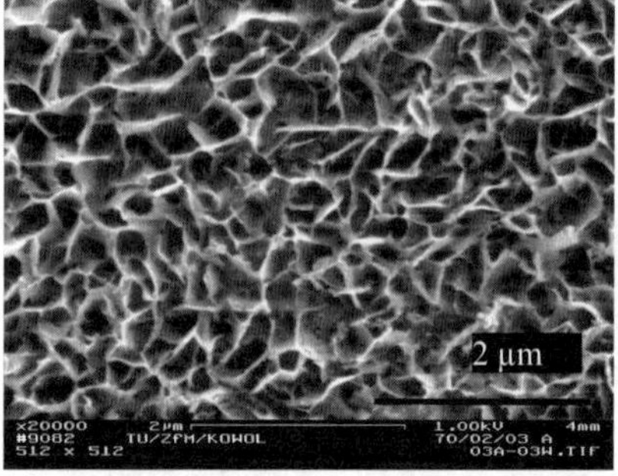

(a)

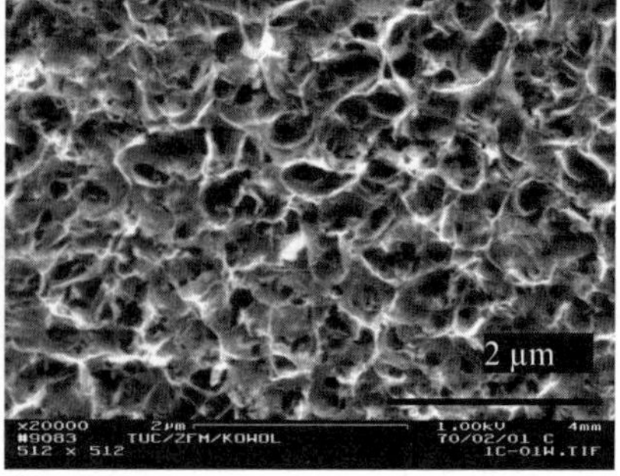

(b)

(c)

Figure 5. SEM images of deposited Zn films for different Zn concentrations: (a) 1 g/l, (b) 5 g/l and (c) 10 g/l

Additional experiments showed that the structure of the deposited film is very coarse grained when no buffer electrolyte is used; in contrary it becomes finer with the addition of such a buffer (e.g. $(NH_4)_2SO_4$) electrolyte. The influence of the temperature is little for the investigated values of 25°C and 50°C, respectively, compared to the influence of the other parameters. The current density was found to achieve best results in terms of film coherence and smoothness for values in the range of 2-3 A/dm^2.

The parameter set that achieved the best results of this experiment were used to investigate the deposition characteristic of the alloy bath. The results from the EDX measurements for three different concentrations of the noble metal Pd are shown in Fig. 6. The diagram shows the ratio of the normalized weight percents of Pd to Zn versus the applied current density. The curves for three different concentrations of Pd show all a strong decay with increasing current density. The degree of the decay is higher for larger concentrations of Pd. Above a certain value of the current density (i.e. 2 ASD) the mass concentration ratio undergoes no remarkable change; moreover there is no remarkable difference of that ratio for the three investigated concentrations of Pd.

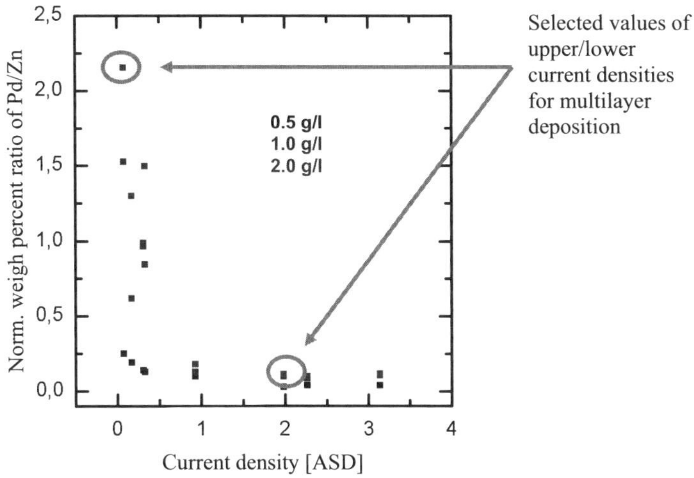

Figure 6. Normalized weigh percent ratio of Pd compared to Zn in dependence on the current density for three concentrations of Pd (highlighted: parameter values for further experiments)

Multilayer depositions were then carried out for the highest concentration of Pd, since hereby the highest mass concentration ratio at low current densities was obtained. Moreover the highlighted current density values in Fig. 6 were used for the deposition. The SEM/TEM images in Fig. 7 show first deposition results. The multilayer clearly shows the separation between two components. The 6 darker thicker layers represent Pd-rich films and the 5 brighter layers the Zn-rich films, respectively. Single layer thicknesses could be identified as 25-70 nm for Pd and 11-18 nm for Zn; the overall thickness of the multilayer was determined as to 320 nm.

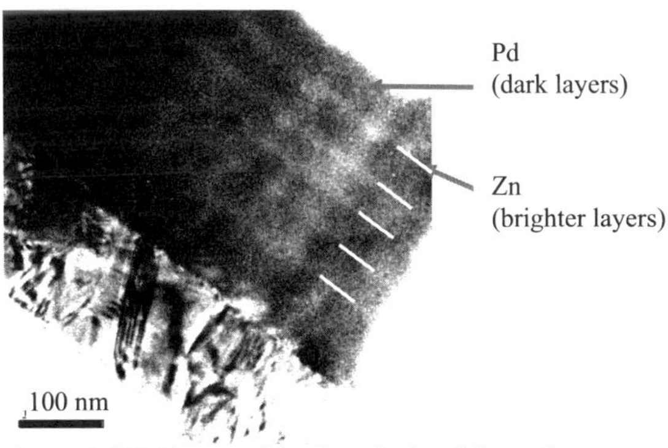

Figure 7. SEM image of the deposited multilayer for a concentration of 2.0 g/l of Pd

For further analysis of the direct deposition by ECD, the deposition of multilayers with a higher number of single layers is momentarily under investigation. Moreover, an optimization of the process parameters including the use of a pulsed current source is necessary to achieve thinner single layers and to achieve a specific mole fraction between the two components. The reactive behavior will be investigated after the deposition of optimized multilayers.

CONCLUSIONS

In contrary to the hitherto usage of reactive multilayer foils, we showed a new approach of direct deposition of reactive multilayers on Si substrates. The principle of this method was demonstrated for sputter deposition with the illustration of functional reactive systems. Improvement of the sputter process will be carried out to achieve Ti/Si multilayers with smaller intermixing zones. These improvements in combination with the use of a better thermal isolation between the multilayer and the substrate, should lead to self propagating directions without any external heating of the substrate. As more promising better method of multilayer deposition the ECD technique was described. First deposition experiments were carried out with this method. By means of this technique multilayer were deposited, that consisted of Pd in combination with Zn. The multilayer system had an overall thickness of 320 nm with single layer thicknesses of 11-18 nm and 25-70 nm, respectively. Further multilayers with a higher number of single layers will be fabricated and will be analyzed regarding their film composition, and the intermixing of the single layers. After the optimization of the deposition the reactive behavior of the deposited multilayers will be tested. In the face of future application for 3D systems integration issues such as material mismatch have to be investigated and other material combinations have to be found if necessary.

ACKNOWLEDGMENTS

The authors like to thank the student assistant Cindy David for carrying out the designed experiments. Moreover, we are indebted to Dr. Schulze's team at the Institute of Physics at the Chemnitz University of Technology for the TEM analysis.

REFERENCES

1. C. K. Tsang, P. S. Andry, E. J. Sprogis, C. S. Patel, B. C. Webb, D. G. Manzer and J. U. Knickerbocker in Enabling Technologies for 3-D Integration, edited by C. A. Bower, P. E. Garrou, P. Ramm and K. Takahashi, (Mater. Res. Soc. Symp. Proc 970, Warrendale, PA, 2007), p. 145-153
2. L. Hofmann, M. Kuechler, T. Gumprecht, R. Ecke, S.E. Schulz, T. Gessner in Advanced Metallization Conference 2007 (AMC 2007), edited by A. J. McKerrow, Y. Shacham-Diamand, S. Shingubara and Y. Shimogaki, (Mater. Res. Soc. Conf. Proc. AMCXXIII, Warrandale, PA, 2008), pp. 623-630
3. M. Wiemer, J. Froemel, J. Bagdahn, R. Knechtel in MicroNanoReliability 2007, edited by B. Michel, (Micromaterials and Nanomaterials 6, Berlin, 2007), p. 264
4. M. Wiemer, J. Froemel, M. Haubold, C. Jia, D. Wuensch and T. Gessner in Semicondurctor Wafer bonding 10: Science, Technology and Applications , edited by T.

Suga, J. Bagdahn, H. Baumgart, C. Collinge, K. Hobart and H. Moriceau (ECS Transactions: 214th Meeting of the ECS, Vol. 16, (8), 2008).

5. T.P. Weihs in Handbook of Thin Film Process Technology, edited by D.A. Glocker and S.I. Shah (Institute of Physics, Philadelphia, 1997), pp. F7:1-F7:13
6. A.J. Gavens, D. Van Heerden, A.B. Mann, M.E. Reiss and T.P. Weihs in Journal of Applied Physics, Vol. 87, 1255-1263 (2000).
7. J. Wang, E. Besnoin, A. Duckham, S. J. Spey, M. E. Reiss, O. M. Knio, M. Powers, M. Whitener and T. P. Weihs in Applied Physics Letters, Vol. 83, 3987-3989 (2003).
8. J. Braeuer, M. Baum, M. Wiemer, T. Gessner in Mikro-Nano-Integration 1. GMM Workshop 2009, edited by VDI Verlag, Berlin-Offenbach, Germany, 2009, ISBN 978-3-8007-3155-8.

Conference Proceedings AMC XXV © 2010 Materials Research Society

Seed Layer Enhancement: An efficient process for the fabrication of 3D interconnects

J. Cuzzocrea[1], E. Deronzier[2], P.H. Haumesser[2], M. Assous[2], J.B. De Labretoigne[3], R. Baskaran[3], E. Chaînet[4]
[1]STMicroelectronics, 850 rue Jean Monnet, Crolles, F-38926, France
[2]CEA-LETI-MINATEC, 17 rue des Martyrs, Grenoble, 38054, France
[3] SEMITOOL, 655 West Reserve Drive, Kalispell, MT 59901, USA
[4]LEPMI, rue de la Piscine, St. Martin d'Hères, 38400, France

ABSTRACT

As two dimensional (2D) scaling of microelectronics devices shows limitations, three dimensional (3D) packaging using Through Silicon Via (TSV) has been investigated as a promising solution to meet tomorrow's electronics needs [1]. One of the key challenges in this research is the capability to fill deep and large vias of high aspect ratios (>5) by electrochemical deposition (ECD) of copper, which is usually limited by discontinuous seeding of sidewalls by PVD. In this work, an electrolytic process called Seed Layer Enhancement (SLE) has been investigated as a solution to improve copper seed continuity. For this purpose, copper nucleation on the resistive barrier material has been optimized. As a result, the SLE process has been straightforwardly inserted in the metallization sequence, as testified by good electrical performances.

INTRODUCTION

Nowadays, 2D integration shows serious limitations when it comes to manufacturing devices with increased functionality and performance. Despite continuous effort spent in scaling 2D devices (copper integration [2], ultra low-k [3],...), physical limits will be reached within the next 10 years.

The continuing demand for smaller device and higher performance lead to the development of 3D integration. Through Silicon Vias (TSV) provide plentiful and short 3-D connections between stacked chips. By providing better performance (less consumption, faster speed, reduction of parasitic loses,...) [4] with low cost, this technology is gaining a lot of interest in the semiconductor industry. However, a critical technological challenge in 3D integration is the fabrication of TSVs with high aspect ratio (>5). In particular, void-free metallization by copper electrodeposition (Cu ECD) is not a straightforward matter, especially when narrow structures (below 5 μm diameter) suited for high density interconnects are concerned.

The ECD process is initiated on a thin copper liner called seed layer. Physical Vapor Deposition (PVD) is traditionally used for seed layer deposition. However, this technique is likely to form discontinuous layers on the sidewalls of narrow and deep structures, due to its limited step coverage. Such seed layers would cause the formation of voids during ECD [5]. To overcome this difficulty (6), a straight forward approach would be to repair the discontinuities of the PVD liner as shown in Fig 1. In previous contributions, we have studied such an economical wet process called Seed Layer Enhancement (SLE) for the fabrication of sub-micrometer interconnects [6]. In this process, copper deposition is performed by electrodeposition (Fig. 1). This solution was shown to provide higher coverage of the conductive layer along the sidewalls, hence leading to void-free structures at significantly lower costs.

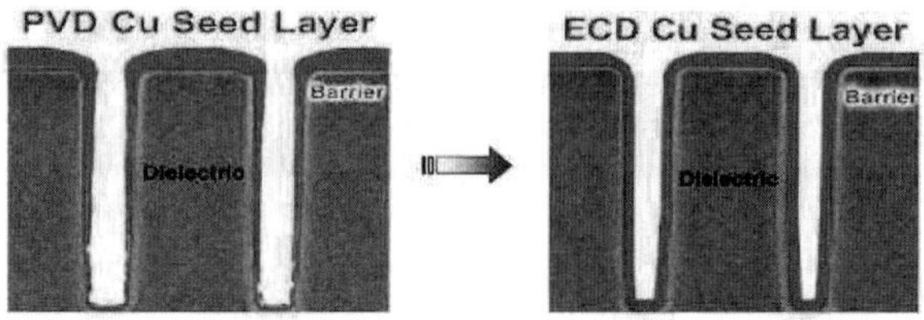

Fig. 1 : *Principle of Seed Layer Enhancement[6]*

In this work, we investigate the capability of this process to efficiently seed narrow and deep TSVs used in high density applications. The impact of potential drops associated with the resistive barrier is examined, and shown to prevent formation of a continuous copper film. Several ways to overcome this effect are investigated, and integration of the SLE process in a conventional metallization sequence is studied.

EXPERIMENTAL

To investigate the benefits of SLE for TSV fabrication, experiments were performed using 200mm wafers patterned with 15μm deep TSVs. The via diameter ranged between 3 μm and 5 μm To fabricate TSVs, RIE etching process was used for dielectric layers and silicon etching. A conformal 200 nm thick SiO_2 film deposited by Sub-Atmospheric Chemical Vapor Deposition (SACVD) was used to isolate the TSVs. A 20nm thick barrier layer of TiN was deposited by metal-organic chemical vapor deposition (MOCVD) in an Endura 5500 tool from Applied Materials. A 50nm thick copper liner was deposited with non collimated PVD in the same tool without air break. The substrates were then ready for the SLE process. A MOCVD copper deposition process was used as an alternative to SLE to grow a continuous seed layer over the PVD liner, and was used as a reference in the present study. In this case, MOCVD copper was performed immediately after Cu PVD, without air break.

All aqueous treatments, including SLE and ECD, were performed in a Raider equipment from Semitool. The incoming wafers were first pre-wetted by DI water in a spray chamber. The SLE process was performed using a specific alkaline electrolyte provided by Rohm and Haas Electronic Materials, in a CFD2 reactor equipped with 4 independent inert anodes. Prior to SLE, a proprietary pretreatment could be applied to restore barrier conductivity. After SLE, copper ECD followed, using a 3-additive medium acid electrolyte in a CFD3m reactor. A hotplate anneal at 400°C was performed in the Raider tool. After chemical-mechanical polishing, an additional metal level was formed using conventional Mx metallization to form Kelvin structures that were tested electrically.

A FEI Expida focus ion beam (FIB) and a HITACHI 5000 scanning electron microscope (SEM) were used to characterize the morphology of SLE films. The electrical resistance of Kelvin via structures was performed following a protocol described elsewhere [6].

RESULTS AND DISCUSSION

An initial filling test was performed using a sample with a 50 nm thick PVD copper liner without any seed repair. Fig. 2a shows that ECD copper growth is essentially blocked below 7 µm from the top of the via. . This experiment confirms that PVD coverage is not sufficient in our experimental conditions, leading to discontinuous seeding of the lower parts of the features. The much higher surface resistance of TiN (137Ω/square) compared to PVD copper (0.75Ω/square) inhibits copper deposition from the conductive ECD electrolyte. This result demonstrates the necessity of a process like Seed Layer Enhancement to form a conformal seed layer which would be continuous all along the via.

With an identical substrate (50 nm PVD copper), such a continuous copper liner is observed after SLE at the bottom of the via (Fig 2a). However, near via bottom, in the same region where ECD copper was not deposited, the deposit is formed from agglomerated copper nodules instead of a smooth film. This might be attributed to a sparser nucleation of electrolytic copper in this region, which is expected to occur on these more resistive portions of the surface. Nucleation rate is probably limited by the high resistance of TiN in competition with the more conductive Cu surface. In addition the availability of ions at this depth (limitations due to mass transfer), and solution resistance drop (IR drop) within the via may hinder nucleation at the bottom of the feature. Moreover, this phenomenon is probably aggravated by superficial oxidation of the barrier during air exposure. This oxide is thus expected to increase surface resistance at the bottom of the via. Therefore, restoring metallic properties of the barrier seems to be a good strategy to improve SLE continuity within the features.

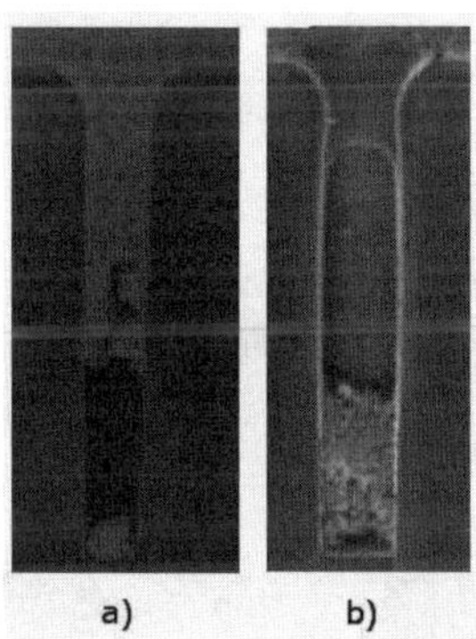

Fig. 2 : *3x15 µm TSVs seeded with 50 nm PVD Cu after a) ECD process and b) SLE process only*

Impact of barrier surface preparation

For this purpose, a proprietary surface pre-treatment was performed prior to SLE. Fig. 3a shows the result of the same SLE process when this surface treatment was employed. A smooth, 80 nm thick copper layer is observed at the top of the via. This thickness decreases to 45nm on the sidewalls of the via, but the copper film remains continuous. This is a significant improvement as compared to the sample without barrier pre-treatment. Fig. 2b

This result is important, as it shows that with appropriate surface treatment of the resistive barrier, efficient copper nucleation is possible directly on barrier with the SLE process. This opens a promising route towards direct on barrier copper plating. Fig. 3

Nevertheless, Fig. 3a shows that the SLE copper film is still discontinuous below 10 µm deep in the TSV in our experimental conditions.

Impact of current density during SLE

To further increase nucleation rate on the more resistive barrier, current density was doubled during deposition. Fig. 3b shows SLE morphology with barrier pre-treatment and higher current density. Comparison with Fig. 3a shows that the higher current during plating step significantly improves seed continuity near the bottom. Nevertheless, SLE copper is still discontinuous in the last 2 µm near via bottom. A first explanation would be that polarization increase induced by higher current density is not sufficient to overcome ohmic drop associated with barrier resistance and improve Cu nucleation. Secondly, the higher current density would deplete copper ions inside the feature rapidly, thus limiting nucleation rate of the electrodeposition due to mass transfer limitations. Thirdly, the increase in current density (higher current) might have increased the solution potential drop (IR drop) with-in the via, leading to poor current distribution and nucleation at the via bottom. These effects are still under investigation, and bath parameters may be optimized to eliminate these limitations and provide a conformal and continuous seed layer.

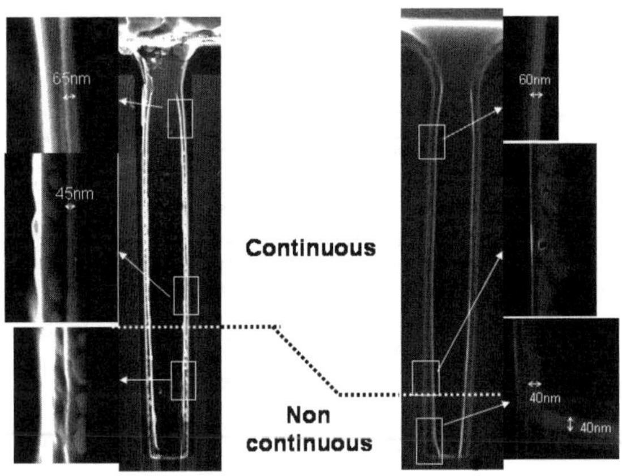

Fig. 3 : *Impact of current density on seed continuity in 3x15µm TSVs. a) nominal current density b) doubled current density*

Integration of SLE in the metallization sequence and electrical performances

Nevertheless, as shown in Fig. 4 the resulting SLE seed layers (with barrier pre-treatment and higher current density during deposition) are suitable to initiate super-filling during the subsequent ECD step as shown in Fig. 4a. After partial -ECD fill, a continuous and smooth copper deposit is obtained inside the via, even at the bottom. Furthermore, the ECD film becomes significantly thicker deep in the via. This is confirmed by the fact that perfectly filled TSVs are obtained after full fill (Fig. 4b).

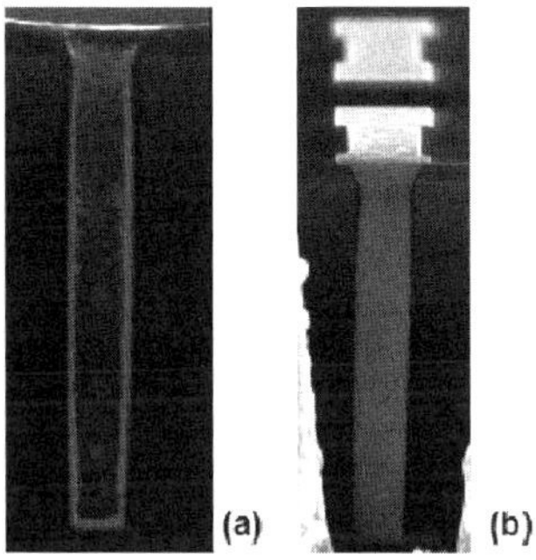

Fig. 4 : *Integration of SLE in the metallization sequence. (a) Efficient initiation of superfill during ECD. (b) Final via structures*

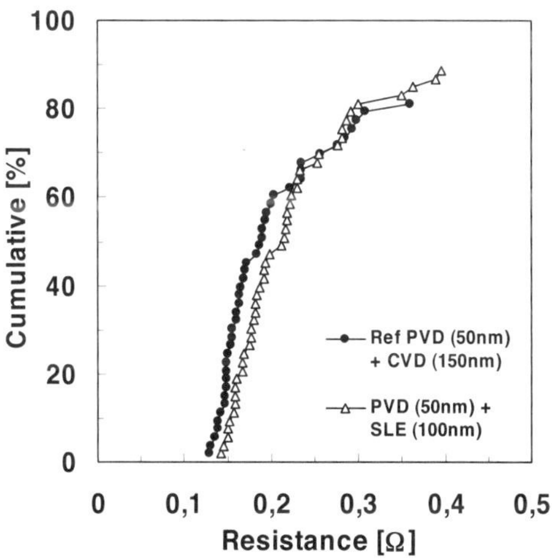

Fig. 5 : *Resistance of TSV (after SLE process) measured with 3x15μm Kelvin structures*

Electrical tests of Kelvin structures confirm these encouraging morphological observations. As shown in Fig. 5, these TSVs are electrically functional with good resistance and yield. These performances are comparable to those obtained when using a MOCVD copper process instead of SLE. This clearly confirms that the SLE process is an efficient solution for TSVs filling by electrodeposition.

CONCLUSIONS

In this study, a seed layer enhancement process (SLE) was evaluated as a suitable solution for copper seeding in the fabrication of 3D-IC TSVs. The ability of this electrolytic process to form continuous copper seed layers from discontinuous PVD Cu liners was demonstrated. In particular, it was shown that barrier oxidation during air exposure can be overcome by appropriate surface pretreatment, which is a key result in the development of direct on barrier plating.

The SLE process was easily and successfully integrated in the metallization sequence, allowing void-free filling of 3x15 μm TSVs and yielding good electrical properties.

ACKNOWLEDGEMENTS

This work has been carried out in the frame of the Leti/STMicroelectronics collaboration, and within the Semitool/Leti Joint Development Program. The authors acknowledge Rohm And Haas Electronic Materials for providing SLE electrolyte. They also wish to thank Céline Jayet and David Bouchu for assistance in FIB/SEM analysis.

REFERENCES

1. Yole Development, 3D IC&TSV Report, http://www.yole.fr/pagesAN/products/Report_sample/3DIC.pdf.
2. C. Charles-Alfred and V. Jousseaume, Surf. Coat. Technol. 201, 9260-9263 (2007).
3. V. Jousseaume, L. Favennec, A. Zenasni and O. Gourhant, Surf. Coat. Technol. 201, 9248-9251 (2007).
4. P. Garrou, S Vitkavage and S. Arkalgud in Handbook of 3D Integration: Technology and Applications of 3D Integrated Circuits, edited by P. Garrou, C. Bower, P. Ramm (Wiley-VCH, 2008), pp. 13-24.
5. R. Beica and P. Siblerud, ECS Trans. 18(1), 689-694 (2009).
6. L.Chen and T. Ritzdorf, Semiconductor Fabtech 12, 267-271 (2000).
7. P. Leduc et al. in Proc. IEEE Very Large Scale Integration (VLSI) Technology System-and-Applications, (IEEE CS Press 2008), pp. 76-78.

Conference Proceedings AMC XXV © 2010 Materials Research Society

Application of Self Assembly Monolayer (SAM) in Cu-Cu Bonding Enhancement at Low Temperature for 3-D Integration

Dau Fatt Lim[1], Shiv Govind Singh[1], Xiao Fang Ang[2], Jun Wei[2], Chee Mang Ng[3], and Chuan Seng Tan[1,*]
[1] School of Electrical and Electronic Engineering, Nanyang Technological University, Singapore
[2] SIMTech, Singapore
[3] Chartered Semiconductor Manufacturing Company, Singapore
*E-mail: tancs@ntu.edu.sg

ABSTRACT

In our work, self assembled monolayer (SAM) of alkane-thiol of 6-carbon (1-hexanethiol, C6) and 12-carbon (1-dodecanethiol, C12) chain length are applied on Cu surface (deposited on Si substrate) and examined carefully. Firstly, the ability of SAM adsorption onto Cu surface is confirmed by the reduction of the surface roughness and the sharp rise of water contact angle (CA) on the surface. Next, SAM stability when it is subjected to various exposure times in ambient environment and annealing in inert ambient are also studied. The CA for both cases decreases as compared to freshly prepared SAM samples. This indicates that SAM degradation or desorption has occurred during exposure to ambient environment and annealing. The desorption behavior of SAM is found to be reversely proportional to the immersion time in SAM solution, whereby longer immersion time shows less degradation and desorption. C12 exhibits a higher resistance to degradation and higher thermal stability as compared to C6 under the above conditions as seen from a slower drop in CA. This is due to the fact that longer chain length has a denser molecular arrangement and hence stronger inter-molecular force which results in better thermal stability. Next, SAM formation is further analyzed by x-ray photoelectron spectroscopy (XPS) to verify its adsorption and desorption properties. It is shown that SAM is able to retard metal oxide formation. The degree of retardation of oxide formation is also found to be proportionate with longer immersion time. The results show that SAM is able to retard copper surface oxidation in the air and the degree of SAM desorption through annealing is proportionate to annealing time and is temperature highly dependent. Finally, bonding experiments are performed to validate effectiveness of SAM in tailoring the Cu surface for bonding enhancement at low temperature.

INTRODUCTION

Thermal budget control by lowering the bonding temperature in three-dimensional integration (3-D IC) flow is critical in order to achieve improved alignment accuracy, lower thermal stress, and lower production cost [1] . However, untailored Cu-Cu diffusion bonding often requires a high temperature bonding (>350 °C) to overcome the metal oxide layer in order to achieve high quality bonding. In this research article, we examine the feasibility of applying a self assembled monolayer (SAM) film to tailor the surface property of Cu for bonding enhancement at low temperature (Figure 1). SAM is readily absorbed onto the Cu surface and organized into a uniform organic layer due to its specific high affinity functional group towards the metal surface. The ability of SAM to retard the Cu surface from oxidation [2-4] and protect Cu surface from contaminants [5] make it an attractive candidate to tailor the Cu surface prior to bonding. Degradation and desorption of SAM behavior is studied here in order to determine the effectiveness of SAM to protect Cu surface. While it is imperative for SAM to protect the Cu surface, the desorption of SAM prior to bonding is as important in order to allow Cu atoms diffusion for successful bonding.

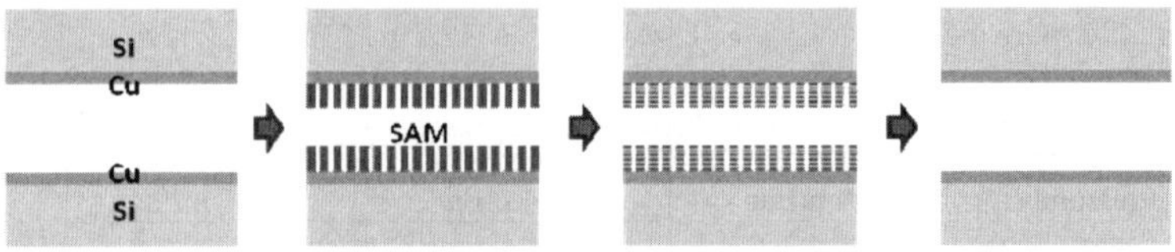

Figure 1: The use of SAM for Cu surface passivation in order to achieve ultra clean surface for bonding enhancement at low temperature.

EXPERIMENT DETAILS

Sample preparation

6 inch *n*-type silicon wafers are first cleaned with standard *RCA* clean prior to the thermal oxidation to form 500 nm thick of silicon dioxide. *Piranha* clean is performed next for pre-metal cleaning. Metal deposition is done consecutively in a sputtering chamber for 50 nm of Ti (as Cu diffusion barrier) and 300 nm of Cu. As-sputtered Cu wafers are separately immersed into two different alkane-thiol solutions: (1) 1mM of 1-hexanethiol, C6, 95% and (2) 1mM of 1-dodecanethiol, C12, ≥98%. Both alkane-tiols are dissolved in ethanol and stored in nitrogen-purged dry box for a specified time interval. Cu wafers are then taken out and rinsed with copious amount of ethanol to remove the unabsorbed molecules and blown dry under nitrogen gas. A series of exposure time in two different ambients are then performed onto the self-assembled monolayer (SAM) coated Cu wafers, namely: (1) room temperature ambient; and (2) high temperature inert ambient (N_2). Various analyses are then performed to characterize the modification of the surface properties.

Surface analysis

Atomic force microscopy (AFM) is utilized in order to determine the modification of surface roughness when SAM is applied onto the Cu surface. Water contact angle measurement is extensively done by static sessile drop method to examine the surface properties change of test samples when exposed in different environments. The surface wetability, as determined by the contact angle profile captured by goniometer, can provide a direct indication of the modification of SAM in different environments. Additionally, XPS with a non-monochromatized Mg Kα radiation is performed to analyze the change in SAM properties more qualitatively.

Wafer bonding

High temperature N_2 ambient annealing and wafer bonding process are performed in a wafer bonder. Different bonding experiments are undertaken for wafer pairs with and without SAM coating as shown in Table I. The bonding process is as described in Figure 1. C-mode scanning acoustic microscopy (c-SAM) is used after bonding process is completed to observe the bonding interface quality. The c-SAM pictures are taken under a c-SAM tool using transducer V390 at 50 MHz. By using c-SAM, bonding interface uniformity and voids can be observed non-destructively.

Table I: Conditions of wafer bonding experiments.

	Figure 6(a)	Figure 6(b)	Figure 6(c)
Chamber Pressure	1E-4 mBar	1E-4 mBar	1E-4 mBar
Tool Pressure	2500 mBar	2500 mBar	2500 mBar
Purging Gas	N_2	N_2	N_2
Days of exposure	5days	5days	5days
Bonding Temperature	250°C	250°C	250°C
Bonding Duration	1 hour	1 hour	1 hour
SAM	No	Yes, C6	Yes, C6
Pre-bonding Annealing	Yes	Yes	No
Annealing	250°C / 30 minutes	250°C / 30 minutes	No

RESULTS AND DISCUSSIONS

Surface analysis

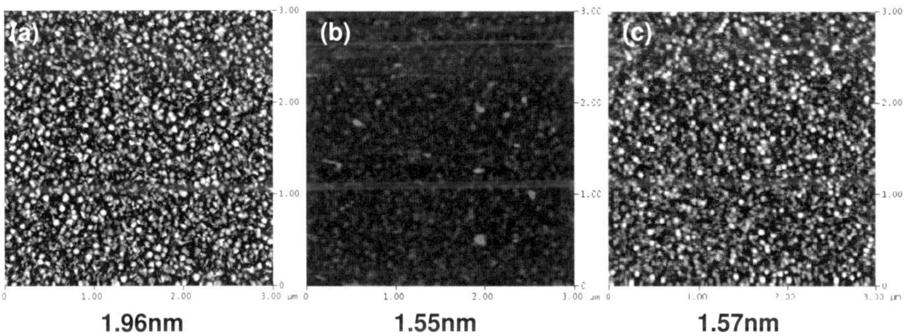

Figure 2: AFM scans (3μm x 3μm) for: (a) blanket Cu surface without SAM, (b) Cu surface after SAM-C6 immersion for 2 hour, and (c) Cu surface after SAM-C12 immersion for 2 hour.

The adsorption of SAM on the Cu surface is first confirmed with the reduction in the surface RMS roughness [5], from 1.96 nm (fresh Cu) to 1.55 nm (C6) and 1.57 nm (C12) respectively after 2 hours of SAM immersion, as shown in Figure 2. In Figure 3, the sharp rise in the contact angle (CA) value whereby Cu surface turns from hydrophilic to hydrophobic also indicates that both SAM are adsorbed onto the Cu surface. The contact angle values in Figure 3 also plateau with SAM solution immersion time. This is because adsorption of SAM has saturated beyond certain immersion duration when the Cu surface is densely populated with SAM and further immersion therefore does not change the CA value appreciably. The changes in the surface properties of SAM-coated Cu surface and the as-sputtered Cu surface over time in room temperature ambient are also shown in Figure 3. Both C6 and C12 SAM are observed to degrade from the decrease in the CA values as exposure time increases in Figure 3. The degradation is likely due to the sulfonation of the thiolate group that is followed by the breaking of the carbon chain. Similar observation has also been reported [3, 6]. C12 shows higher resistance to degradation as compared to C6 because longer carbon chain length in C12 has a stronger barrier against oxidation [3, 7]. The degradation is also slowed down with a longer immersion time. This is because longer immersion time can improve SAM conformality [5] and hence provide more resistance to degradation. In contrast, the contact angle of as-sputtered Cu surface (with no SAM adsorption) increases over time, and this is most likely due to surface oxide formation over time when SAM is absent from the Cu surface.

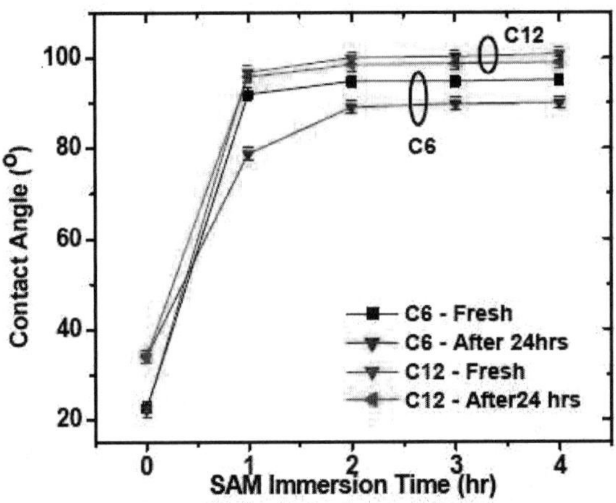

Figure 3: Comparison of SAM-C6 and -C12 degradation in ambient air versus immersion time in SAM solution. Longer immersion time has a higher resistance against degradation. C12 shows lesser degradation as compared to C6.

The adsorption and degradation properties of SAM C6 are studied further using XPS as shown in Figure 4. Figure 4(a) shows the ratios of the relative percentage of atomic concentration of the C6-coated Cu surface with immersion time. The adsorption of SAM onto Cu surface is confirmed by an increase in the S/Cu ratio. Besides, O/Cu ratio is also found to reduce with immersion time and this behavior is described in [8]. Figure 4(b) shows the change in the ratios of the relative percentage of atomic concentration of the C6-coated Cu surface (with 2 hours immersion) with different exposure times. O/Cu ratio shows a slight increase after 1 day of exposure, however this value is still lower than the corresponding value taken from blank Cu surface with no C6 adsorption (Figure 4(a)) and this suggests that oxidation process is retarded on Cu surface coated with C6.

SAM degradation at high temperature inert ambient (N_2) annealing is also studied, as shown in Figure 5. In Figure 5(a), the contact angle is seen to decrease as the duration of 150°C annealing increases. It is also observed the drop in contact angle at higher immersion time (up to 21 hours) is less significant as compared with samples with shorter immersion time. This is because SAM conformality improves with immersion time [5] as described earlier. Since it is crucial to determine the optimal conditions for complete desorption of SAM in order to achieve successful bonding, higher annealing temperature of 200°C and 250°C are also employed. Figure 5(b) shows the contact angle measurement on the samples (with 2 hours of SAM immersion time) after higher temperature annealing (150 - 250°C) for 30 minutes. A drastic drop in contact angle of C6 sample is observed around 200°C, and the contact angle continues to drop further to around 30° after annealing at 250°C. Similarly, contact angle of C12 sample is observed to drop sharply after annealing at temperature > 200°C. A sharp drop in the contact angle measurement as observed in Figure 5(b) implies that SAM can be effectively desorbed from Cu surface via thermal means. As C12 has longer carbon chain than C6, the critical temperature for abrupt desorption is also slightly higher. Similar finding has also been observed by monitoring SAM desorption peaks at 200°C with other means [8].

(a)

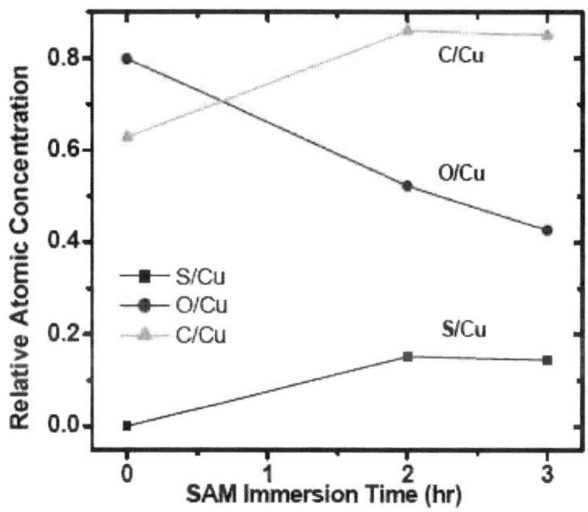

Figure 4 (a): XPS analysis on C6-coated Cu surface showing the relative atomic composition ratio versus immersion time.

(b)

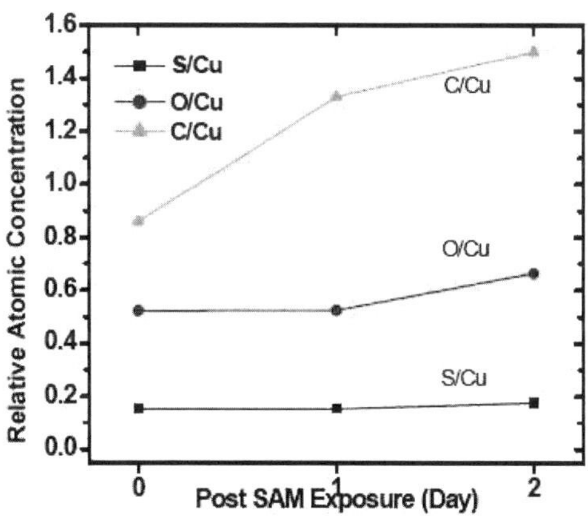

Figure 4(b): XPS analysis on C6-coated Cu surface showing the composition ratio after exposure in ambient air. SAM immersion time is 2 hours.

(a)

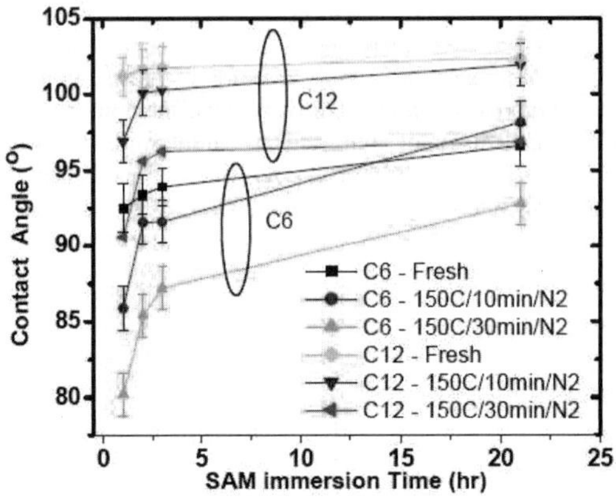

Figure 5(a): Comparison of C6 and C12 SAM degradation in inert ambient after anneal (150°C/N2) versus the immersion time in SAM solution. Longer annealing time results in more degradation for both SAM.

(b)

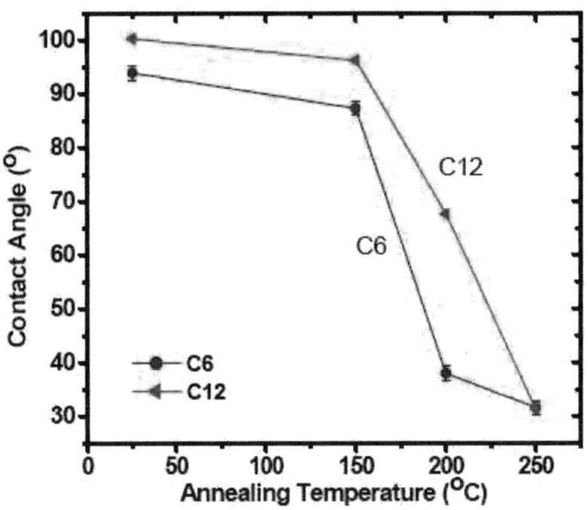

Figure 5(b): Comparison of C6 and C12 SAM desorption in inert ambient with annealing (150°C to 250°C, 30 min) with 2 hours of immersion in SAM solution. SAM desorption occurs at a critical temperature of >200°C.

Wafer Bonding

C-SAM pictures are taken for bonded wafer pairs to verify the ability of SAM to protect the Cu surface from oxidation and contaminants. Figure 6(a) shows the c-SAM image for bonded wafer pair without SAM coating. It is observed that non-uniform bonding and "stain-like" dark patches (especially at the wafer edge) exist at the bonded interface. These dark patches are probably due to Cu oxide formation or surface contaminants during exposure which inhibits uniform bonding. On the other hand, Figure 6(b) shows the c-SAM image for bonded wafer pair with SAM coating that is annealed (i.e. thermal desorption) before bonding taking place. The rationale of the annealing step is to desorb the SAM before Cu bonding process take place. An improved and uniform bonded interface is observed with no observable "stain-like" patches. The "bubble" like voids in Figures 6(a) and 6(b) at the center of bonded wafers are most likely due to inherent wafer bow which deter uniform bonding across the entire wafer.

Figure 6(c) highlights the importance of SAM desorption in order to achieve a high quality bonding interface. Without the additional annealing step, only rim of the wafers is bonded and a large void is observed at the center of the bonded wafers. This is because the non-desorbed SAM acts as a barrier layer that inhibits atomic diffusion of contacted Cu-Cu interface from bonding. However, Cu-Cu is observed to bond together at the rim of the wafer pair, and this is most likely due to the fact that SAM are able to desorb more readily at this area as they are under higher contact pressure due to the wafer bow and SAM are mainly desorbed via mechanical means. Besides, SAM at this area may also be able to escape due to its close proximity to the non-contacted open space.

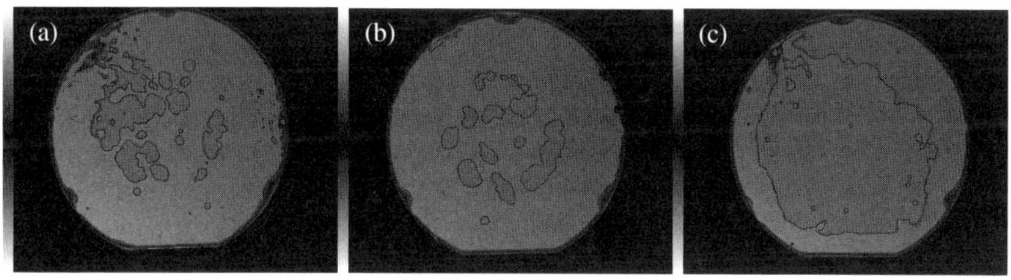

Figure 6: (a) c-SAM image for bonded wafer pair without SAM. Non uniform bonding and interface voids are observed. (b) c-SAM image of bonded wafer pair with SAM and pre-bonding annealing. The bonding interface improves with SAM passivation. (c) c-SAM image of bonded wafer pair with SAM but without pre-bonding annealing. A large void area is observed at the center of the bonded wafer pair implies that pre-bonding annealing for SAM desorption is vital to achieve uniform bonding.

CONCLUSIONS

SAM of 1-hexanethiol and 1-dodecanethiol have been shown to absorb onto Cu surface and its degradation and desorption property in two different environments are studied. When it is exposed to the clean room ambient, SAM degradation is inversely proportionate to the immersion time. 1-Dodecanethiol SAM shows a larger degradation resistance due to more crystalline nature of longer carbon chain length. Significant desorption for both SAM is observed after annealing at 200°C and higher temperature. From the bonding experiments using 1-exanethiol SAM, bonding quality is improved. The complete desorption of SAM before bonding take place via annealing is also found to be critical in order to achieve successful and large area bonding. In conclusion, adsorption of SAM is able to protect or retard Cu surface from contaminants and oxidation and it can be thermally desorbed to provide an ultraclean surface for low temperature (~250°C) bonding of Cu. This process finds potential application in realizing 3-D integration of integrated circuits.

ACKNOWLEDGEMENTS

Authors are grateful with funding support from Nanyang Technological University through an award of Nanyang Assistant Professorship and the Semiconductor Research Corporation (SRC) through a subcontract (R0286-S3) from Georgia Institute of Technology.

REFERENCES

1. Tan, C.S., Gutmann, R.J. and Reif, L.R., Overview of Wafer-Level 3D ICs. (Springer, New York, 2008) p. 1.
2. Hutt, D.A. and Liu, C., *Oxidation protection of copper surfaces using self-assembled monolayers of octadecanethiol.* Applied Surface Science, 2005. **252**(2): p. 400-411.
3. Laibinis, P.E. and Whitesides, G.M., *Self-assembled monolayers of n-alkanethiolates on copper are barrier films that protect the metal against oxidation by air.* Journal of the American Chemical Society, 2002. **114**(23): p. 9022-9028.
4. Metikos-Hukovic, M., et al., *Copper protection by a self-assembled monolayer of alkanethiol.* Journal of the Electrochemical Society, 2007. **154**(2): p. 138-43.
5. Love, J.C., et al., *Self-assembled monolayers of thiolates on metals as a form of nanotechnology.* Chemical Reviews, 2005. **105**(4): p. 1103-1169.
6. Sung, M.M., Yun, W.J., Lee, S.S., Kim, Y., *Thermal Decomposition of Octanethiolate Self-Assembled Monolayers on Cu(111) in UHV* Bull. Korean Chem. Soc., 2003. **24** (5).
7. Srivastava, P., Chapman, W.G. and Laibinis, P.E., *Molecular dynamics simulation of oxygen transport through n-alkanethiolate self-assembled monolayers on gold and copper.* Journal of Physical Chemistry B, 2009. **113**(2): p. 456-464.
8. Carbonell, L., et al., *A thermal stability study of alkane and aromatic thiolate self-assembled monolayers on copper surfaces.* Superlattices and Microstructures, 2004. **36**(1-3): p. 149-160.

Emerging Applications

Conference Proceedings AMC XXV © 2010 Materials Research Society

Metallic Conduction in ZnO/Cu/ZnO Thin Films

K. Sivaramakrishnan[1], N. D. Theodore[2], J. F. Moulder[3] and T. L. Alford[1,a)]

[1]School of Materials and Flexible Display Center at ASU, Arizona State University, Tempe, Arizona 85287, U.S.A.

[2]Freescale Semiconductor Inc. and Arizona State University, Tempe, Arizona 85284, U.S.A.

[3]Physical Electronics, Chanhassen, Minnesota 55317, U.S.A.

ABSTRACT

ZnO/Cu/ZnO multilayer films have been obtained with carrier concentration of 1.2×10^{22} cm^{-3} and resistivity of 6.9×10^{-5} Ω-cm at the optimum copper layer thickness. The conduction mechanism involves metal to oxide carrier injection at low copper layer thickness prior to the formation of a continuous metal conduction pathway at higher copper thicknesses. The peak transmittance, photopic averaged transmittance, and Haacke figure of merit are 88%, 75%, and 8.7×10^{-3} Ω^{-1}, respectively. Optical transmission is elucidated in terms of copper's absorption due to *d*-band to Fermi surface transitions at short wavelengths and reflectance combined with scattering losses at long wavelengths. A Burstein-Moss shift in the band gap of the films is seen to take place with increase in thickness of the copper layer. Annealing at 150 °C for up to 24 hours in a reducing ambient(Ar + 5%H_2) was found to have negligible effect on the properties of the films adding to the significance of the room temperature results obtained.

INTRODUCTION

Transparent conducting oxides (TCOs) exhibit simultaneous conduction and optical transmission, a pair of properties which are usually mutually exclusive.[1] ZnO is a well-known transparent conducting oxide. It is a useful Ohmic contact for several systems including GaN based LEDs[2], GaAs based room temperature radiation and particle detectors[3], and AlGaN/GaN high electron mobility transistors used in developing high-frequency and high-power devices for microwave and millimeter wave applications.[4] The key property of an Ohmic contact is high conductivity but good transmission properties of the contact can help improve efficiency in optical applications like LEDs. For instance, it has been observed that the light out power of an LED with a doped zinc oxide based contact improved 34% compared to that of an LED using a Ni/Au film. [2] Doped zinc oxide (ZnO) films are known to exhibit good electrical conduction and optical transmission in the visible region.[5, 6] but are limited by impurity scattering at high carrier concentrations. As a consequence, TCOs with sufficiently low resistivity and transparency are still unavailable for many applications. Dielectric-metal-dielectric thin films have been studied in the hope of obtaining near metallic conductivity

a)Electronic mail: TA@asu.edu

without significant losses in transmission.[7,8] Several studies have been carried out using a silver sandwiched layer.[5,8,9] Copper, with a comparable resistivity has also been tried. [10,11] However, there is a lack of understanding of the role of copper in affecting the conduction and transmission behavior. In this study, we prepare ZnO/Cu/ZnO films on flexible polyethylene naphthalate (PEN) substrates with the highest conductivity of dielectric-metal-dielectric thin film structures reported in literature, while retaining transmittance values acceptable for optoelectronics applications. We have investigated the role of copper and its thickness in affecting the film properties.

EXPERIMENTAL DETAILS

Thin film layers of ZnO were deposited by rf magnetron sputtering onto PEN substrates at 10 mTorr pressure and 100 W power. Copper was deposited by dc magnetron sputtering at 10 mTorr and 40 W. The base pressure prior to deposition was approximately $1x10^{-7}$ Torr. Both the top and bottom ZnO layers were 30 nm thick; while, the copper thicknesses varied between 3 nm and 8 nm. Bare ZnO layers were prepared under identical conditions for comparison of properties.

Rutherford backscattering spectrometry (RBS) was used to determine the thickness of the films. The analysis was performed in a vacuum of 10^{-6} Torr using a 2.0 MeV He^{++} ion beam. Energy spectra were analyzed using the RUMP computer simulation program.[12] Transmission electron microscopy (TEM) using a Philips CM200-FEG TEM operating at 200kV was used to confirm the thickness of the Cu midlayer and study the continuity of the metal layer.

Electrical resistivity, Hall mobility, and carrier concentration of the films were measured by Hall measurements in the van der Pauw geometry using an Ecopia HMS-3000 instrument. Four-point-probe analysis, to determine the sheet resistance, was carried out using a Keithley 2700 data acquisition system.

Optical transmittance and reflectivity were measured using an Ocean Optics double channel spectrometer (model DS200) in the wavelength range 300-900 nm with an aluminum mirror as the reference for reflectivity and a bare PEN substrate for transmittance.

X-ray photoelectron spectroscopy (XPS) sputter depth profiles obtained using a Physical Electronics Quantera XPS microprobe were used to determine the chemical state of the buried metal layers. XPS spectra were acquired using a focused, monochromatic, Al K_α x-ray beam that was adjusted to be 200 μm in diameter. The sputter etch rate for an SiO_2 thin film reference sample was approximately 1.25 nm/min.

X-ray diffraction (XRD) in the Bragg-Brentano configuration (θ-2θ geometry) was used to determine crystallinity while pole figure analysis using a Philips X'pert multipurpose diffractometer system was used to examine the evolution of (002) ZnO texture in the films. The sample tilt (ψ) was varied from 0° to 85°. Cu K_α radiation was used for the incident X-ray beam.

Annealing of the samples was done in a reducing atmosphere (Ar+5% H_2) at 150 °C for up to 24 hours.

DISCUSSION

Figure 1 shows a glancing angle XRD profile from one of the ZnO/Cu/ZnO multilayer films. The tiny peak at 34.4° corresponds to the ZnO (002) set of planes.[13] The low intensity of this peak indicates that the ZnO layers are mostly amorphous.

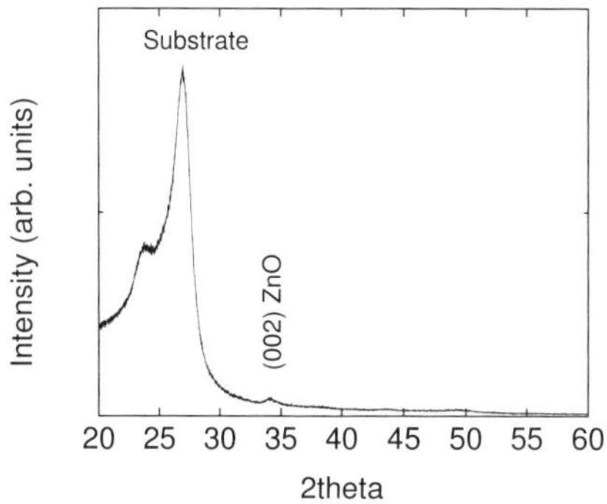

Figure 1. Glancing angle XRD profile showing the largely amorphous structure of ZnO in the ZnO/Cu/ZnO multilayers.

The variation of resistivity and sheet resistance of the ZnO/Cu/ZnO films as a function of the thickness of the copper layer is shown in figure 2. It is well-known that as-deposited ZnO films can vary considerably in their resistivity.[14-16]

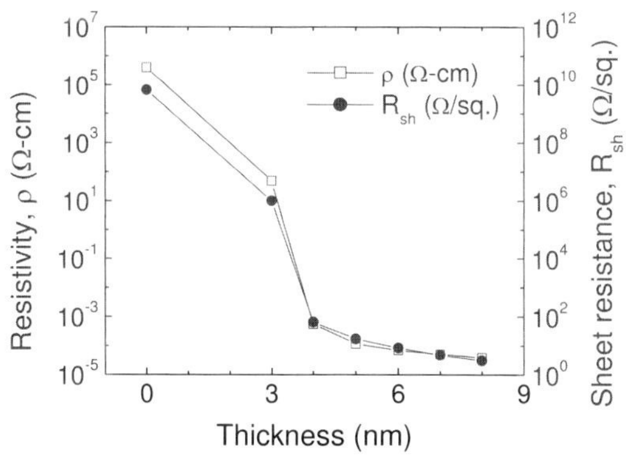

Figure 2. Resistivity and sheet resistance of the ZnO/Cu/ZnO thin films as a function of copper layer thickness.

The resistivity of the as-deposited ZnO films decreases considerably from 0.46 MΩ-cm for 0 nm of Cu thickness to 3.87×10^{-5} Ω-cm at 8 nm of Cu thickness while the sheet resistance decreases dramatically from 6000MΩ/sq. to 3.06 Ω/sq. with increase in thickness of the copper midlayer. The variation of resistivity and sheet resistance is a consequence of changes in carrier concentration and mobility. Figure 3 shows the variation of carrier concentration in the ZnO/Cu/ZnO films as a function of copper layer thickness.

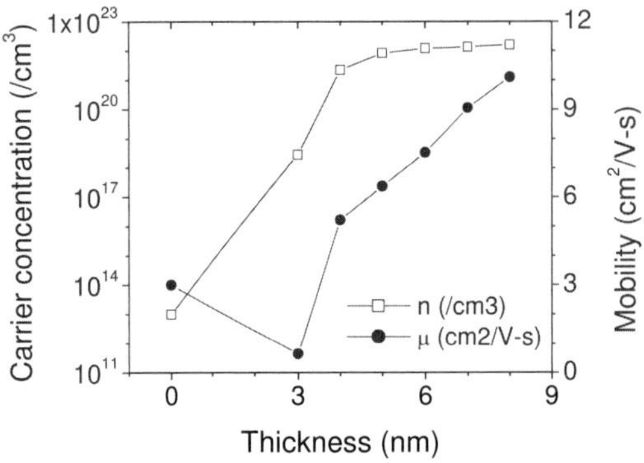

Figure 3. Carrier concentration and mobility of the ZnO/Cu/ZnO thins films as a function of copper layer thickness.

The carrier concentration increases with increasing copper midlayer thickness from 3×10^{13} cm^{-3} for undoped ZnO on PEN by nine orders of magnitude to 1.59×10^{22} cm^{-3}. However, it can be seen that most of the increase happens before the Cu midlayer reaches 5 nm in thickness. Beyond that there is a small rise in carrier concentration up to 8 nm of Cu thickness. Variation of Hall mobility with thickness of the copper layer is also shown in figure 3. The mobility of the as-deposited ZnO films is 2.8 cm^2/V-s. The mobility drops to a low 0.64 cm^2/V-s for films with a 3 nm copper midlayer. Thereafter, the mobility increases with increasing copper midlayer thickness and reaches 10.1 cm^2/V-s at 8 nm of Cu thickness. To understand this behavior it is essential to look at the energy band diagrams for Cu and ZnO. Copper has a work function of ϕ = 4.5 eV [17] and ZnO has a work function of ϕ=5.4 eV.[18] The presence of a copper metal layer of lower work function with respect to the semiconducting ZnO film leads to the formation of an accumulation layer at the interface due to transfer of electrons from the metal to the semiconductor across the interface in order for the Fermi levels to align. As the work function difference is large, there is significant injection of carriers into the ZnO layer. This is seen in the substantial decrease in both resistivity and sheet resistance of the multilayer structure even with 3 nm of copper film in between the ZnO layers. However at this Cu thickness, the ZnO/Cu/ZnO multilayer structure has a lower Hall mobility than the ZnO film with no Cu, as shown in figure

3. This suggests that most of the current passes through the amorphous ZnO layer with Cu islands acting as discontinuous scattering sites that reduce the mobility below that of a bare ZnO layer. As the copper midlayer thickness increases, the carrier concentration increases by another 3 orders of magnitude from 2.9×10^{18} cm^{-3} to 2.2×10^{21} cm^{-3}. The mobility is seen to increase from 0.64 cm^2/V-s to 5.22 cm^2/V-s which is higher than for bare ZnO. This suggests that the copper layer becomes near-continuous to electron conduction and allows a significant amount of current to pass through it. Beyond this thickness, the carrier concentration increases only slowly from 1.2×10^{22} cm^{-3} at 6 nm of Cu midlayer thickness to 1.6×10^{22} cm^{-3} at 8 nm of Cu thickness. So it can be seen that the rate of decrease in resistivity drops as the copper layer thickness increases, particularly when the copper layer is contiguous. This behavior is likely due to the resistivity values approaching closer to the bulk resistivity value of copper.

The Haacke figure of merit, φ_{TC}, is a commonly used standard for estimating the quality of transparent conductors defined as[19]:

$$\varphi_{TC} = \frac{T_{av}^{10}}{R_{sh}} \quad (1)$$

where T_{av} is the transmittance and R_{sh} is the sheet resistance. This figure of merit is used in stead of a simple ratio of T_{av}/R_{sh} because for a given thickness we get a maximum figure of merit at 90% average transmittance by this approach in stead of a maximum φ_{TC} at 37% average transmittance when using the *T/R* ratio, thus making it a better measure of the quality of a transparent conductor. The average transmittance was calculated as:

$$T_{av} = \frac{\int V(\lambda)T(\lambda)d\lambda}{\int V(\lambda)d\lambda} \quad (2)$$

where $T(\lambda)$ is the transmittance and $V(\lambda)$ is the photopic luminous efficiency function defining the standard observer for photometry.[20] The average transmittance T_{av} has been calculated by the above method because the photopic function is a measure of the sensitivity of the human eye to color. By taking a weighted average of the transmittance values with the sensitivity of the eye, in stead of taking a linearly averaged value of transmittance or using the peak value of transmittance, a more accurate measure of the average transmittance is obtained keeping in mind the application of these films to flexible display devices. The Haacke figures of merit obtained are 0.3×10^{-3}, 3.4×10^{-3}, 6.8×10^{-3}, 8.7×10^{-3}, and 7.9×10^{-3} Ω^{-1} for the 4, 5, 6, 7, and 8 nm films, respectively. The best figure of merit is obtained when the copper layer is just continuous.

Figure 4 shows optical transmittance spectra relative to the PEN substrate for ZnO/Cu/ZnO thin films with different copper thicknesses. The optical transmittance drops from about 90% over the visible range of wavelengths (400 nm – 700 nm) for undoped ZnO films to between 70 % and 75 % upon insertion of a copper layer. However, the variation in transmittance with copper thickness is different in the short and long wavelengths. In the short wavelengths near the blue part of the visible spectrum, as the thickness of the copper layer increases the transmission decreases because the reflectance of copper is low and transmission is mainly affected by absorption of light due to interband electronic transitions, particularly, due to the excitation of electrons from the *d*-band to the Fermi surface.[21] The number of bound electrons available for excitation increases with thickness leading to decreasing transmission. In the red part of the visible spectrum, copper films are reflective.[22] For the film with the 8 nm copper midlayer, the transmittance is low due to high reflectance in red part of the spectrum. As the copper layer thickness decreases from 8 nm to 6 nm transmission increases; however, below this thickness, the transmittance decreases again suggesting that the presence of discontinuous copper islands leads to significant scattering losses. The optical transmittance results therefore substantiate the inference of formation of a contiguous copper conduction layer at 6 nm of Cu thickness arrived at based on conduction behavior. Free carrier absorption is significant at longer wavelengths.[23] However, the screening effect of the bound electrons leads to an increase in transmission with decreasing wavelength in the wavelengths beyond 580 nm, which is known to be the crossover point.[21]

The change in absorption coefficient with incident energy for the multilayer films with varying copper thicknesses is shown in figure 5. The energy gap E_g is determined as:[24]

$$\alpha(h\nu) \propto (h\nu - E_g)^{1/2} \quad (3)$$

where $h\nu$ is the photon energy and α is the absorption coefficient.

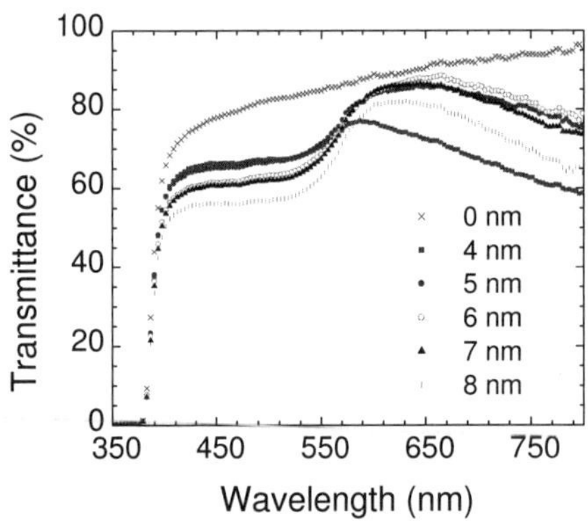

Figure 4. Optical transmittance spectra relative to the PEN substrate for ZnO/Cu/ZnO thin films as a function of Cu.

The optical band gap is deduced from the absorption edge by linear extrapolation back to zero. ZnO is reported to have an optical band gap of ~3.2-3.3 eV.[25, 26] The optical band gap obtained from figure 5 is ~3.21 eV. Though this value is within the range for ZnO band gap, it is more advantageous to utilize the combined band gap of the ZnO+PEN structure because PEN is known to have a band gap of ~3.2 eV.[6] From figure 5 it is evident that the band gap of the ZnO films undergo a blue shift with increasing thickness of the copper layer. The increase in carrier concentration with increased copper thickness leads to filling of the small density of states of ZnO near the conduction band minimum.[27] This results in a shift in the absorption edge to higher energies, known as the Burstein-Moss effect.[28]

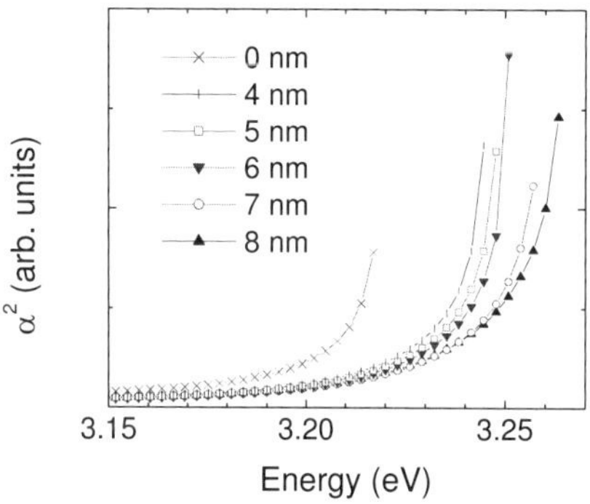

Figure 5. Absorption coefficient versus energy of photons showing the Burstein-Moss effect in the ZnO/Cu/ZnO multilayer films with increasing copper thickness.

Figure 6 (a) is a three dimensional representation of XPS intensity of the Cu LMM line for the metallic state of copper(binding energy = 568.2 eV)[29] versus depth of sputter for the multilayer film with 8 nm copper thickness. Figure 6(b) is a 2-D representation of the same information. The position of the Cu LMM peak can be seen to be unchanged across the thickness of the copper layer. This confirms that the copper is in metallic state in the films, supporting the theory of metallic conduction presented earlier.

Pole figure profile of ZnO (002) texture with intensity plotted against angle of tilt (ψ) is shown in figure 7. The two thicker films of 7 and 8 nm Cu thickness have been plotted along with the two thinner films of 3 and 4 nm thickness. It is interesting to note that there is a marginal decrease in texturing with increase in thickness of the copper midlayer. The result suggests that the copper layers being thin and discontinuous act more as an interference to alignment between the bottom and top ZnO layers than as seed layers for further texturing. This confirms that the increase in mobility seen at higher Cu thicknesses occurs through metal layer conduction and not due to increased texturing of the ZnO layers.

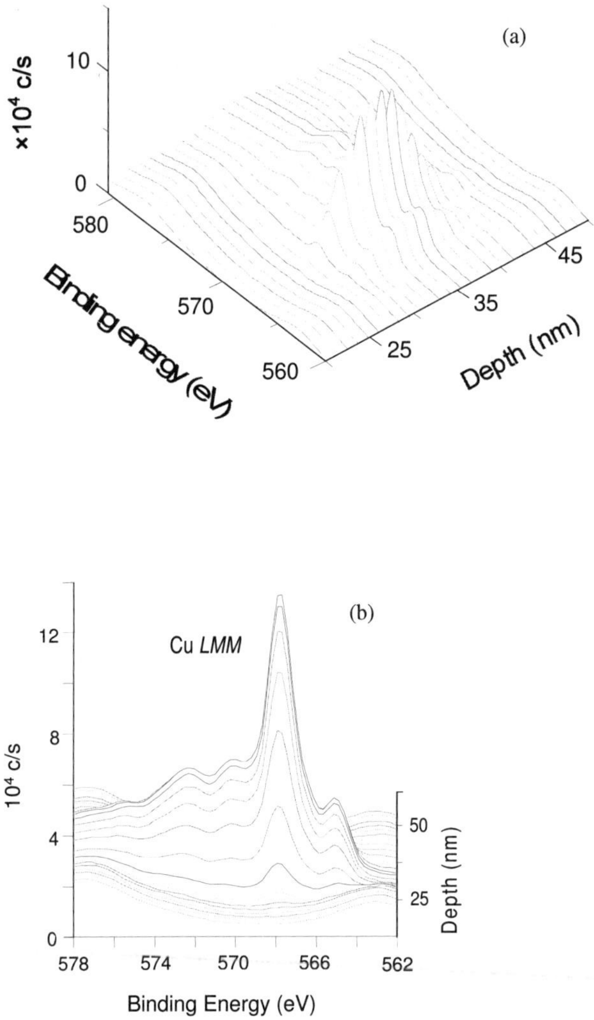

Figure 6. XPS profile of the ZnO/Cu/ZnO samples showing intensity of Cu LMM line with depth of sputter for 8 nm thick Cu layer structure in (a) 3-D and (b) 2-D.

Fig. 8 (a) shows a cross-sectional TEM micrograph from the multilayer structure with a 3 nm copper midlayer. The Cu layer is seen to be in the form of poorly connected islands. Fig. 8 (b) shows the formation of a contiguous layer at 6 nm of copper layer thickness and Fig. 8 (c) shows a TEM micrograph from the multilayer structure with an 8 nm copper midlayer. The islands are contiguous thus providing a continuous path for electron conduction as was expected based on Hall conductivity results. The TEM images help to affirm the theory of conduction and transmission hypothesized in this paper.

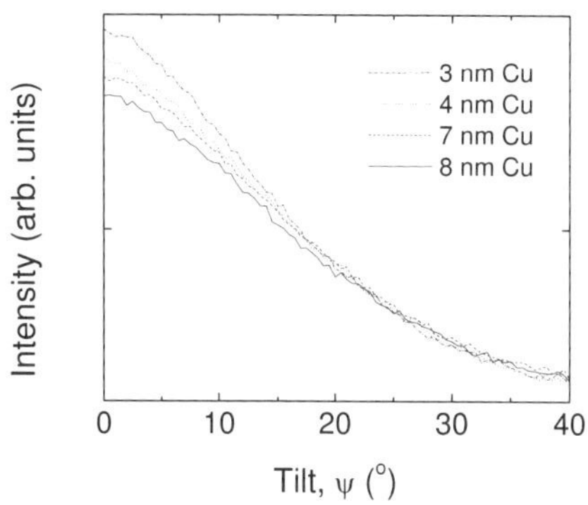

Figure 7. Pole figure profile of ZnO (002) texture showing intensity plotted against angle of tilt, ψ.

Annealing of the films for up to 24 hours in a reducing ambient at 150 °C didn't result in any significant change in conduction and transmittance properties of the films. However, PEN cannot be taken to higher temperatures and held for long times without risk of damaging the substrate.

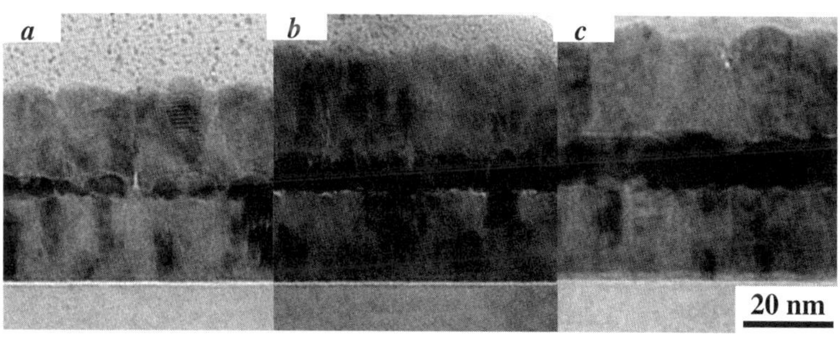

Figure 8. Cross-section TEM micrographs of ZnO/Cu/ZnO films with varying Cu thickness: (a) 3 nm, (b) 6 nm, and (c) 8 nm.

CONCLUSIONS

In conclusion, very high conductivity ZnO/Cu/ZnO films with good transmittance for optoelectronic device applications have been obtained by using the magnetron sputtering technique. For structures with 6 nm copper layer, the carrier concentration is up to 1.2×10^{22} cm^{-3} and resistivity is 6.92×10^{-5} Ω-cm; while, the peak transmittance is still as high as 88%. The film also had a Haacke figure of merit [19] of 6.7×10^{-3} Ω^{-1}. The room temperature properties of these films make them a better choice than metallic contacts for use in photovoltaic and optoelectronic devices.

ACKNOWLEDGMENTS

This work was partially supported by the National Science Foundation (L. Hess, Grant No. DMR-0602716), to whom the authors are greatly indebted. Research was sponsored by the Army Research Laboratory (ARL) and was accomplished under Cooperative Agreement NO. W911NG-04-2-0005. The views and conclusions contained in this document are those of the authors and should not be interpreted as representing the official policies, either expressed or implied, of the ARL or the U.S. Government. The U.S. Government is authorized to reproduce and distribute reprints for Government purposes notwithstanding any copyright notation hereon.

REFERENCES

1. G. J. Exarhos, and X. D. Zhou, Thin Solid Films **515**, 7025 (2007)
2. J. H. Lim, D. K. Hwang, H. S. Kim, J. Y. Oh, J. H. Yang, R. Navamathavan, and S. J. Park, Appl. Phys. Lett. **85**, 6191 (2004)
3. V. Palumbo, A. Valentini, A. Cola, and F. M. Quaranta, Solid-State Elec. **43**, 2021 (1999)
4. K. Nishizono, M. Okada, M. Kamei, D. Kikuta, K. Tominaga, Y. Ohno, and J. P. Ao, Appl. Phys. Lett. **84**, 3996 (2004)
5. S. Kim, Y. Moon, D. Moon, M. Hong, Y. Jeon, and J. Park, J. Korean Phys. Soc. **49**, 1256 (2006)
6. D.C. Look, Semicond. Sci. Technol. **20**, S55 (2005)
7. C. Guillen, and J. Herrero, Solar Energy Mat. & Solar cells **92**, 938 (2008)
8. M. Bender, W. Seelig, C. Daube, H. Frankenberger, B. Ocker, and J. Stollenwerk, Thin Solid Films **326**, 72 (1998)
9. D.R. Sahu, S. Lin, and J. Huang, Appl. Surf. Sci. **252**, 7509 (2006)
10. D. R. Sahu, J. Huang, Appl. Surf. Sci. **253**, 827 (2006)
11. D.R. Sahu, and J. Huang, Thin Solid Films **516**, 208 (2007)
12. L. R. Doolittle, Nucl. Instrum. Methods Phys. Res. B **9**, 344 (1985)
13. JCPDS card no. 89-1397
14. H. Han, N.D. Theodore, and T.L. Alford, J. Appl. Phys. **103**, 013708 (2008)
15. S. Baik, J. H. Jang, C. H. Lee, W. Y Cho, and K. S. Lim, Appl. Phys. Lett. **70**, 3516 (1997)

16. G. Z. Xing, B. Yao, C. X. Cong, T. Yang, Y. P. Xie, B. H. Li and D. Z. Shen and J. of Alloys and Compounds **457**, 36 (2008)
17. V. Dose, W. Altmann, A. Goldmann, U. Kolac, and J. Rogozik, Phys. Rev. Lett. **52**, 1919 (1984)
18. Y. W. Zhu, H. Z. Zhang, X. C. Sun, S. Q. Feng, J. Xu, Q. Zhao, B. Xiang, R. M. Wang, and D. P. Yu, Appl. Phys. Lett. **83**, 144 (2003)
19. G. Haacke, J. Appl. Phys. **47**, 4086 (1976)
20. W. G. Driscoll, and W. Vaughan, *Handbook of Optics* (McGraw-Hill, USA, 1978)
21. H. Ehrenreich, and H. R. Philipp, Phys. Rev. **128**, 1622 (1962)
22. D. R. Lide, *CRC handbook of chemistry and physics* (CRC Press, Boca Raton, 1996)
23. K Sivaramakrishnan, T. L. Alford, Applied Physics Letters **94**, 052104 (2009)
24. H. Han, J. W. Mayer, and T. L. Alford, J. Appl. Phys. **100**, 083715 (2006)
25. T. Minemoto, T. Negami, S. Nishiwaki, H. Takakura, and Y. Hamakawa, Thin Solid Films **372**, 173 (2000)
26. S. Choopun, R. D. Vispute, W. Yang, R. P. Sharma, and T. Venkatesan, Appl. Phys. Lett. **80**, 1529 (2002)
27. K. J. Kim, and Y. R. Park, Appl. Phys. Lett. **78**, 475 (2001)
28. V. Bhosle, A. Tiwari, and J. Narayan, Appl. Phys. Lett. **88**, 032106 (2006)
29. S. Poulston, P. M. Parlett, P. Stone, and M. Bowker, Surf. and Inter. Anal. **24**, 811 (1996)

Conference Proceedings AMC XXV © 2010 Materials Research Society

Effect of thermal processing on silver contacts for zinc oxide and indium tin oxide

K. Sivaramakrishnan[1], S. Iyer[2], N. D. Theodore[3], and T. L. Alford[1]

[1]School of Materials and Flexible Display Center at ASU, Arizona State University, Tempe, Arizona 85287, U.S.A.

[2]Department of Electrical and Computer Engineering, North Carolina Agricultural and Technical State University, Greensboro, NC 27411, U.S.A.

[3]Freescale Semiconductor Inc. and Arizona State University, Tempe, Arizona 85284, U.S.A.

ABSTRACT

Silver thin film contacts of varying thickness (25, 45, and 60 nm) were deposited on indium tin oxide (ITO) on silicon and zinc oxide (ZnO) on silicon. The films were annealed in vacuum for 1 hour at different temperatures (300 - 650 ^{o}C). The resistivity of the films was determined using four-point-probe measurements. The films underwent an abrupt change in resistivity beyond an onset temperature that varied with thickness. Rutherford backscattering spectrometry measurements suggest agglomeration of the Ag films upon annealing caused the resistivity breakdown. X-ray pole figure analysis reveals that the annealed films took on a preferential <111> texturing; however, the degree of texturing was significantly higher in Ag on ZnO than on ITO. This observation was accounted for by interface energy minimization. Atomic force microscopy (AFM) measurements revealed an increasing surface roughness of the annealed films with temperature. The resistivity behavior is accounted for by the increasing crystallinity countered by an increasing surface roughness. Average surface roughness obtained from AFM measurements were also used to model the agglomeration of Ag based on Ostwald ripening theory.

INTRODUCTION

Transparent conducting oxides (TCOs) are materials which can simultaneously display transparency and conductivity. [1] TCOs find industrial application in flat-panel displays (*e.g.*, liquid crystal displays), solar cells, and electromagnetic shielding of cathode ray tubes used for video display terminals.[2, 3] Zinc oxide and indium tin oxide (ITO) are the most popular choice of TCOs. [4]

Thin films of silver are frequently used as contacts for photovoltaic devices on flexible substrates.[5] The resistance of these films is dependent upon scattering of electrons at grain boundaries, defects, and surfaces. In order to minimize these effects, the films must possess large aligned grains and smooth surfaces.[6] It is important to understand the annealing behavior of the films as the silver layer could be subject to high temperatures in the processing stages for these applications. [7] In this study, different thicknesses of Ag are deposited on ZnO and ITO to compare the surface roughness and texturing of the Ag films on the two substrates and to understand the effects of annealing on the films.

EXPERIMENT

The films of Ag on ZnO and ITO were prepared by magnetron sputtering. The oxides were deposited on Si wafers by rf sputtering while dc sputtering was used to deposit silver at a pressure of 10 mTorr and 40 W power. The sputtering targets used in this experiment were 99.999% In_2O_3 containing 10 wt% SnO_2, 99.999% pure ZnO, and 99.99% Ag. The ZnO and ITO thicknesses were kept fixed at 20 nm and 40 nm, respectively. The Ag thicknesses were 25 nm, 45 nm, and 60 nm. After deposition, the samples were annealed in a vacuum (base pressure ~ 10^{-8} Torr) for 1 hr at various temperatures (300, 400, 500, 600, and 650 °C) in order to investigate the changes in Ag film behavior. Resistivity of the Ag films was measured using an in-line, four-point probe configuration.

Rutherford backscattering spectrometry (RBS) was used to determine the thickness of the as-deposited silver films and to study the effect of annealing. RBS analysis was performed in a vacuum of 10^{-6} Torr using a 2.0 MeV He^{++} ion beam. The samples were tilted to 60° off normal incidence to increase the depth resolution. Energy spectra were were analyzed using the RUMP computer simulation program.[8]

Atomic force microscopy (AFM) in the tapping mode was used to obtain information on surface morphology and roughness.

Phase analysis of the films was performed by x-ray diffraction (XRD) using a Philips X'pert multipurpose diffractometer system with Cu K_α radiation incident on the sample at a 1° glancing angle. Texture evolution of the Ag films was investigated by pole figure analysis. Texture along the (111) and (200) poles was measured with the sample tilt ψ varied from 0° to 85°.

DISCUSSION

The change in resistivity as a function of annealing temperature is shown in figure 1. A thicker silver film results in a reduced resistivity on both ITO and ZnO likely due to reduced interface scattering. The resistivity decrease in the as-deposited samples however is lesser with increasing Ag film thickness because the values approach the value of resistivity for bulk silver(1.6 μm-cm[9]). Among the annealed samples, the 25 nm film of Ag on ITO and ZnO is found to show a sudden increase in resistivity after annealing at 300 °C. This is probably due to rapid agglomeration of the Ag in order to reduce its free energy.[10] The process of agglomeration reduces the surface-to-volume ratio at the same time leading to the formation of a discontinuous film. The higher resistivity of the as-deposited 25 nm film of Ag is suggestive of a discontinuous film which leads to potential barriers in the electron conduction path, thus increasing the resistivity. Unlike the 25 nm films, the 45 nm films show a significant change in resistivity only at 600 °C. The greater thickness ensures a continuous path for conduction of the electrons up to 500 °C prior to agglomeration induced discontinuity effects. Similarly, the 60 nm films of Ag were stable up to 600 °C and began degrading only at 650 °C. In the Ag films of 45 nm and 60 nm thickness there is also the possibility of degradation by sublimation at the high temperature of 600 °C, apart from the process of agglomeration.[11]

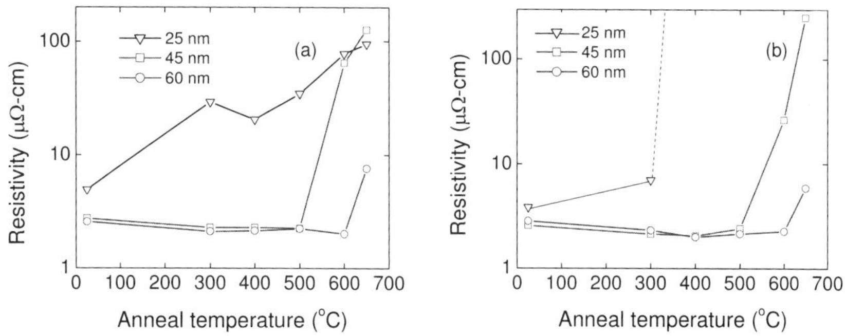

Figure 1. Electrical resistivity of Ag of various thicknesses on: (a) ITO and, (b) ZnO as a function of annealing temperature.

Figure 2 shows the backscattering results for 25 and 45 nm silver films on ITO and ZnO. In Figure 2(a) and Figure 2(c), there is a substantial drop in intensity of the Ag signal upon annealing at 300 °C. This observation confirms the inference of agglomeration of Ag films arrived at based on resistivity data. Agglomeration can lead to the formation of exposed regions in thin films, which translates to a drop in intensity of the RBS signal. At a temperature of 300 °C, which is considerably less than the melting point (961 °C[12]) of Ag, the effect of the sublimation is negligible and the drop in intensity of the RBS signal can be assumed to be primarily due to agglomeration. The early onset of agglomeration makes extremely thin silver films unviable for low resistivity applications. By contrast, the 45 nm films of Ag on ITO and ZnO show a drop in intensity of the Ag peak from agglomeration only at 600 °C as shown in Figure 2(b) and Figure 2(d) respectively. However, the possibility of sublimation cannot be completely ruled out due to the high temperature of anneal.[11] Further, the peaks of Ag, ITO and the Si interfaces are broadened in this spectrum suggesting more inter-diffusion than for Ag films of 25 nm thickness annealed at 300 °C. The 60 nm films show reduced intensity only at 650 °C. Hence, the application temperatures of interest will determine the optimized Ag film thickness to be used.

Pole figure analysis helps in determining the presence of texture in films. Figure 3 shows the pole figure profile of (111) texture with the intensity plotted against the tilt angle ψ for 45 nm Ag films on ITO and ZnO: as deposited, and for samples annealed at 400 °C and 600 °C.

The as-deposited samples show highest intensity for the (111) set of planes at $\psi = 0^{\circ}$ suggesting preferential orientation of the silver. This is to be expected based one energy considerations for silver surface.[13] The intensity peak at $\psi = 67^{\circ}$ corresponds to the {111} planes too which are at 70.5° to the surface normal. The shift in (111) peak position observed from $\psi = 70.5^{\circ}$ to $\psi = 67^{\circ}$ is due to an inaccurate defocusing correction at large tilts.[14] Monitoring the [111] Ag peak is a useful method of quantification of the extent of (111) fiber texture. This information is shown in Table I for all the samples. The drastic increase in (111) intensity for the 400 °C annealed samples compared to the as-deposited samples is due to thermal energy of the Ag atoms causing diffusion to attain lowest free energy configuration for the system. Although the as-deposited films of silver on ITO and ZnO have comparable intensities,

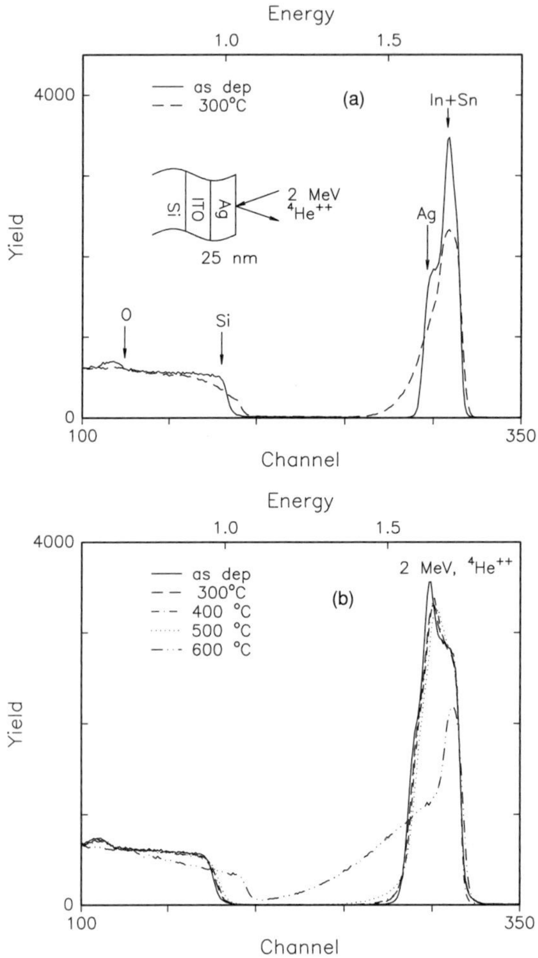

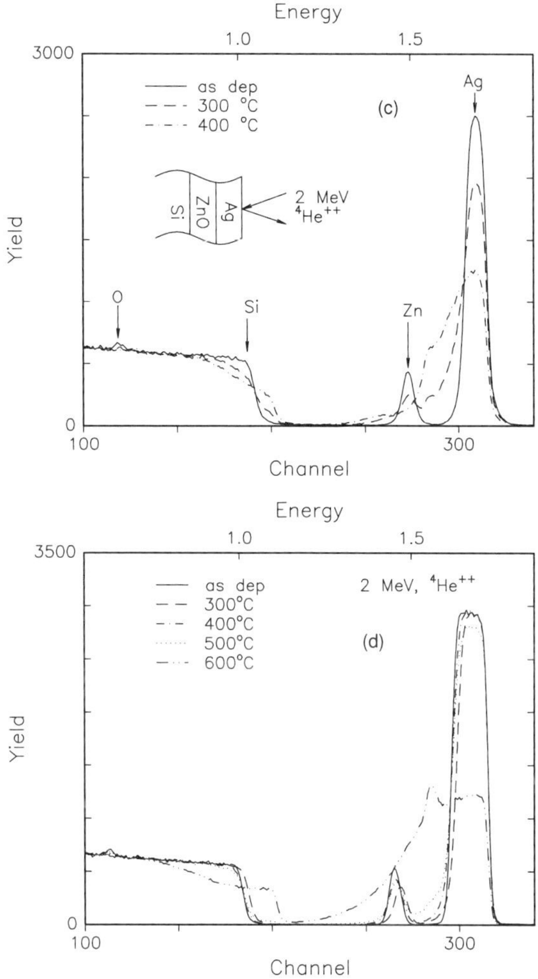

Figure 2. 2.0 MeV RBS spectra for: (a) 25 nm, and (b) 45 nm Ag films on ITO, and (c) 25 nm, and (d) 45 nm Ag films on ZnO.

in the case of the 400 °C annealed samples, the silver films on ZnO seem to have a greater (111) intensity and hence greater texture (see Table 1). This might be because crystalline ZnO, which has a strong preference for (002) orientation with the hexagonal lattice, enhances (111)

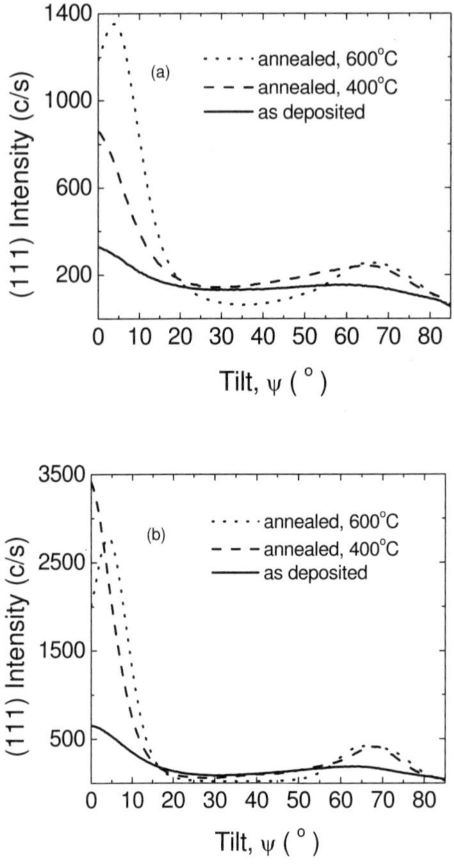

Figure 3. (111) pole figure profiles for 45 nm film of Ag on: (a) ITO and (b) ZnO as deposited, and annealed at 400 °C and 600 °C for 1 h in vacuum.

nucleation and growth of silver. [15] In contrast, ITO crystallizes into a BCC structure that does not particularly favour the enhancement of (111) silver texture. The background intensity in the plot is a measure of the degree of random orientation in the film, which has been quantified by the minimum in the ψ plot[10] tabulated in Table I.

It is worth noting that for silver films on ITO there is an increase in intensity at all ψ values for the 400 °C annealed sample when compared to the as-deposited sample. This indicates an overall increase in crystallinity. On the other hand, for silver on ZnO, the degree of random orientation is less in the 400 °C annealed sample than in the as-deposited sample. For the 600 °C

Table I. [111] peak parameters from [111] pole figure texture analysis on Ag/ITO and Ag/ZnO samples.

Sample	Background intensity (arb. units)	Height above random (arb. units)	HWHM (arb. units)
Ag/ITO as-dep	89	313	15
Ag/ITO 400 °C	112	798	10.5
Ag/ITO 600 °C	63	1290	7.5
Ag/ZnO as-dep	42	611	11.5
Ag/ZnO 400 °C	36	3383	6
Ag/ZnO 600 °C	19	2796	5.5

annealed samples on both ITO and ZnO, one can observe a decrease in background intensity when compared to the as-deposited samples, due to the formation of enhanced (111) texture. Further, the silver films have the peak slightly shifted from $\psi = 0^o$ to about 7°. This could be either due to the formation of a near-<111> fiber texture [16,17] or the agglomerated islands having an uneven surface. Due to the lack of motivation for the formation of a near-<111> Ag texture on top of BCC structured ITO it is speculated that the shift in the position of the [111] peak must be due to the Ag island formation. [18]

Pole figure profiles of (200) texture (not shown) for Ag on both substrates, revealed that the greatest increase in pole figure intensity is in the <111> directions at $\psi = 54.7^o$. For the 600 °C annealed samples, there is seen to be a clear increase in <111> intensity and a simultaneous drop in <200> intensity. It is also seen that the drop in intensity of random is much greater for the Ag on ZnO than for the Ag on ITO. These results further confirm an increase in the degree of randomly oriented grain growth in the case of Ag on ITO. However for Ag deposited on ZnO, despite an increase in crystallinity in specific directions such as [111] at $\psi = 54.7^o$, the overall degree of random orientation seems to have reduced. This suggests that the ZnO substrate favors Ag growth in [111] direction; whereas, ITO merely serves as a passive substrate. Thus the texturing of Ag on ZnO is determined by the following two factors: i) the tendency for Ag to favor (111) orientation based on surface energy considerations, and ii) the tendency for ZnO to favor (111) orientation of Ag based on interface energy consideration. In the case of Ag on ITO only the former plays a role and hence the degree of texturing is reduced.

The as-deposited samples of Ag on both ZnO and ITO substrates have an average surface roughness that varies between 0 to 2 nm for different thicknesses. Immediately upon annealing however, there is at least a five-fold increase in roughness. The 3-D AFM image shown in figure 4 of the 300 °C annealed 25 nm film of Ag on ITO clearly shows the formation of Ag islands upon annealing, characteristic of a discontinuous film.[16] The resistivity of the 25 nm films of Ag increases upon annealing due to severe agglomeration; however in all the other films, the resistivity improves initially upon annealing, and suggests an improved crystallinity.

Figure 4. 3D AFM image of 25 nm of Ag film on ITO substrates.

Figure 5 shows an Arrhenius plot of agglomeration of the 45 nm films of Ag on ITO and ZnO. The agglomeration has been modeled using the Ostwald ripening theory.[19] This theory can be used to predict both the power-law dependence and the temperature dependence of the growth rate.

The temperature dependence for island growth can be described by:

$$K = (K_o / T)\exp(-E_A / k_B T) \quad (3)$$

where K is the rate of change in the mean island size with time and E_A represents the activation energy for diffusion. The applicability of the Ostwald theory is tested by plotting the growth rate dependence on temperature. The term $\ln(T \times R_{av}^3 / t)$ on the ordinate is a logarithmic measure of the volume growth rate of the Ag islands as approximated using the cube of the root-mean-square roughness (R_{av}). The additional temperature factor on the ordinate is due to the Ostwald

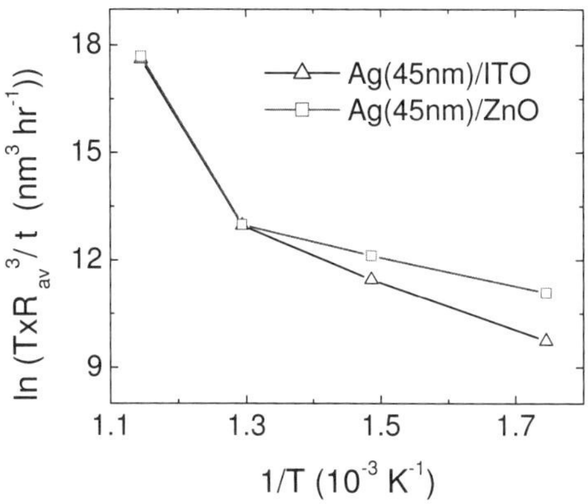

Figure 5. Arrhenius plot of agglomeration of the 45 nm films of Ag on ITO and ZnO

approach which predicts an $[\exp(-E_A / k_B T)]/T$ dependence for the island growth.[20] Figure 5 shows that the third-power dependence of the average roughness of the films (as measured using AFM) varies linearly with time up to 500 °C before increasing at a greater rate. Using equation 3, activation energy was determined from the slope for the linear part of the curve which extends up to 500 °C. The activation energies for diffusion of Ag on ITO and ZnO films have been calculated to be 0.62 ± 0.08 eV and 0.36 ± 0.06 eV, respectively. Beyond 500 °C there is an increase in the rate of agglomeration. This might be due to the onset of lattice diffusion as it is known that initial agglomeration happens by surface diffusion related grooving in Ag [21] and that volume diffusion has a larger onset time.[22]

The resistivity data, x-ray pole figure analysis of texturing, and AFM measurements combined suggest that the initial decrease in resistivity of the Ag films on both ITO and ZnO substrates with anneal temperature is due to increased silver texturing. However, increasing surface roughness (see figure 4) leads to increased surface scattering thus leading to a decrease in resistivity that is not in proportion to the increase in texture. It has been found that beyond a certain anneal temperature, the surface roughness effect starts to dominate over texturing effects and the resistivity rises with temperature.

Conclusion

Silver metal contacts have been deposited on to ITO and ZnO. Island formation of the films at high anneal temperature has been inferred from RBS and resistivity data. AFM roughness measurements combined with pole figure texture measurements can be used to explain the resistivity data in terms of the counterbalancing effects of increased texturing and increased surface roughness with temperature of anneal. Pole figure data also reveal a greater texture enhancement for the Ag films on ZnO than on ITO due to minimization of the surface and interface energies. Ostwald ripening theory is used to explain the agglomeration behavior of the Ag films. In conclusion, film thickness if found to have a significant effect on high temperature stability of the contacts. This understanding of the high temperature processing behavior of Ag films on transparent conducting oxides is advantageous for the use of Ag and its alloys in optoelectronic applications including solid-state photovoltaic devices on flexible substrates.

Acknowledgements

This work was partially supported by National Science Foundation (L. Hess, Grant No. DMR-0602716), to whom the authors are greatly indebted. Research was sponsored by the Army Research Laboratory (ARL) and was accomplished under Cooperative Agreement No. W911NG-04-2-0005. The views and conclusions contained in this document are those of the authors and should not be interpreted as representing the official policies, either expressed or implied, of the ARL or the U.S. Government. The U.S. Government is authorized to reproduce and distribute reprints for Government purposes notwithstanding any copyright notation hereon.

References

1. K. Sivaramakrishnan, and T. L. Alford, Appl. Phys. Lett. **94**, 052104 (2009)
2. C. G. Granqvist, and A. Hultaker, Thin Solid Films **411,** 1 (2002)
3. Roy G. Gordon, MRS Bulletin **25**, 52 (2000)
4. T. Minami, Semicond. Sci. Technol. **20**, S35 (2005)
5. X. Fang, T. Ma, M. Akiyama, G. Guan, S. Tsunematsu, and E. Abe, Thin Solid Films, **472**, 242 (2006)
6. A. Anders, E. Byon, D. Kim, K. Fukuda, and S. H. N. Lim, Solid State Comm. **140**, 225 (2006)
7. K. Fukuda, S. H. N. Lim, and A. Anders, Thin Solid Films, **516,** 4546 (2008)
8. L. R. Doolittle, Nucl. Instrum. Methods Phys. Res. B **9**, 344 (1985)
9. J. R. Greer, and R. A. Street, Acta Materialia **55**, 6345 (2007)
10. H. C. Kim, and T. L. Alford, J. Appl. Phys. **94**, 5393 (2003)
11. Y. Zoo, H. Han, and T. L. Alford, J. Appl. Phys. **102**, 083548 (2007)
12. D. Soh, Y. Shan, J. Park, Y. Li, and Y. Cho, Physica C **337**, 44 (2000)
13. Y. S. Jung, Appl. Surf. Sci. **221**, 281 (2004)
14. T. J. Klemmer, V. R. Inturi, and J. A. Barnard, J. Vac. Sci. Technol. A **15**, 1190 (1997)
15. N. Fujimara, T. Nishikara, S. Goto, J. Xu, and T. Ito, J. Crystal Growth **130**, 269 (1993)
16. A. E. Lita, and J. E. Sanchez, J. Elec. Matls., **31**, 55 (2002)
17. D. R. Waryoba, and P. N. Kalu, Microsc. Microanal. **11** (Suppl 2), 1694 (2005)
18. M. Birkholz, *Thin Film Analysis by X-Ray Scattering* (Wiley-VCH, Germany, 2006)
19. M. Zinke-Allmang, L. C. Feldman, and S. Nakahara, Appl. Phys. Lett. **51**, 975 (1987)
20. K. Tu, J. W. Mayer, and L. C. Feldman, *Electronic Thin Film Science (*Macmillan, USA, 1992)
21. H. Han, Y. Zoo, J. W. Mayer, and T. L. Alford, J. Appl. Phys. **102**, 036101 (2007)
22. D.J. Srolovitz, and M. G. Goldiner, JOM **47**, 31 (1995)

AUTHOR INDEX

SUBJECT INDEX